Fundamentals of Physics

Fundamentals of Physics

Edited by
Leonel Ware

Larsen & Keller
www.larsen-keller.com

Fundamentals of Physics
Edited by Leonel Ware
ISBN: 978-1-63549-038-1 (Hardback)

Larsen & Keller

Published by Larsen and Keller Education,
5 Penn Plaza,
19th Floor,
New York, NY 10001, USA

Cataloging-in-Publication Data

Fundamentals of physics / edited by Leonel Ware.
    p. cm.
Includes bibliographical references and index.
ISBN 978-1-63549-038-1
1. Physics. I. Ware, Leonel.
QC23.2 .F66 2017
530--dc23

The publisher's policy is to use permanent paper from mills that operate a sustainable forestry policy. Furthermore, the publisher ensures that the text paper and cover boards used have met acceptable environmental accreditation standards.

Printed and bound in the United States of America.

For more information regarding Larsen and Keller Education and its products, please visit the publisher's website www.larsen-keller.com

Table of Contents

Permissions

Index

Preface

This book provides comprehensive insights into the field of physics. It picks up individual branches and explains their need and contribution in the context of the growth of this field. Physics is a vast subject and has multiple subfields. This textbook attempts to provide complete information about the basic principles and fundamental concepts of this huge subject area. It presents this complex subject in the most comprehensible and easy to understand language. For all those who are interested in physics, this textbook can prove to be an essential guide. With its detailed analyses and data, this book on physics will prove immensely beneficial to students involved in this area at various levels.

To facilitate a deeper understanding of the contents of this book a short introduction of every chapter is written below:

Chapter 1- Physics is a vast subject of science that majorly concerns itself with the study of matter. The basic aim of physics is to understand the functionality of the universe. The chapter on physics offers an insightful focus, keeping in mind the complex subject matter.

Chapter 2- Physics has a number of branches; some of these are particle physics, condensed matter physics, electromagnetism and quantum mechanics. The branch of physics that studies the nature of particles is termed as particle physics whereas astrophysics is the branch that studies heavenly bodies and their positions in space. This text is a compilation of the various branches of physics that form an integral part of the broader subject matter.

Chapter 3- Classical physics studies the widely applicable theories of physics. Hamiltonian mechanics, Archimedes' principle, physical law, conservation law, inverse-square law, Stokes' law etc. have been explained in this section. The chapter serves as a source to understand the major theories and principles of physics.

Chapter 4- The movement of any object from one point to another is because of motion. Velocity of an object is the rate of this change from one point to another. Momentum, frame of reference and Newton's laws of motion have also been explained in the following section. Physics is best understood in confluence with the major topics listed in the following chapter.

Chapter 5- In physics, force can change the velocity of any object. It has both magnitude and direction. The section on work and energy also covers themes such as friction, power, energy, forms of energy, conservation of energy etc. The major concepts of work and energy are explained in this chapter.

Chapter 6- Matter is observed on a daily basis in the form of solid, liquid, gas and plasma. The other states of matter exist only in extreme conditions. The topics discussed in the section are of great importance to broaden the existing on fundamental states of matter.

I would like to share the credit of this book with my editorial team who worked tirelessly on this book. I owe the completion of this book to the never-ending support of my family, who supported me throughout the project.

Editor

Introduction to Physics

Physics is a vast subject of science that majorly concerns itself with the study of matter. The basic aim of physics is to understand the functionality of the universe. The chapter on physics offers an insightful focus, keeping in mind the complex subject matter.

Physics is the natural science that involves the study of matter and its motion and behavior through space and time, along with related concepts such as energy and force. One of the most fundamental scientific disciplines, the main goal of physics is to understand how the universe behaves.

Various examples of physical phenomena

Physics is one of the oldest academic disciplines, perhaps the oldest through its inclusion of astronomy. Over the last two millennia, physics was a part of natural philosophy along with chemistry, biology, and certain branches of mathematics, but during the scientific revolution in the 17th century, the natural sciences emerged as unique research programs in their own right. Physics intersects with many interdisciplinary areas of research, such as biophysics and quantum chemistry, and the boundaries of physics are not rigidly defined. New ideas in physics often explain the fundamental mechanisms of other sciences while opening new avenues of research in areas such as mathematics and philosophy.

Physics also makes significant contributions through advances in new technologies that arise from theoretical breakthroughs. For example, advances in the understanding of electromagnetism or nuclear physics led directly to the development of new products that have dramatically transformed modern-day society, such as television, computers, domestic appliances, and nuclear weapons; advances in thermodynamics led to the development of industrialization, and advances in mechanics inspired the development of calculus.

The United Nations named 2005 the World Year of Physics.

History

Ancient Astronomy

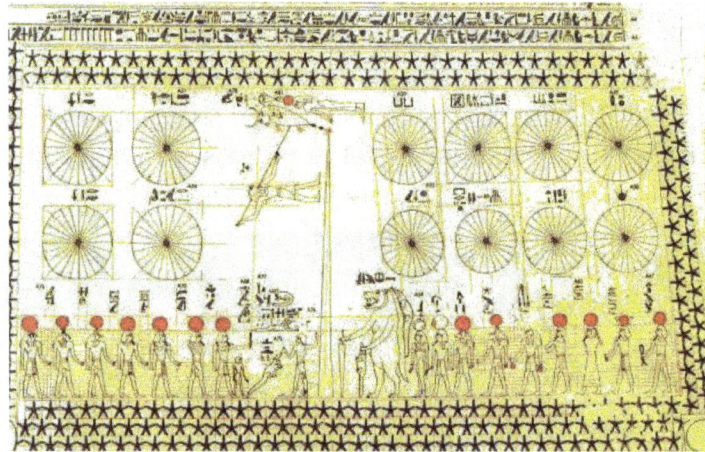

Ancient Egyptian astronomy is evident in monuments like the ceiling of Senemut's tomb from the Eighteenth Dynasty of Egypt.

Astronomy is the oldest of the natural sciences. The earliest civilizations dating back to beyond 3000 BCE, such as the Sumerians, ancient Egyptians, and the Indus Valley Civilization, all had a predictive knowledge and a basic understanding of the motions of the Sun, Moon, and stars. The stars and planets were often a target of worship, believed to represent their gods. While the explanations for these phenomena were often unscientific and lacking in evidence, these early observations laid the foundation for later astronomy.

According to Asger Aaboe, the origins of Western astronomy can be found in Mesopotamia, and all Western efforts in the exact sciences are descended from late Babylonian astronomy. Egyptian astronomers left monuments showing knowledge of the constellations and the motions of the celestial bodies, while Greek poet Homer wrote of various celestial objects in his *Iliad* and *Odyssey*; later Greek astronomers provided names, which are still used today, for most constellations visible from the northern hemisphere.

Natural Philosophy

Natural philosophy has its origins in Greece during the Archaic period, (650 BCE – 480 BCE), when pre-Socratic philosophers like Thales rejected non-naturalistic explanations for natural phenomena and proclaimed that every event had a natural cause. They proposed ideas verified by reason and observation, and many of their hypotheses proved successful in experiment; for example, atomism was found to be correct approximately 2000 years after it was first proposed by Leucippus and his pupil Democritus.

Physics in the Medieval Islamic World

Islamic scholarship had inherited Aristotelian physics from the Greeks and during the Islamic Golden Age developed it further, especially placing emphasis on observation and *a priori* reasoning, developing early forms of the scientific method.

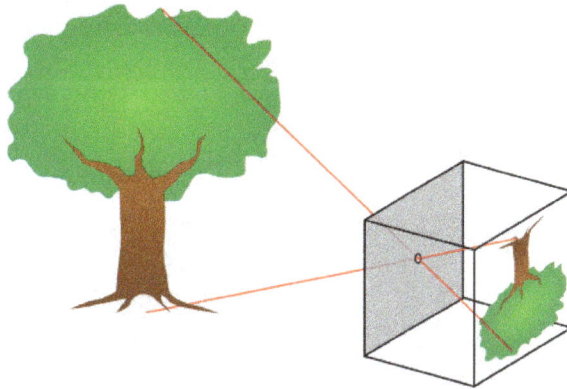

The basic way a pinhole camera works

The most notable innovations were in the field of optics and vision, which came from the works of many scientists like Ibn Sahl, Al-Kindi, Ibn Al-Haitham, Al-Farisi and Avicenna. The most notable work was *The Book of Optics* (also known as Kitāb al-Manāẓir), written by Ibn Al-Haitham, in which he was not only the first to disprove the ancient Greek idea about vision, but also came up with a new theory. In the book, he was also the first to study the phenomenon of the pinhole camera and delved further into the way the eye itself works. Using dissections and the knowledge of previous scholars, he was able to begin to explain how light enters the eye, is focused, and is projected to the back of the eye: and built then the world's first camera obscura hundreds of years before the modern development of photography.

Ibn Al Haitham, the pioneer of optics

The seven-volume *Book of Optics* (*Kitab al-Manathir*) hugely influenced thinking across disciplines from the theory of visual perception to the nature of perspective in medieval art, in both the

East and the West, for more than 600 years. Many later European scholars and fellow polymaths, from Robert Grosseteste and Leonardo da Vinci to René Descartes, Johannes Kepler and Isaac Newton, were in his debt. Indeed, the influence of Ibn al-Haytham's Optics ranks alongside that of Newton's work of the same title, published 700 years later.

The translation of *The Book of Optics* had a huge impact on Europe. From it, later European scholars were able to build the same devices as what Ibn Al Haytham did, and understand the way light works. From this, such important things as eyeglasses, magnifying glasses, telescopes, and cameras were developed.

Classical Physics

Sir Isaac Newton (1643–1727), whose laws of motion and universal gravitation were major milestones in classical physics

Physics became a separate science when early modern Europeans used experimental and quantitative methods to discover what are now considered to be the laws of physics.

Major developments in this period include the replacement of the geocentric model of the solar system with the heliocentric Copernican model, the laws governing the motion of planetary bodies determined by Johannes Kepler between 1609 and 1619, pioneering work on telescopes and observational astronomy by Galileo Galilei in the 16th and 17th Centuries, and Isaac Newton's discovery and unification of the laws of motion and universal gravitation that would come to bear his name. Newton also developed calculus, the mathematical study of change, which provided new mathematical methods for solving physical problems.

The discovery of new laws in thermodynamics, chemistry, and electromagnetics resulted from greater research efforts during the Industrial Revolution as energy needs increased. The laws comprising classical physics remain very widely used for objects on everyday scales travelling at non-relativistic speeds, since they provide a very close approximation in such situations, and theories such as quantum mechanics and the theory of relativity simplify to their classical equivalents at such scales. However, inaccuracies in classical mechanics for very small objects and very high velocities led to the development of modern physics in the 20th century.

Modern Physics

Modern physics began in the early 20th century with the work of Max Planck in quantum theory and Albert Einstein's theory of relativity. Both of these theories came about due to inaccuracies in classical mechanics in certain situations. Classical mechanics predicted a varying speed of light, which could not be resolved with the constant speed predicted by Maxwell's equations of electromagnetism; this discrepancy was corrected by Einstein's theory of special relativity, which replaced classical mechanics for fast-moving bodies and allowed for a constant speed of light. Black body radiation provided another problem for classical physics, which was corrected when Planck proposed that the excitation of material oscillators is possible only in discrete steps proportional to their frequency; this, along with the photoelectric effect and a complete theory predicting discrete energy levels of electron orbitals, led to the theory of quantum mechanics taking over from classical physics at very small scales.

Albert Einstein (1879–1955), whose work on the photoelectric effect and
the theory of relativity led to a revolution in 20th century physics

Max Planck (1858–1947), the originator of the theory of quantum mechanics

Quantum mechanics would come to be pioneered by Werner Heisenberg, Erwin Schrödinger and Paul Dirac. From this early work, and work in related fields, the Standard Model of particle physics was derived. Following the discovery of a particle with properties consistent with the Higgs boson at CERN in 2012, all fundamental particles predicted by the standard model, and no others, appear to exist; however, physics beyond the Standard Model, with theories such as supersymmetry, is an active area of research. Areas of mathematics in general are important to this field, such as the study of probabilities.

Philosophy

In many ways, physics stems from ancient Greek philosophy. From Thales' first attempt to characterise matter, to Democritus' deduction that matter ought to reduce to an invariant state, the Ptolemaic astronomy of a crystalline firmament, and Aristotle's book *Physics* (an early book on physics, which attempted to analyze and define motion from a philosophical point of view), various Greek philosophers advanced their own theories of nature. Physics was known as natural philosophy until the late 18th century.

By the 19th century, physics was realised as a discipline distinct from philosophy and the other sciences. Physics, as with the rest of science, relies on philosophy of science and its "scientific method" to advance our knowledge of the physical world. The scientific method employs *a priori reasoning* as well as *a posteriori* reasoning and the use of Bayesian inference to measure the validity of a given theory.

The development of physics has answered many questions of early philosophers, but has also raised new questions. Study of the philosophical issues surrounding physics, the philosophy of physics, involves issues such as the nature of space and time, determinism, and metaphysical outlooks such as empiricism, naturalism and realism.

Many physicists have written about the philosophical implications of their work, for instance Laplace, who championed causal determinism, and Erwin Schrödinger, who wrote on quantum mechanics. The mathematical physicist Roger Penrose has been called a Platonist by Stephen Hawking, a view Penrose discusses in his book, *The Road to Reality*. Hawking refers to himself as an "unashamed reductionist" and takes issue with Penrose's views.

Core Theories

Though physics deals with a wide variety of systems, certain theories are used by all physicists. Each of these theories were experimentally tested numerous times and found to be an adequate approximation of nature. For instance, the theory of classical mechanics accurately describes the motion of objects, provided they are much larger than atoms and moving at much less than the speed of light. These theories continue to be areas of active research today. Chaos theory, a remarkable aspect of classical mechanics was discovered in the 20th century, three centuries after the original formulation of classical mechanics by Isaac Newton (1642–1727).

These central theories are important tools for research into more specialised topics, and any physicist, regardless of their specialisation, is expected to be literate in them. These include classical mechanics, quantum mechanics, thermodynamics and statistical mechanics, electromagnetism, and special relativity.

Classical Physics

Classical physics includes the traditional branches and topics that were recognised and well-developed before the beginning of the 20th century—classical mechanics, acoustics, optics, thermodynamics, and electromagnetism. Classical mechanics is concerned with bodies acted on by forces and bodies in motion and may be divided into statics (study of the forces on a body or bodies not subject to an acceleration), kinematics (study of motion without regard to its causes), and dynamics (study of motion and the forces that affect it); mechanics may also be divided into solid mechanics and fluid mechanics (known together as continuum mechanics), the latter include such branches as hydrostatics, hydrodynamics, aerodynamics, and pneumatics. Acoustics is the study of how sound is produced, controlled, transmitted and received. Important modern branches of acoustics include ultrasonics, the study of sound waves of very high frequency beyond the range of human hearing; bioacoustics, the physics of animal calls and hearing, and electroacoustics, the manipulation of audible sound waves using electronics.

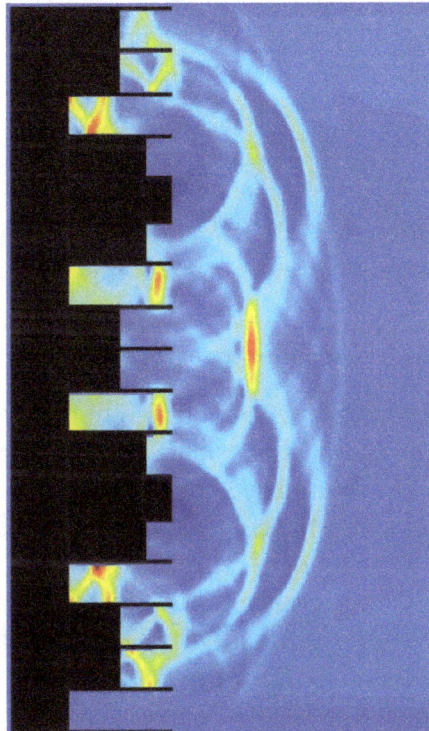

Classical physics implemented in an acoustic engineering model of sound reflecting from an acoustic diffuser

Optics, the study of light, is concerned not only with visible light but also with infrared and ultraviolet radiation, which exhibit all of the phenomena of visible light except visibility, e.g., reflection, refraction, interference, diffraction, dispersion, and polarization of light. Heat is a form of energy, the internal energy possessed by the particles of which a substance is composed; thermodynamics deals with the relationships between heat and other forms of energy. Electricity and magnetism have been studied as a single branch of physics since the intimate connection between them was discovered in the early 19th century; an electric current gives rise to a magnetic field, and a changing magnetic field induces an electric current. Electrostatics deals with electric charges at rest, electrodynamics with moving charges, and magnetostatics with magnetic poles at rest.

Modern Physics

Classical physics is generally concerned with matter and energy on the normal scale of observation, while much of modern physics is concerned with the behavior of matter and energy under extreme conditions or on a very large or very small scale. For example, atomic and nuclear physics studies matter on the smallest scale at which chemical elements can be identified. The physics of elementary particles is on an even smaller scale since it is concerned with the most basic units of matter; this branch of physics is also known as high-energy physics because of the extremely high energies necessary to produce many types of particles in particle accelerators. On this scale, ordinary, commonsense notions of space, time, matter, and energy are no longer valid.

Solvay Conference of 1927, with prominent physicists such as Albert Einstein, Werner Heisenberg, Max Planck, Hendrik Lorentz, Niels Bohr, Marie Curie, Erwin Schrödinger and Paul Dirac

The two chief theories of modern physics present a different picture of the concepts of space, time, and matter from that presented by classical physics. Classical mechanics approximates nature as continuous, while quantum theory is concerned with the discrete nature of many phenomena at the atomic and subatomic level and with the complementary aspects of particles and waves in the description of such phenomena. The theory of relativity is concerned with the description of phenomena that take place in a frame of reference that is in motion with respect to an observer; the special theory of relativity is concerned with relative uniform motion in a straight line and the general theory of relativity with accelerated motion and its connection with gravitation. Both quantum theory and the theory of relativity find applications in all areas of modern physics.

Difference Between Classical and Modern Physics

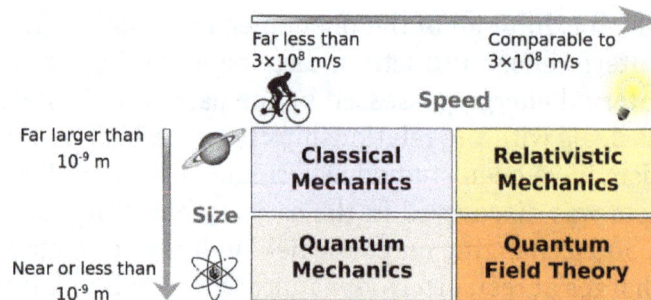

The basic domains of physics

While physics aims to discover universal laws, its theories lie in explicit domains of applicability. Loosely speaking, the laws of classical physics accurately describe systems whose important length scales are greater than the atomic scale and whose motions are much slower than the speed of light. Outside of this domain, observations do not match predictions provided by classical mechanics. Albert Einstein contributed the framework of special relativity, which replaced notions of absolute time and space with spacetime and allowed an accurate description of systems whose components have speeds approaching the speed of light. Max Planck, Erwin Schrödinger, and others introduced quantum mechanics, a probabilistic notion of particles and interactions that allowed an accurate description of atomic and subatomic scales. Later, quantum field theory unified quantum mechanics and special relativity. General relativity allowed for a dynamical, curved spacetime, with which highly massive systems and the large-scale structure of the universe can be well-described. General relativity has not yet been unified with the other fundamental descriptions; several candidate theories of quantum gravity are being developed.

Relation to Other Fields

This parabola-shaped lava flow illustrates the application of mathematics in physics—in this case, Galileo's law of falling bodies.

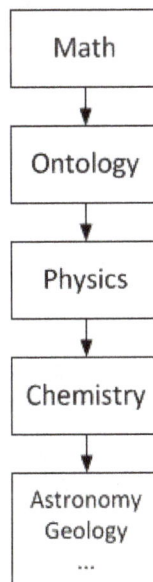

Mathematics and ontology are used in physics. Physics is used in chemistry and cosmology.

Prerequisites

Mathematics provides a compact and exact language used to describe of the order in nature. This was noted and advocated by Pythagoras, Plato, Galileo, and Newton.

Physics uses mathematics to organise and formulate experimental results. From those results, precise or estimated solutions, quantitative results from which new predictions can be made and experimentally confirmed or negated. The results from physics experiments are numerical measurements. Technologies based on mathematics, like computation have made computational physics an active area of research.

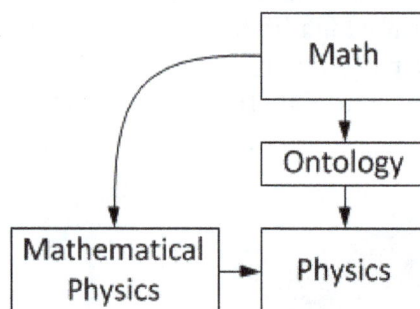

The distinction between mathematics and physics is clear-cut, but not always obvious,
especially in mathematical physics.

Ontology is a prerequisite for physics, but not for mathematics. It means physics is ultimately concerned with descriptions of the real world, while mathematics is concerned with abstract patterns, even beyond the real world. Thus physics statements are synthetic, while mathematical statements are analytic. Mathematics contains hypotheses, while physics contains theories. Mathematics statements have to be only logically true, while predictions of physics statements must match observed and experimental data.

The distinction is clear-cut, but not always obvious. For example, mathematical physics is the application of mathematics in physics. Its methods are mathematical, but its subject is physical. The problems in this field start with a "mathematical model of a physical situation" (system) and a "mathematical description of a physical law" that will be applied to that system. Every mathematical statement used for solving has a hard-to-find physical meaning. The final mathematical solution has an easier-to-find meaning, because it is what the solver is looking for.

Physics is a branch of fundamental science, not practical science. Physics is also called "the fundamental science" because the subject of study of all branches of natural science like chemistry, astronomy, geology, and biology are constrained by laws of physics, similar to how chemistry is often called the central science because of its role in linking the physical sciences. For example, chemistry studies properties, structures, and reactions of matter (chemistry's focus on the atomic scale distinguishes it from physics). Structures are formed because particles exert electrical forces on each other, properties include physical characteristics of given substances, and reactions are bound by laws of physics, like conservation of energy, mass, and charge.

Physics is applied in industries like engineering and medicine.

Application and Influence

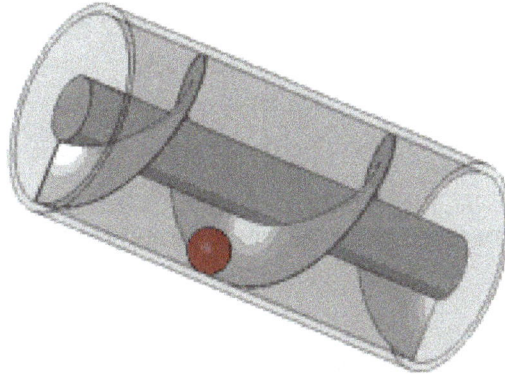

Archimedes' screw, a simple machine for lifting

The application of physical laws in lifting liquids

Applied physics is a general term for physics research which is intended for a particular use. An applied physics curriculum usually contains a few classes in an applied discipline, like geology or electrical engineering. It usually differs from engineering in that an applied physicist may not be designing something in particular, but rather is using physics or conducting physics research with the aim of developing new technologies or solving a problem.

The approach is similar to that of applied mathematics. Applied physicists use physics in scientific research. For instance, people working on accelerator physics might seek to build better particle detectors for research in theoretical physics.

Physics is used heavily in engineering. For example, statics, a subfield of mechanics, is used in the building of bridges and other static structures. The understanding and use of acoustics results in sound control and better concert halls; similarly, the use of optics creates better optical devices. An understanding of physics makes for more realistic flight simulators, video games, and movies, and is often critical in forensic investigations.

With the standard consensus that the laws of physics are universal and do not change with time, physics can be used to study things that would ordinarily be mired in uncertainty. For example, in the study of the origin of the earth, one can reasonably model earth's mass, temperature, and rate of rotation, as a function of time allowing one to extrapolate forward or backward in time and so predict future or prior events. It also allows for simulations in engineering which drastically speed up the development of a new technology.

But there is also considerable interdisciplinarity in the physicist's methods, so many other important fields are influenced by physics (e.g., the fields of econophysics and sociophysics).

Research

Scientific Method

Physicists use the scientific method to test the validity of a physical theory. By using a methodical approach to compare the implications of a theory with the conclusions drawn from its related experiments and observations, physicists are better able to test the validity of a theory in a logical, unbiased, and repeatable way. To that end, experiments are performed and observations are made in order to determine the validity or invalidity of the theory.

A scientific law is a concise verbal or mathematical statement of a relation which expresses a fundamental principle of some theory, such as Newton's law of universal gravitation.

Theory and Experiment

Theorists seek to develop mathematical models that both agree with existing experiments and successfully predict future experimental results, while experimentalists devise and perform experiments to test theoretical predictions and explore new phenomena. Although theory and experiment are developed separately, they are strongly dependent upon each other. Progress in physics frequently comes about when experimentalists make a discovery that existing theories cannot explain, or when new theories generate experimentally testable predictions, which inspire new experiments.

The astronaut and Earth are both in free-fall

Lightning is an electric current

Physicists who work at the interplay of theory and experiment are called phenomenologists, who study complex phenomena observed in experiment and work to relate them to a fundamental theory.

Theoretical physics has historically taken inspiration from philosophy; electromagnetism was unified this way. Beyond the known universe, the field of theoretical physics also deals with hypothetical issues, such as parallel universes, a multiverse, and higher dimensions. Theorists invoke these ideas in hopes of solving particular problems with existing theories. They then explore the consequences of these ideas and work toward making testable predictions.

Experimental physics expands, and is expanded by, engineering and technology. Experimental physicists involved in basic research design and perform experiments with equipment such as particle accelerators and lasers, whereas those involved in applied research often work in industry developing technologies such as magnetic resonance imaging (MRI) and transistors. Feynman has noted that experimentalists may seek areas which are not well-explored by theorists.

Scope and Aims

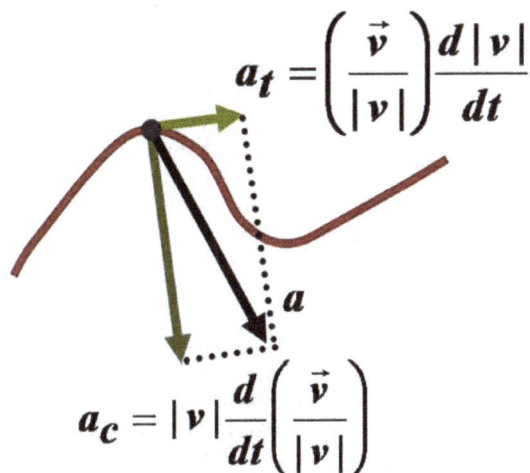

$$a_t = \left(\frac{\vec{v}}{|v|}\right)\frac{d|v|}{dt}$$

$$a$$

$$a_c = |v|\frac{d}{dt}\left(\frac{\vec{v}}{|v|}\right)$$

Physics involves modeling the natural world with theory, usually quantitative. Here, the path of a particle is modeled with the mathematics of calculus to explain its behavior: the purview of the branch of physics known as mechanics.

Physics covers a wide range of phenomena, from elementary particles (such as quarks, neutrinos, and electrons) to the largest superclusters of galaxies. Included in these phenomena are the most basic objects composing all other things. Therefore, physics is sometimes called the "fundamental science". Physics aims to describe the various phenomena that occur in nature in terms of simpler phenomena. Thus, physics aims to both connect the things observable to humans to root causes, and then connect these causes together.

For example, the ancient Chinese observed that certain rocks (lodestone and magnetite) were attracted to one another by an invisible force. This effect was later called magnetism, which was first rigorously studied in the 17th century. But even before the Chinese discovered magnetism, the ancient Greeks knew of other objects such as amber, that when rubbed with fur would cause a similar invisible attraction between the two. This was also first studied rigorously in the 17th century and came to be called electricity. Thus, physics had come to understand two observations of nature in terms of some root cause (electricity and magnetism). However, further work in the 19th century revealed that these two forces were just two different aspects of one force—electromagnetism. This process of "unifying" forces continues today, and electromagnetism and the weak nuclear force are now considered to be two aspects of the electroweak interaction.

Research Fields

Contemporary research in physics can be broadly divided into particle physics; condensed matter physics; atomic, molecular, and optical physics; astrophysics; and applied physics. Some physics departments also support physics education research and physics outreach.

Since the 20th century, the individual fields of physics have become increasingly specialised, and today most physicists work in a single field for their entire careers. "Universalists" such as Albert Einstein (1879–1955) and Lev Landau (1908–1968), who worked in multiple fields of physics, are now very rare.

The major fields of physics, along with their subfields and the theories and concepts they employ, are shown in the following table.

Field	Subfields	Major theories	Concepts
Particle physics	Nuclear physics, Nuclear astrophysics, Particle astrophysics, Particle physics phenomenology	Standard Model, Quantum field theory, Quantum electrodynamics, Quantum chromodynamics, Electroweak theory, Effective field theory, Lattice field theory, Lattice gauge theory, Gauge theory, Supersymmetry, Grand unification theory, Superstring theory, M-theory	Fundamental force (gravitational, electromagnetic, weak, strong), Elementary particle, Spin, Antimatter, Spontaneous symmetry breaking, Neutrino oscillation, Seesaw mechanism, Brane, String, Quantum gravity, Theory of everything, Vacuum energy
Atomic, molecular, and optical physics	Atomic physics, Molecular physics, Atomic and Molecular astrophysics, Chemical physics, Optics, Photonics	Quantum optics, Quantum chemistry, Quantum information science	Photon, Atom, Molecule, Diffraction, Electromagnetic radiation, Laser, Polarization (waves), Spectral line, Casimir effect

Condensed matter physics	Solid state physics, High pressure physics, Low-temperature physics, Surface Physics, Nanoscale and Mesoscopic physics, Polymer physics	BCS theory, Bloch wave, Density functional theory, Fermi gas, Fermi liquid, Many-body theory, Statistical Mechanics	Phases (gas, liquid, solid), Bose-Einstein condensate, Electrical conduction, Phonon, Magnetism, Self-organization, Semiconductor, superconductor, superfluid, Spin,
Astrophysics	Astronomy, Astrometry, Cosmology, Gravitation physics, High-energy astrophysics, Planetary astrophysics, Plasma physics, Solar physics, Space physics, Stellar astrophysics	Big Bang, Cosmic inflation, General relativity, Newton's law of universal gravitation, Lambda-CDM model, Magnetohydrodynamics	Black hole, Cosmic background radiation, Cosmic string, Cosmos, Dark energy, Dark matter, Galaxy, Gravity, Gravitational radiation, Gravitational singularity, Planet, Solar System, Star, Supernova, Universe
Applied Physics	Accelerator physics, Acoustics, Agrophysics, Biophysics, Chemical Physics, Communication Physics, Econophysics, Engineering physics, Fluid dynamics, Geophysics, Laser Physics, Materials physics, Medical physics, Nanotechnology, Optics, Optoelectronics, Photonics, Photovoltaics, Physical chemistry, Physics of computation, Plasma physics, Solid-state devices, Quantum chemistry, Quantum electronics, Quantum information science, Vehicle dynamics		

Particle Physics

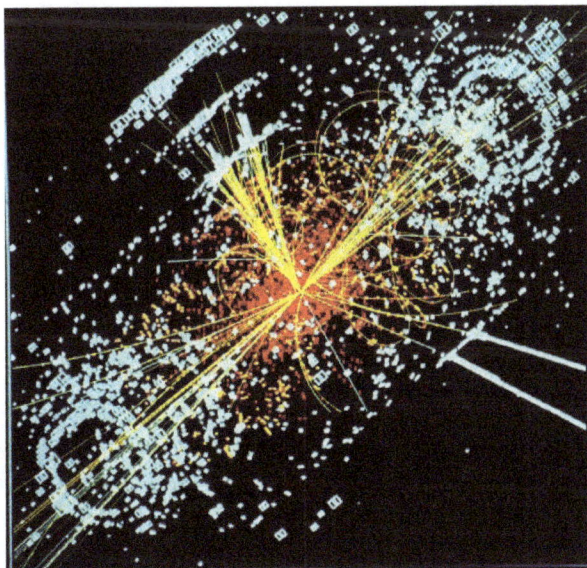

A simulated event in the CMS detector of the Large Hadron Collider, featuring a possible appearance of the Higgs boson.

Particle physics is the study of the elementary constituents of matter and energy and the interactions between them. In addition, particle physicists design and develop the high energy accelerators, detectors, and computer programs necessary for this research. The field is also called "high-energy physics" because many elementary particles do not occur naturally but are created only during high-energy collisions of other particles.

Currently, the interactions of elementary particles and fields are described by the Standard Model. The model accounts for the 12 known particles of matter (quarks and leptons) that interact via the strong, weak, and electromagnetic fundamental forces. Dynamics are described in terms of matter particles exchanging gauge bosons (gluons, W and Z bosons, and photons, respectively). The Standard Model also predicts a particle known as the Higgs boson. In July 2012 CERN, the European laboratory for particle physics, announced the detection of a particle consistent with the Higgs boson, an integral part of a Higgs mechanism.

Nuclear physics is the field of physics that studies the constituents and interactions of atomic nuclei. The most commonly known applications of nuclear physics are nuclear power generation and nuclear weapons technology, but the research has provided application in many fields, including those in nuclear medicine and magnetic resonance imaging, ion implantation in materials engineering, and radiocarbon dating in geology and archaeology.

Atomic, Molecular, and Optical Physics

Atomic, molecular, and optical physics (AMO) is the study of matter–matter and light–matter interactions on the scale of single atoms and molecules. The three areas are grouped together because of their interrelationships, the similarity of methods used, and the commonality of their relevant energy scales. All three areas include both classical, semi-classical and quantum treatments; they can treat their subject from a microscopic view (in contrast to a macroscopic view).

Atomic physics studies the electron shells of atoms. Current research focuses on activities in quantum control, cooling and trapping of atoms and ions, low-temperature collision dynamics and the effects of electron correlation on structure and dynamics. Atomic physics is influenced by the nucleus, but intra-nuclear phenomena such as fission and fusion are considered part of high-energy physics.

Molecular physics focuses on multi-atomic structures and their internal and external interactions with matter and light. Optical physics is distinct from optics in that it tends to focus not on the control of classical light fields by macroscopic objects but on the fundamental properties of optical fields and their interactions with matter in the microscopic realm.

Condensed Matter Physics

Condensed matter physics is the field of physics that deals with the macroscopic physical properties of matter. In particular, it is concerned with the "condensed" phases that appear whenever the number of particles in a system is extremely large and the interactions between them are strong.

The most familiar examples of condensed phases are solids and liquids, which arise from the bonding by way of the electromagnetic force between atoms. More exotic condensed phases include the superfluid and the Bose–Einstein condensate found in certain atomic systems at very low temperature, the superconducting phase exhibited by conduction electrons in certain materials, and the ferromagnetic and antiferromagnetic phases of spins on atomic lattices.

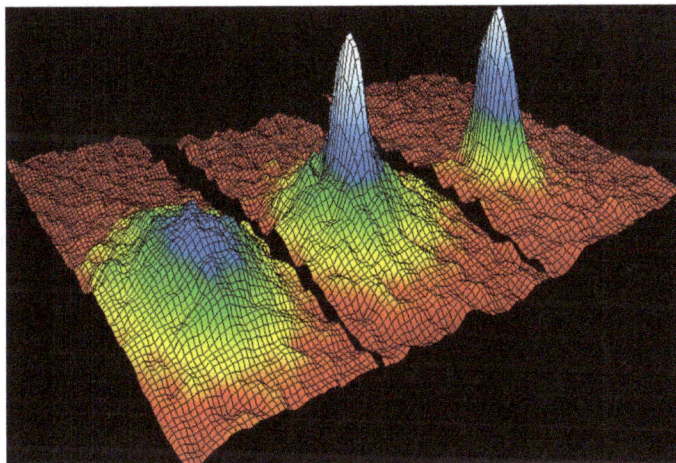

Velocity-distribution data of a gas of rubidium atoms, confirming the discovery of a new phase of matter, the Bose–Einstein condensate

Condensed matter physics is the largest field of contemporary physics. Historically, condensed matter physics grew out of solid-state physics, which is now considered one of its main subfields. The term *condensed matter physics* was apparently coined by Philip Anderson when he renamed his research group—previously *solid-state theory*—in 1967. In 1978, the Division of Solid State Physics of the American Physical Society was renamed as the Division of Condensed Matter Physics. Condensed matter physics has a large overlap with chemistry, materials science, nanotechnology and engineering.

Astrophysics

The deepest visible-light image of the universe, the Hubble Ultra Deep Field

Astrophysics and astronomy are the application of the theories and methods of physics to the study of stellar structure, stellar evolution, the origin of the Solar System, and related problems of cosmology. Because astrophysics is a broad subject, astrophysicists typically apply many disciplines of physics, including mechanics, electromagnetism, statistical mechanics, thermodynamics, quantum mechanics, relativity, nuclear and particle physics, and atomic and molecular physics.

The discovery by Karl Jansky in 1931 that radio signals were emitted by celestial bodies initiated the science of radio astronomy. Most recently, the frontiers of astronomy have been expanded by space exploration. Perturbations and interference from the earth's atmosphere make space-based observations necessary for infrared, ultraviolet, gamma-ray, and X-ray astronomy.

Physical cosmology is the study of the formation and evolution of the universe on its largest scales. Albert Einstein's theory of relativity plays a central role in all modern cosmological theories. In the early 20th century, Hubble's discovery that the universe is expanding, as shown by the Hubble diagram, prompted rival explanations known as the steady state universe and the Big Bang.

The Big Bang was confirmed by the success of Big Bang nucleosynthesis and the discovery of the cosmic microwave background in 1964. The Big Bang model rests on two theoretical pillars: Albert Einstein's general relativity and the cosmological principle. Cosmologists have recently established the ΛCDM model of the evolution of the universe, which includes cosmic inflation, dark energy, and dark matter.

Numerous possibilities and discoveries are anticipated to emerge from new data from the Fermi Gamma-ray Space Telescope over the upcoming decade and vastly revise or clarify existing models of the universe. In particular, the potential for a tremendous discovery surrounding dark matter is possible over the next several years. Fermi will search for evidence that dark matter is composed of weakly interacting massive particles, complementing similar experiments with the Large Hadron Collider and other underground detectors.

IBEX is already yielding new astrophysical discoveries: "No one knows what is creating the ENA (energetic neutral atoms) ribbon" along the termination shock of the solar wind, "but everyone agrees that it means the textbook picture of the heliosphere—in which the Solar System's enveloping pocket filled with the solar wind's charged particles is plowing through the onrushing 'galactic wind' of the interstellar medium in the shape of a comet—is wrong."

Current Research

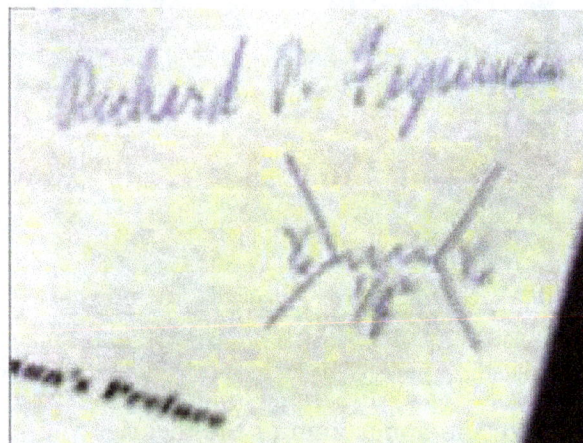

Feynman diagram signed by R. P. Feynman.

A typical event described by physics: a magnet levitating above a superconductor demonstrates the Meissner effect.

Research in physics is continually progressing on a large number of fronts.

In condensed matter physics, an important unsolved theoretical problem is that of high-temperature superconductivity. Many condensed matter experiments are aiming to fabricate workable spintronics and quantum computers.

In particle physics, the first pieces of experimental evidence for physics beyond the Standard Model have begun to appear. Foremost among these are indications that neutrinos have non-zero mass. These experimental results appear to have solved the long-standing solar neutrino problem, and the physics of massive neutrinos remains an area of active theoretical and experimental research. Large Hadron Collider had already found the Higgs Boson. Future research aims to prove or disprove the supersymmetry, which extends the Standard Model of particle physics. The research on dark matter and dark energy is also on the agenda.

Theoretical attempts to unify quantum mechanics and general relativity into a single theory of quantum gravity, a program ongoing for over half a century, have not yet been decisively resolved. The current leading candidates are M-theory, superstring theory and loop quantum gravity.

Many astronomical and cosmological phenomena have yet to be satisfactorily explained, including the existence of ultra-high energy cosmic rays, the baryon asymmetry, the acceleration of the universe and the anomalous rotation rates of galaxies.

Although much progress has been made in high-energy, quantum, and astronomical physics, many everyday phenomena involving complexity, chaos, or turbulence are still poorly understood. Complex problems that seem like they could be solved by a clever application of dynamics and mechanics remain unsolved; examples include the formation of sandpiles, nodes in trickling water, the shape of water droplets, mechanisms of surface tension catastrophes, and self-sorting in shaken heterogeneous collections.

These complex phenomena have received growing attention since the 1970s for several reasons, including the availability of modern mathematical methods and computers, which enabled complex systems to be modeled in new ways. Complex physics has become part of increasingly interdisciplinary research, as exemplified by the study of turbulence in aerodynamics and the observation of pattern formation in biological systems. In the 1932 *Annual Review of Fluid Mechanics*, Horace Lamb said:

I am an old man now, and when I die and go to heaven there are two matters on which I hope for enlightenment. One is quantum electrodynamics, and the other is the turbulent motion of fluids. And about the former I am rather optimistic.

References

- Weinberg, S. (1993). Dreams of a Final Theory: The Search for the Fundamental Laws of Nature. Hutchinson Radius. ISBN 0-09-177395-4.

- "Phenomenology". Max Planck Institute for Physics. Archived from the original on 7 March 2016. Retrieved 22 October 2016.

- Gibney, E. (2015). "LHC 2.0: A new view of the Universe". Nature. 519 (7542): 142–143. Bibcode:2015Natur.519..142G. doi:10.1038/519142a.

- "Applications of Mathematics to the Sciences". 25 January 2000. Archived from the original on 2015-05-10. Retrieved 30 January 2012.

- Ellis, G.; Silk, J. (16 December 2014). "Scientific method: Defend the integrity of physics". Nature. 516 (7531): 321–323. Bibcode:2014Natur.516..321E. doi:10.1038/516321a.

- "Physics is an experimental science. Physicists observe the phenomena of nature and try to find patterns that relate these phenomena."Young & Freedman 2014, p. 2

Branches of Physics

Physics has a number of branches; some of these are particle physics, condensed matter physics, electromagnetism and quantum mechanics. The branch of physics that studies the nature of particles is termed as particle physics whereas astrophysics is the branch that studies heavenly bodies and their positions in space. This text is a compilation of the various branches of physics that form an integral part of the broader subject matter.

Particle Physics

Particle physics (also high energy physics) is the branch of physics that studies the nature of the particles that constitute *matter* (particles with mass) and *radiation* (massless particles). Although the word "particle" can refer to various types of very small objects (e.g. protons, gas particles, or even household dust), "particle physics" usually investigates the irreducibly smallest detectable particles and the irreducibly fundamental force fields necessary to explain them. By our current understanding, these elementary particles are excitations of the quantum fields that also govern their interactions. The currently dominant theory explaining these fundamental particles and fields, along with their dynamics, is called the Standard Model. Thus, modern particle physics generally investigates the Standard Model and its various possible extensions, e.g. to the newest "known" particle, the Higgs boson, or even to the oldest known force field, gravity.

Subatomic Particles

The particle content of the Standard Model of Physics

Elementary Particles					
	Types	**Generations**	**Antiparticle**	**Colours**	**Total**
Quarks	2	3	Pair	3	36
Leptons			Pair	None	12
Gluons	1	1	Own	8	8
Photon	Own				1
Z Boson			Own	None	1
W Boson			Pair		2
Higgs			Own		1
Total number of (known) elementary particles:					61

Modern particle physics research is focused on subatomic particles, including atomic constituents such as electrons, protons, and neutrons (protons and neutrons are composite particles called baryons, made of quarks), produced by radioactive and scattering processes, such as photons, neutrinos, and muons, as well as a wide range of exotic particles. Dynamics of particles is also governed by quantum mechanics; they exhibit wave–particle duality, displaying particle-like behaviour under certain experimental conditions and wave-like behaviour in others. In more technical terms, they are described by quantum state vectors in a Hilbert space, which is also treated in quantum field theory. Following the convention of particle physicists, the term *elementary particles* is applied to those particles that are, according to current understanding, presumed to be indivisible and not composed of other particles.

All particles and their interactions observed to date can be described almost entirely by a quantum field theory called the Standard Model. The Standard Model, as currently formulated, has 61 elementary particles. Those elementary particles can combine to form composite particles, accounting for the hundreds of other species of particles that have been discovered since the 1960s. The Standard Model has been found to agree with almost all the experimental tests conducted to date. However, most particle physicists believe that it is an incomplete description of nature and that a more fundamental theory awaits discovery. In recent years, measurements of neutrino mass have provided the first experimental deviations from the Standard Model.

History

The idea that all matter is composed of elementary particles dates to at least the 6th century BC. In the 19th century, John Dalton, through his work on stoichiometry, concluded that each element of nature was composed of a single, unique type of particle. The word *atom*, after the Greek word *atomos* meaning "indivisible", denotes the smallest particle of a chemical element since then, but physicists soon discovered that *atoms* are not, in fact, the fundamental particles of nature, but conglomerates of even smaller particles, such as the electron. The early 20th-century explorations of nuclear physics and quantum physics culminated in proofs of nuclear fission in 1939 by Lise Meitner (based on experiments by Otto Hahn), and nuclear fusion by Hans Bethe in that same year; both discoveries also led to the development of nuclear weapons. Throughout the 1950s and 1960s, a bewildering variety of particles were found in scattering experiments. It was referred to as the "particle zoo". That term was deprecated after the formulation of the Standard Model during the 1970s in which the large number of particles was explained as combinations of a (relatively) small number of fundamental particles.

Standard Model

The current state of the classification of all elementary particles is explained by the Standard Model. It describes the strong, weak, and electromagnetic fundamental interactions, using mediating gauge bosons. The species of gauge bosons are the gluons, W−, W+and Zbosons, and the photons. The Standard Model also contains 24 fundamental particles, (12 particles and their associated anti-particles), which are the constituents of all matter. Finally, the Standard Model also predicted the existence of a type of boson known as the Higgs boson. Early in the morning on 4 July 2012, physicists with the Large Hadron Collider at CERN announced they had found a new particle that behaves similarly to what is expected from the Higgs boson.

Experimental Laboratories

In particle physics, the major international laboratories are located at the:

- Brookhaven National Laboratory (Long Island, United States). Its main facility is the Relativistic Heavy Ion Collider (RHIC), which collides heavy ions such as gold ions and polarized protons. It is the world's first heavy ion collider, and the world's only polarized proton collider.

- Budker Institute of Nuclear Physics (Novosibirsk, Russia). Its main projects are now the electron-positron colliders VEPP-2000, operated since 2006, and VEPP-4, started experiments in 1994. Earlier facilities include the first electron-electron beam-beam collider VEP-1, which conducted experiments from 1964 to 1968; the electron-positron colliders VEPP-2, operated from 1965 to 1974; and, its successor VEPP-2M, performed experiments from 1974 to 2000.

- CERN, (Conseil Européen pour la Recherche Nucléaire) (Franco-Swiss border, near Geneva). Its main project is now the Large Hadron Collider (LHC), which had its first beam circulation on 10 September 2008, and is now the world's most energetic collider of protons. It also became the most energetic collider of heavy ions after it began colliding lead ions. Earlier facilities include the Large Electron–Positron Collider (LEP), which was stopped on 2 November 2000 and then dismantled to give way for LHC; and the Super Proton Synchrotron, which is being reused as a pre-accelerator for the LHC.

- DESY (Deutsches Elektronen-Synchrotron) (Hamburg, Germany). Its main facility is the Hadron Elektron Ring Anlage (HERA), which collides electrons and positrons with protons.

- Fermi National Accelerator Laboratory (Fermilab), (Batavia, United States). Its main facility until 2011 was the Tevatron, which collided protons and antiprotons and was the highest-energy particle collider on earth until the Large Hadron Collider surpassed it on 29 November 2009.

- Institute of High Energy Physics (IHEP), (Beijing, China). IHEP manages a number of China's major particle physics facilities, including the Beijing Electron Positron Collider (BEPC), the Beijing Spectrometer (BES), the Beijing Synchrotron Radiation Facility (BSRF), the International Cosmic-Ray Observatory at Yangbajing in Tibet, the Daya Bay Reactor Neutrino Experiment, the China Spallation Neutron Source, the Hard X-ray Modulation Telescope (HXMT), and the Accelerator-driven Sub-critical System (ADS) as well as the Jiangmen Underground Neutrino Observatory (JUNO).

- KEK, (Tsukuba, Japan). It is the home of a number of experiments such as the K2K experiment, a neutrino oscillation experiment and Belle, an experiment measuring the CP violation of B mesons.

- SLAC National Accelerator Laboratory, (Menlo Park, United States). Its 2-mile-long linear particle accelerator began operating in 1962 and was the basis for numerous electron and positron collision experiments until 2008. Since then the linear accelerator is being used for the Linac Coherent Light Source X-ray laser as well as advanced accelerator design research. SLAC staff continue to participate in developing and building many particle physics experiments around the world.

Many other particle accelerators do exist.

The techniques required to do modern, experimental, particle physics are quite varied and complex, constituting a sub-specialty nearly completely distinct from the theoretical side of the field.

Theory

Theoretical particle physics attempts to develop the models, theoretical framework, and mathematical tools to understand current experiments and make predictions for future experiments. There are several major interrelated efforts being made in theoretical particle physics today. One important branch attempts to better understand the Standard Model and its tests. By extracting the parameters of the Standard Model, from experiments with less uncertainty, this work probes the limits of the Standard Model and therefore expands our understanding of nature's building blocks. Those efforts are made challenging by the difficulty of calculating quantities in quantum chromodynamics. Some theorists working in this area refer to themselves as phenomenologists and they may use the tools of quantum field theory and effective field theory. Others make use of lattice field theory and call themselves *lattice theorists*.

Another major effort is in model building where model builders develop ideas for what physics may lie beyond the Standard Model (at higher energies or smaller distances). This work is often motivated by the hierarchy problem and is constrained by existing experimental data. It may involve work on supersymmetry, alternatives to the Higgs mechanism, extra spatial dimensions (such as the Randall-Sundrum models), Preon theory, combinations of these, or other ideas.

A third major effort in theoretical particle physics is string theory. *String theorists* attempt to construct a unified description of quantum mechanics and general relativity by building a theory based on small strings, and branes rather than particles. If the theory is successful, it may be considered a "Theory of Everything", or "TOE".

There are also other areas of work in theoretical particle physics ranging from particle cosmology to loop quantum gravity.

This division of efforts in particle physics is reflected in the names of categories on the arXiv, a preprint archive: hep-th (theory), hep-ph (phenomenology), hep-ex (experiments), hep-lat (lattice gauge theory).

Practical Applications

In principle, all physics (and practical applications developed there from) can be derived from the study of fundamental particles. In practice, even if "particle physics" is taken to mean only "high-energy atom smashers", many technologies have been developed during these pioneering investigations that later find wide uses in society. Cyclotrons are used to produce medical isotopes for research and treatment (for example, isotopes used in PET imaging), or used directly for certain cancer treatments. The development of Superconductors has been pushed forward by their use in particle physics. The World Wide Web and touchscreen technology were initially developed at CERN.

Additional applications are found in medicine, national security, industry, computing, science, and workforce development, illustrating a long and growing list of beneficial practical applications with contributions from particle physics.

Future

The primary goal, which is pursued in several distinct ways, is to find and understand what physics may lie beyond the standard model. There are several powerful experimental reasons to expect new physics, including dark matter and neutrino mass. There are also theoretical hints that this new physics should be found at accessible energy scales.

Much of the effort to find this new physics are focused on new collider experiments. The Large Hadron Collider (LHC) was completed in 2008 to help continue the search for the Higgs boson, supersymmetric particles, and other new physics. An intermediate goal is the construction of the International Linear Collider (ILC), which will complement the LHC by allowing more precise measurements of the properties of newly found particles. In August 2004, a decision for the technology of the ILC was taken but the site has still to be agreed upon.

In addition, there are important non-collider experiments that also attempt to find and understand physics beyond the Standard Model. One important non-collider effort is the determination of the neutrino masses, since these masses may arise from neutrinos mixing with very heavy particles. In addition, cosmological observations provide many useful constraints on the dark matter, although it may be impossible to determine the exact nature of the dark matter without the colliders. Finally, lower bounds on the very long lifetime of the proton put constraints on Grand Unified Theories at energy scales much higher than collider experiments will be able to probe any time soon.

In May 2014, the Particle Physics Project Prioritization Panel released its report on particle physics funding priorities for the United States over the next decade. This report emphasized continued U.S. participation in the LHC and ILC, and expansion of the Long Baseline Neutrino Experiment, among other recommendations.

Astrophysics

Astrophysics is the branch of astronomy that employs the principles of physics and chemistry "to ascertain the nature of the heavenly bodies, rather than their positions or motions in space." Among the objects studied are the Sun, other stars, galaxies, extrasolar planets, the interstellar medium and the cosmic microwave background. Their emissions are examined across all parts of the electromagnetic spectrum, and the properties examined include luminosity, density, temperature, and chemical composition. Because astrophysics is a very broad subject, *astrophysicists* typically apply many disciplines of physics, including mechanics, electromagnetism, statistical mechanics, thermodynamics, quantum mechanics, relativity, nuclear and particle physics, and atomic and molecular physics.

In practice, modern astronomical research often involves a substantial amount of work in the realms of theoretical and observational physics. Some areas of study for astrophysicists include their attempts to determine: the properties of dark matter, dark energy, and black holes; whether or not time travel is possible, wormholes can form, or the multiverse exists; and the origin and ultimate fate of the universe. Topics also studied by theoretical astrophysicists include: Solar System formation and evolution; stellar dynamics and evolution; galaxy formation and evolution; magnetohydrodynamics; large-scale structure of matter in the universe; origin of cosmic rays; general relativity and physical cosmology, including string cosmology and astroparticle physics.

History

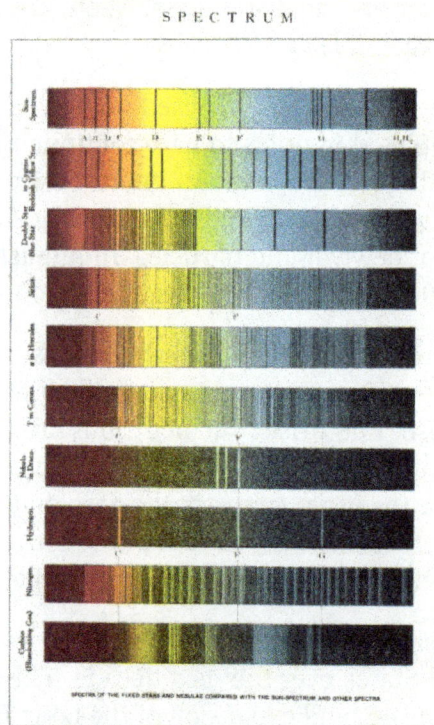

Early 20th-century comparison of elemental, solar, and stellar spectra

Although astronomy is as ancient as recorded history itself, it was long separated from the study of terrestrial physics. In the Aristotelian worldview, bodies in the sky appeared to be unchanging spheres whose only motion was uniform motion in a circle, while the earthly world was the realm which underwent growth and decay and in which natural motion was in a straight line and ended when the moving object reached its goal. Consequently, it was held that the celestial region was made of a fundamentally different kind of matter from that found in the terrestrial sphere; either Fire as maintained by Plato, or Aether as maintained by Aristotle. During the 17th century, natural philosophers such as Galileo, Descartes, and Newton began to maintain that the celestial and terrestrial regions were made of similar kinds of material and were subject to the same natural laws. Their challenge was that the tools had not yet been invented with which to prove these assertions.

For much of the nineteenth century, astronomical research was focused on the routine work of measuring the positions and computing the motions of astronomical objects. A new astronomy, soon to be called astrophysics, began to emerge when William Hyde Wollaston and Joseph von Fraunhofer independently discovered that, when decomposing the light from the Sun, a multitude of dark lines (regions where there was less or no light) were observed in the spectrum. By 1860 the physicist, Gustav Kirchhoff, and the chemist, Robert Bunsen, had demonstrated that the dark lines in the solar spectrum corresponded to bright lines in the spectra of known gases, specific lines corresponding to unique chemical elements. Kirchhoff deduced that the dark lines in the solar spectrum are caused by absorption by chemical elements in the Solar atmosphere. In this way it was proved that the chemical elements found in the Sun and stars were also found on Earth.

Among those who extended the study of solar and stellar spectra was Norman Lockyer, who in 1868 detected bright, as well as dark, lines in solar spectra. Working with the chemist, Edward Frankland, to investigate the spectra of elements at various temperatures and pressures, he could not associate a yellow line in the solar spectrum with any known elements. He thus claimed the line represented a new element, which was called helium, after the Greek Helios, the Sun personified.

In 1885, Edward C. Pickering undertook an ambitious program of stellar spectral classification at Harvard College Observatory, in which a team of woman computers, notably Williamina Fleming, Antonia Maury, and Annie Jump Cannon, classified the spectra recorded on photographic plates. By 1890, a catalog of over 10,000 stars had been prepared that grouped them into thirteen spectral types. Following Pickering's vision, by 1924 Cannon expanded the catalog to nine volumes and over a quarter of a million stars, developing the Harvard Classification Scheme which was accepted for worldwide use in 1922.

In 1895, George Ellery Hale and James E. Keeler, along with a group of ten associate editors from Europe and the United States, established. It was intended that the journal would fill the gap between journals in astronomy and physics, providing a venue for publication of articles on astronomical applications of the spectroscope; on laboratory research closely allied to astronomical physics, including wavelength determinations of metallic and gaseous spectra and experiments on radiation and absorption; on theories of the Sun, Moon, planets, comets, meteors, and nebulae; and on instrumentation for telescopes and laboratories.

In 1925 Cecilia Helena Payne (later Cecilia Payne-Gaposchkin) wrote an influential doctoral dissertation at Radcliffe College, in which she applied ionization theory to stellar atmospheres to relate the spectral classes to the temperature of stars. Most significantly, she discovered that hydrogen and helium were the principal components of stars. This discovery was so unexpected that her dissertation readers convinced her to modify the conclusion before publication. However, later research confirmed her discovery.

By the end of the 20th century, further study of stellar and experimental spectra advanced, particularly as a result of the advent of quantum physics.

Observational Astrophysics

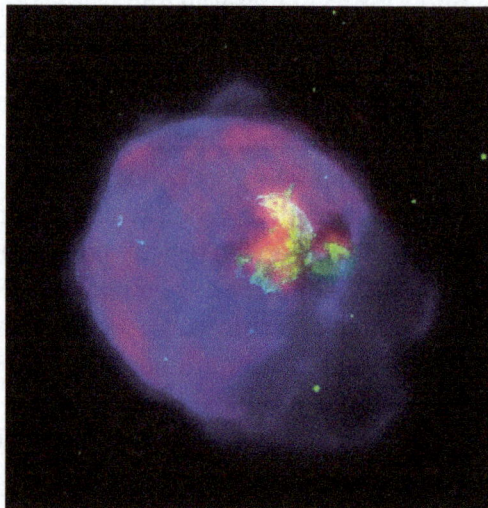

Supernova remnant LMC N 63A imaged in x-ray (blue), optical (green) and radio (red) wavelengths. The X-ray glow is from material heated to about ten million degrees Celsius by a shock wave generated by the supernova explosion.

Observational astronomy is a division of the astronomical science that is concerned with recording data, in contrast with theoretical astrophysics, which is mainly concerned with finding out the measurable implications of physical models. It is the practice of observing celestial objects by using telescopes and other astronomical apparatus.

The majority of astrophysical observations are made using the electromagnetic spectrum.

- Radio astronomy studies radiation with a wavelength greater than a few millimeters. Example areas of study are radio waves, usually emitted by cold objects such as interstellar gas and dust clouds; the cosmic microwave background radiation which is the redshifted light from the Big Bang; pulsars, which were first detected at microwave frequencies. The study of these waves requires very large radio telescopes.

- Infrared astronomy studies radiation with a wavelength that is too long to be visible to the naked eye but is shorter than radio waves. Infrared observations are usually made with telescopes similar to the familiar optical telescopes. Objects colder than stars (such as planets) are normally studied at infrared frequencies.

- Optical astronomy is the oldest kind of astronomy. Telescopes paired with a charge-coupled device or spectroscopes are the most common instruments used. The Earth's atmosphere interferes somewhat with optical observations, so adaptive optics and space telescopes are used to obtain the highest possible image quality. In this wavelength range, stars are highly visible, and many chemical spectra can be observed to study the chemical composition of stars, galaxies and nebulae.

- Ultraviolet, X-ray and gamma ray astronomy study very energetic processes such as binary pulsars, black holes, magnetars, and many others. These kinds of radiation do not penetrate the Earth's atmosphere well. There are two methods in use to observe this part of the electromagnetic spectrum—space-based telescopes and ground-based imaging air Cherenkov telescopes (IACT). Examples of Observatories of the first type are RXTE, the Chandra

X-ray Observatory and the Compton Gamma Ray Observatory. Examples of IACTs are the High Energy Stereoscopic System (H.E.S.S.) and the MAGIC telescope.

Other than electromagnetic radiation, few things may be observed from the Earth that originate from great distances. A few gravitational wave observatories have been constructed, but gravitational waves are extremely difficult to detect. Neutrino observatories have also been built, primarily to study our Sun. Cosmic rays consisting of very high energy particles can be observed hitting the Earth's atmosphere.

Observations can also vary in their time scale. Most optical observations take minutes to hours, so phenomena that change faster than this cannot readily be observed. However, historical data on some objects is available, spanning centuries or millennia. On the other hand, radio observations may look at events on a millisecond timescale (millisecond pulsars) or combine years of data (pulsar deceleration studies). The information obtained from these different timescales is very different.

The study of our very own Sun has a special place in observational astrophysics. Due to the tremendous distance of all other stars, the Sun can be observed in a kind of detail unparalleled by any other star. Our understanding of our own Sun serves as a guide to our understanding of other stars.

The topic of how stars change, or stellar evolution, is often modeled by placing the varieties of star types in their respective positions on the Hertzsprung–Russell diagram, which can be viewed as representing the state of a stellar object, from birth to destruction.

Theoretical Astrophysics

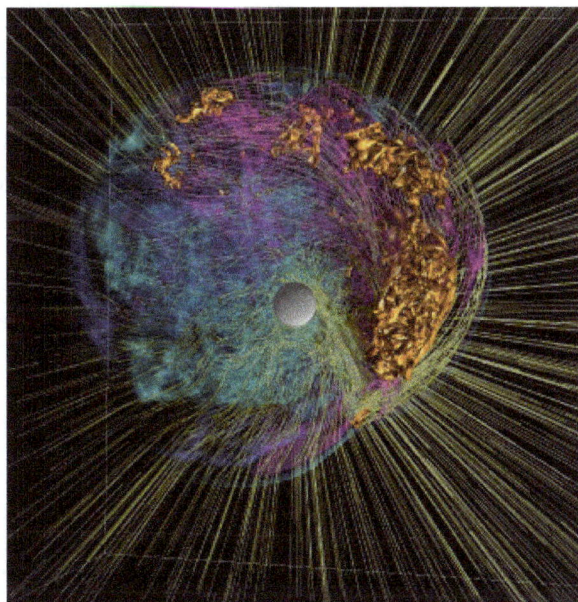

Stream lines on this simulation of a supernova show the flow of matter behind
the shock wave giving clues as to the origin of pulsars

Theoretical astrophysicists use a wide variety of tools which include analytical models (for example, polytropes to approximate the behaviors of a star) and computational numerical simulations. Each has some advantages. Analytical models of a process are generally better for giving insight into the heart of what is going on. Numerical models can reveal the existence of phenomena and effects that would otherwise not be seen.

Theorists in astrophysics endeavor to create theoretical models and figure out the observational consequences of those models. This helps allow observers to look for data that can refute a model or help in choosing between several alternate or conflicting models.

Theorists also try to generate or modify models to take into account new data. In the case of an inconsistency, the general tendency is to try to make minimal modifications to the model to fit the data. In some cases, a large amount of inconsistent data over time may lead to total abandonment of a model.

Topics studied by theoretical astrophysicists include: stellar dynamics and evolution; galaxy formation and evolution; magnetohydrodynamics; large-scale structure of matter in the universe; origin of cosmic rays; general relativity and physical cosmology, including string cosmology and astroparticle physics. Astrophysical relativity serves as a tool to gauge the properties of large scale structures for which gravitation plays a significant role in physical phenomena investigated and as the basis for black hole (*astro*)physics and the study of gravitational waves.

Some widely accepted and studied theories and models in astrophysics, now included in the Lambda-CDM model, are the Big Bang, cosmic inflation, dark matter, dark energy and fundamental theories of physics. Wormholes are examples of hypotheses which are yet to be proven (or disproven).

Popularization

The roots of astrophysics can be found in the seventeenth century emergence of a unified physics, in which the same laws applied to the celestial and terrestrial realms. There were scientists who were qualified in both physics and astronomy who laid the firm foundation for the current science of astrophysics. In modern times, students continue to be drawn to astrophysics due to its popularization by the Royal Astronomical Society and notable educators such as prominent professors Subrahmanyan Chandrasekhar, Stephen Hawking, Hubert Reeves, Carl Sagan and Neil deGrasse Tyson. The efforts of the early, late, and present scientists continue to attract young people to study the history and science of astrophysics.

Condensed Matter Physics

Condensed matter physics is a branch of physics that deals with the physical properties of condensed phases of matter. Condensed matter physicists seek to understand the behavior of these phases by using physical laws. In particular, they include the laws of quantum mechanics, electromagnetism and statistical mechanics.

The most familiar condensed phases are solids and liquids while more exotic condensed phases include the superconducting phase exhibited by certain materials at low temperature, the ferromagnetic and antiferromagnetic phases of spins on atomic lattices, and the Bose–Einstein condensate found in cold atomic systems. The study of condensed matter physics involves measuring various material properties via experimental probes along with using techniques of theoretical physics to develop mathematical models that help in understanding physical behavior.

The diversity of systems and phenomena available for study makes condensed matter physics the most active field of contemporary physics: one third of all American physicists identify themselves as condensed matter physicists, and the Division of Condensed Matter Physics is the largest division at the American Physical Society. The field overlaps with chemistry, materials science, and nanotechnology, and relates closely to atomic physics and biophysics. Theoretical condensed matter physics shares important concepts and techniques with theoretical particle and nuclear physics.

A variety of topics in physics such as crystallography, metallurgy, elasticity, magnetism, etc., were treated as distinct areas until the 1940s, when they were grouped together as *solid state physics*. Around the 1960s, the study of physical properties of liquids was added to this list, forming the basis for the new, related specialty of condensed matter physics. According to physicist Philip Warren Anderson, the term was coined by him and Volker Heine, when they changed the name of their group at the Cavendish Laboratories, Cambridge from "Solid state theory" to "Theory of Condensed Matter" in 1967, as they felt it did not exclude their interests in the study of liquids, nuclear matter and so on. Although Anderson and Heine helped popularize the name "condensed matter", it had been present in Europe for some years, most prominently in the form of a journal published in English, French, and German by Springer-Verlag titled *Physics of Condensed Matter*, which was launched in 1963. The funding environment and Cold War politics of the 1960s and 1970s were also factors that lead some physicists to prefer the name "condensed matter physics", which emphasized the commonality of scientific problems encountered by physicists working on solids, liquids, plasmas, and other complex matter, over "solid state physics", which was often associated with the industrial applications of metals and semiconductors. The Bell Telephone Laboratories was one of the first institutes to conduct a research program in condensed matter physics.

References to "condensed" state can be traced to earlier sources. For example, in the introduction to his 1947 book "Kinetic Theory of Liquids", Yakov Frenkel proposed that "The kinetic theory of liquids must accordingly be developed as a generalization and extension of the kinetic theory of solid bodies. As a matter of fact, it would be more correct to unify them under the title of 'condensed bodies'".

History

Classical Physics

One of the first studies of condensed states of matter was by English chemist Humphry Davy, in the first decades of the nineteenth century. Davy observed that of the forty chemical elements known at the time, twenty-six had metallic properties such as lustre, ductility and high electrical and thermal conductivity. This indicated that the atoms in Dalton's atomic theory were not indivisible as Dalton claimed, but had inner structure. Davy further claimed that elements that were then believed to be gases, such as nitrogen and hydrogen could be liquefied under the right conditions and would then behave as metals.

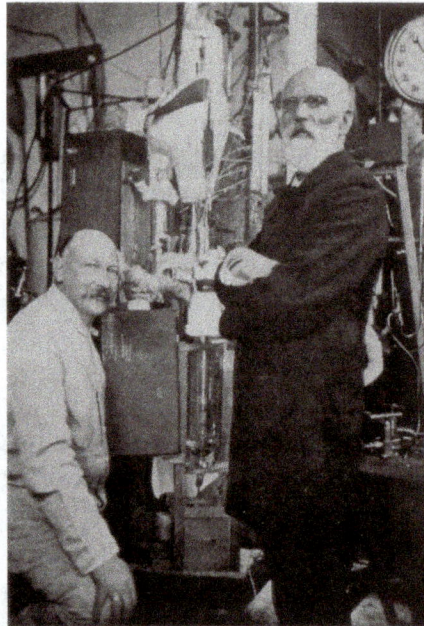

Heike Kamerlingh Onnes and Johannes van der Waals with the helium "liquefactor" in Leiden (1908)

In 1823, Michael Faraday, then an assistant in Davy's lab, successfully liquefied chlorine and went on to liquefy all known gaseous elements, with the exception of nitrogen, hydrogen and oxygen. Shortly after, in 1869, Irish chemist Thomas Andrews studied the phase transition from a liquid to a gas and coined the term critical point to describe the condition where a gas and a liquid were indistinguishable as phases, and Dutch physicist Johannes van der Waals supplied the theoretical framework which allowed the prediction of critical behavior based on measurements at much higher temperatures. By 1908, James Dewar and H. Kamerlingh Onnes were successfully able to liquefy hydrogen and then newly discovered helium, respectively.

Paul Drude in 1900 proposed the first theoretical model for a classical electron moving through a metallic solid. Drude's model described properties of metals in terms of a gas of free electrons, and was the first microscopic model to explain empirical observations such as the Wiedemann–Franz law. However, despite the success of Drude's free electron model, it had one notable problem, in that it was unable to correctly explain the electronic contribution to the specific heat and magnetic properties of metals, as well as the temperature dependence of resistivity at low temperatures.

In 1911, three years after helium was first liquefied, Onnes working at University of Leiden discovered superconductivity in mercury, when he observed the electrical resistivity of mercury to vanish at temperatures below a certain value. The phenomenon completely surprised the best theoretical physicists of the time, and it remained unexplained for several decades. Albert Einstein, in 1922, said regarding contemporary theories of superconductivity that "with our far-reaching ignorance of the quantum mechanics of composite systems we are very far from being able to compose a theory out of these vague ideas".

Advent of Quantum Mechanics

Drude's classical model was augmented by Wolfgang Pauli, Arnold Sommerfeld, Felix Bloch and other physicists. Pauli realized that the free electrons in metal must obey the Fermi–Dirac statistics. Using this idea, he developed the theory of paramagnetism in 1926. Shortly after, Sommerfeld incorporated the Fermi–Dirac statistics into the free electron model and made it better able to explain the heat capacity. Two years later, Bloch used quantum mechanics to describe the motion of a quantum electron in a periodic lattice. The mathematics of crystal structures developed by Auguste Bravais, Yevgraf Fyodorov and others was used to classify crystals by their symmetry group, and tables of crystal structures were the basis for the series *International Tables of Crystallography*, first published in 1935. Band structure calculations was first used in 1930 to predict the properties of new materials, and in 1947 John Bardeen, Walter Brattain and William Shockley developed the first semiconductor-based transistor, heralding a revolution in electronics.

A replica of the first point-contact transistor in Bell labs

In 1879, Edwin Herbert Hall working at the Johns Hopkins University discovered the development of a voltage across conductors transverse to an electric current in the conductor and magnetic field perpendicular to the current. This phenomenon arising due to the nature of charge carriers in the conductor came to be known as the Hall effect, but it was not properly explained at the time, since the electron was experimentally discovered 18 years later. After the advent of quantum mechanics, Lev Landau in 1930 developed the theory of landau quantization and laid the foundation for the theoretical explanation for the quantum hall effect discovered half a century later.

Magnetism as a property of matter has been known in China since 4000BC. However, the first modern studies of magnetism only started with the development of electrodynamics by Faraday, Maxwell and others in the nineteenth century, which included the classification of materials as ferromagnetic, paramagnetic and diamagnetic based on their response to magnetization. Pierre Curie studied the dependence of magnetization on temperature and discovered the Curie point phase transition in ferromagnetic materials. In 1906, Pierre Weiss introduced the concept of magnetic domains to explain the main properties of ferromagnets. The first attempt at a microscopic description of magnetism was by Wilhelm Lenz and Ernst Ising through the Ising model that described magnetic materials as consisting of a periodic lattice of spins that collectively acquired magnetization. The Ising model was solved exactly to show that spontaneous magnetization cannot occur in one dimension but is possible in higher-dimensional lattices. Further research such as by Bloch on spin waves and Néel on antiferromagnetism led to the development of new magnetic materials with applications to magnetic storage devices.

Modern Many-body Physics

The Sommerfeld model and spin models for ferromagnetism illustrated the successful application of quantum mechanics to condensed matter problems in the 1930s. However, there still were several unsolved problems, most notably the description of superconductivity and the Kondo effect. After World War II, several ideas from quantum field theory were applied to condensed matter problems. These included recognition of collective modes of excitation of solids and the important notion of a quasiparticle. Russian physicist Lev Landau used the idea for the Fermi liquid theory wherein low energy properties of interacting fermion systems were given in terms of what are now known as Landau-quasiparticles. Landau also developed a mean field theory for continuous phase transitions, which described ordered phases as spontaneous breakdown of symmetry. The theory also introduced the notion of an order parameter to distinguish between ordered phases. Eventually in 1965, John Bardeen, Leon Cooper and John Schrieffer developed the so-called BCS theory of superconductivity, based on the discovery that arbitrarily small attraction between two electrons of opposite spin mediated by phonons in the lattice can give rise to a bound state called a Cooper pair.

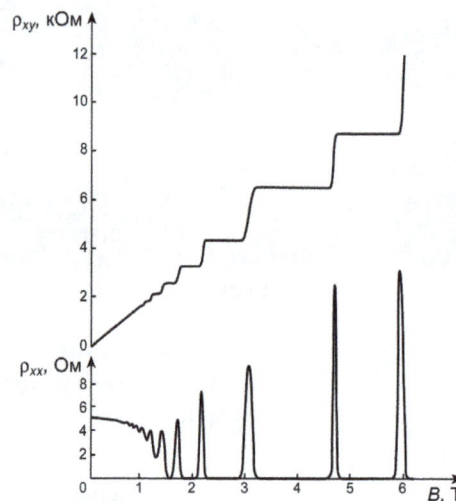

The quantum Hall effect: Components of the Hall resistivity as a function of the external magnetic field.

The study of phase transition and the critical behavior of observables, known as critical phenomena, was a major field of interest in the 1960s. Leo Kadanoff, Benjamin Widom and Michael Fisher developed the ideas of critical exponents and scaling. These ideas were unified by Kenneth Wilson in 1972, under the formalism of the renormalization group in the context of quantum field theory.

The Landau quasiparticles possessed their second wind in the field of strongly correlated materials with specific quantum critical point represented by fermion condensation quantum phase transition that supports quasiparticles. This phase transition does preserve the Pomeranchuk stability conditions. These unique properties of the phase transition allow one to explain both the scaling and the non-Fermi liquid behaviour observed in strongly correlated materials.

The quantum Hall effect was discovered by Klaus von Klitzing in 1980 when he observed the Hall conductance to be integer multiples of a fundamental constant e^2/h. The effect was observed to be independent of parameters such as the system size and impurities. In 1981, theorist Robert Laughlin proposed a theory explaining the unanticipated precision of the integral plateau. It also implied that the Hall conductance can be characterized in terms of a topological invariable called Chern number. Shortly after, in 1982, Horst Störmer and Daniel Tsui observed the fractional quantum Hall effect where the conductance was now a rational multiple of a constant. Laughlin, in 1983, realized that this was a consequence of quasiparticle interaction in the Hall states and formulated a variational solution, known as the Laughlin wavefunction. The study of topological properties of the fractional Hall effect remains an active field of research.

In 1986, Karl Müller and Johannes Bednorz discovered the first high temperature superconductor, a material which was superconducting at temperatures as high as 50 Kelvin. It was realized that the high temperature superconductors are examples of strongly correlated materials where the electron–electron interactions play an important role. A satisfactory theoretical description of high-temperature superconductors is still not known and the field of strongly correlated materials continues to be an active research topic.

In 2009, David Field and researchers at Aarhus University discovered spontaneous electric fields when creating prosaic films of various gases. This has more recently expanded to form the research area of spontelectrics.

In 2012 several groups released preprints which suggest that samarium hexaboride has the properties of a topological insulator in accordance with the earlier theoretical predictions. Since samarium hexaboride is an established Kondo insulator, i.e. a strongly correlated electron material, the existence of a topological surface state in this material would lead to a topological insulator with strong electronic correlations.

Theoretical

Theoretical condensed matter physics involves the use of theoretical models to understand properties of states of matter. These include models to study the electronic properties of solids, such as the Drude model, the Band structure and the density functional theory. Theoretical models have also been developed to study the physics of phase transitions, such as the Ginzburg–Landau theory, critical exponents and the use of mathematical techniques of quantum field theory and the renormalization group. Modern theoretical studies involve the use of numerical computation of electronic structure and mathematical tools to understand phenomena such as high-temperature superconductivity, topological phases and gauge symmetries.

Emergence

Theoretical understanding of condensed matter physics is closely related to the notion of emergence, wherein complex assemblies of particles behave in ways dramatically different from their individual constituents. For example, a range of phenomena related to high temperature superconductivity are not well understood, although the microscopic physics of individual electrons and lattices is well known. Similarly, models of condensed matter systems have been studied where collective excitations behave like photons and electrons, thereby describing electromagnetism as an emergent phenomenon. Emergent properties can also occur at the interface between materials: one example is the lanthanum-aluminate-strontium-titanate interface, where two non-magnetic insulators are joined to create conductivity, superconductivity, and ferromagnetism.

Electronic Theory of Solids

The metallic state has historically been an important building block for studying properties of solids. The first theoretical description of metals was given by Paul Drude in 1900 with the Drude model, which explained electrical and thermal properties by describing a metal as an ideal gas of then-newly discovered electrons. He was able to derive the empirical Wiedemann-Franz law and get results in close agreement with the experiments. This classical model was then improved by Arnold Sommerfeld who incorporated the Fermi–Dirac statistics of electrons and was able to explain the anomalous behavior of the specific heat of metals in the Wiedemann–Franz law. In 1912, The structure of crystalline solids was studied by Max von Laue and Paul Knipping, when they observed the X-ray diffraction pattern of crystals, and concluded that crystals get their structure from periodic lattices of atoms. In 1928, Swiss physicist Felix Bloch provided a wave function solution to the Schrödinger equation with a periodic potential, called the Bloch wave.

Calculating electronic properties of metals by solving the many-body wavefunction is often computationally hard, and hence, approximation techniques are necessary to obtain meaningful predictions. The Thomas–Fermi theory, developed in the 1920s, was used to estimate system energy and electronic density by treating the local electron density as a variational parameter. Later in the 1930s, Douglas Hartree, Vladimir Fock and John Slater developed the so-called Hartree–Fock wavefunction as an improvement over the Thomas–Fermi model. The Hartree–Fock method accounted for exchange statistics of single particle electron wavefunctions. In general, it's very difficult to solve the Hartree–Fock equation. Only the free electron gas case can be solved exactly. Finally in 1964–65, Walter Kohn, Pierre Hohenberg and Lu Jeu Sham proposed the density functional theory which gave realistic descriptions for bulk and surface properties of metals. The density functional theory (DFT) has been widely used since the 1970s for band structure calculations of variety of solids.

Symmetry Breaking

Certain states of matter exhibit symmetry breaking, where the relevant laws of physics possess some symmetry that is broken. A common example is crystalline solids, which break continuous translational symmetry. Other examples include magnetized ferromagnets, which break rotational symmetry, and more exotic states such as the ground state of a BCS superconductor, that breaks U(1) phase rotational symmetry.

Goldstone's theorem in quantum field theory states that in a system with broken continuous symmetry, there may exist excitations with arbitrarily low energy, called the Goldstone bosons. For example, in crystalline solids, these correspond to phonons, which are quantized versions of lattice vibrations.

Phase Transition

Phase transition refers to the change of phase of a system, which is brought about by change in an external parameter such as temperature. Classical phase transition occurs at finite temperature when the order of the system was destroyed. For example, when ice melts and becomes water, the ordered crystal structure is destroyed. In quantum phase transitions, the temperature is set to absolute zero, and the non-thermal control parameter, such as pressure or magnetic field, causes the phase transitions when order is destroyed by quantum fluctuations originating from the Heisenberg uncertainty principle. Here, the different quantum phases of the system refer to distinct ground states of the Hamiltonian. Understanding the behavior of quantum phase transition is important in the difficult tasks of explaining the properties of rare-earth magnetic insulators, high-temperature superconductors and other substances.

There are two classes of phase tranisitions: first-order transitions and continuous transitions. For the continuous transitions, the two phases involved do not co-exist at the transition temperature, also called critical point. Near the critical point, systems undergoes displays critical behavior, wherein several of their properties such as correlation length, specific heat and susceptibility diverge exponentially. These critical phenomena poses serious challenges to physicists because normal macroscopic laws are no longer valid in the region and novel ideas and methods has to be invented to find the new laws that can describe the system.

The simplest theory that can describe continuous phase transitions is the Ginzburg–Landau theory, which works in the so-called mean field approximation. However, it can only roughly explain continuous phase transition for ferroelectrics and type I superconductors which invoves long range microscopic interactions. For other types of systems that involves short range interactions near the critical point, a better theory is needed.

Near the critical point, the fluctuations happen over broad range of size scales while the feature of the whole system is scale invariant. Renormalization group techniques successively average out the shortest wavelength fluctuations in stages while retaining their effects into the next stage. Thus, the changes of a physical system as viewed at different size scales can be investigated systematically. The techniques, together with powerful computer simulation, contribute greatly to the explanation of the critical phenomena associated with continuous phase transition.

Experimental

Experimental condensed matter physics involves the use of experimental probes to try to discover new properties of materials. Experimental probes include effects of electric and magnetic fields, measurement of response functions, transport properties and thermometry. Commonly used experimental techniques include spectroscopy, with probes such as X-rays, infrared light and inelastic neutron scattering; study of thermal response, such as specific heat and measurement of transport via thermal and heat conduction.

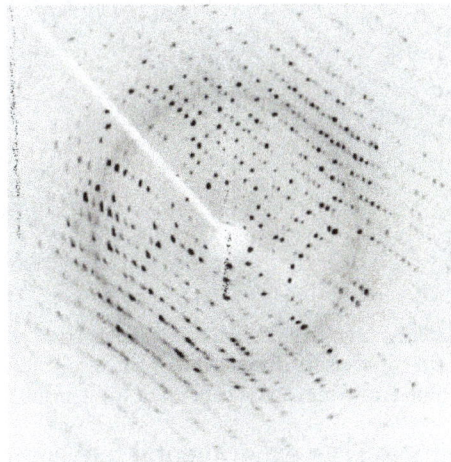

Image of X-ray diffraction pattern from a protein crystal.

Scattering

Several condensed matter experiments involve scattering of an experimental probe, such as X-ray, optical photons, neutrons, etc., on constituents of a material. The choice of scattering probe depends on the observation energy scale of interest. Visible light has energy on the scale of 1 eV and is used as a scattering probe to measure variations in material properties such as dielectric constant and refractive index. X-rays have energies of the order of 10 keV and hence are able to probe atomic length scales, and are used to measure variations in electron charge density.

Neutrons can also probe atomic length scales and are used to study scattering off nuclei and electron spins and magnetization (as neutrons themselves have spin but no charge). Coulomb and Mott scattering measurements can be made by using electron beams as scattering probes. Similarly, positron annihilation can be used as an indirect measurement of local electron density. Laser spectroscopy is an excellent tool for studying the microscopic properties of a medium, for example, to study forbidden transitions in media with nonlinear optical spectroscopy.

External Magnetic Fields

In experimental condensed matter physics, external magnetic fields act as thermodynamic variables that control the state, phase transitions and properties of material systems. Nuclear magnetic resonance (NMR) is a technique by which external magnetic fields can be used to find resonance modes of individual electrons, thus giving information about the atomic, molecular and bond structure of their neighborhood. NMR experiments can be made in magnetic fields with strengths up to 60 Tesla. Higher magnetic fields can improve the quality of NMR measurement data. Quantum oscillations is another experimental technique where high magnetic fields are used to study material properties such as the geometry of the Fermi surface. High magnetic fields will be useful in experimentally testing of the various theoretical predictions such as the quantized magnetoelectric effect, image magnetic monopole, and the half-integer quantum Hall effect.

Cold Atomic Gases

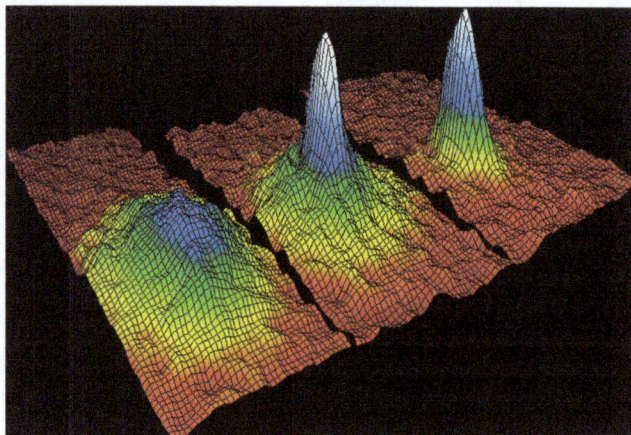

The first Bose–Einstein condensate observed in a gas of ultracold rubidium atoms.
The blue and white areas represent higher density.

Cold atom trapping in optical lattices is an experimental tool commonly used in condensed matter as well as atomic, molecular, and optical physics. The technique involves using optical lasers to create an interference pattern, which acts as a "lattice", in which ions or atoms can be placed at very low temperatures. Cold atoms in optical lattices are used as "quantum simulators", that is, they act as controllable systems that can model behavior of more complicated systems, such as frustrated magnets. In particular, they are used to engineer one-, two- and three-dimensional lattices for a Hubbard model with pre-specified parameters, and to study phase transitions for antiferromagnetic and spin liquid ordering.

In 1995, a gas of rubidium atoms cooled down to a temperature of 170 nK was used to experimentally realize the Bose–Einstein condensate, a novel state of matter originally predicted by S. N. Bose and Albert Einstein, wherein a large number of atoms occupy a single quantum state.

Applications

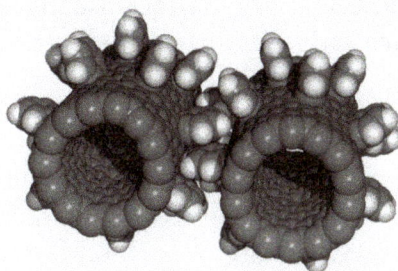

Computer simulation of "nanogears" made of fullerene molecules.
It is hoped that advances in nanoscience will lead to machines working on the molecular scale.

Research in condensed matter physics has given rise to several device applications, such as the development of the semiconductor transistor, and laser technology. Several phenomena studied in the context of nanotechnology come under the purview of condensed matter physics. Techniques such as scanning-tunneling microscopy can be used to control processes at the nanometer scale, and have given rise to the study of nanofabrication.

In quantum computation, information is represented by quantum bits, or qubits. The qubits may decohere quickly before useful computation is completed. This serious problem must be solved before quantum computation may be realized. The superconducting Josephson junction qubits, the spintronic qubits using the spin orientation of magnetic materials, or the topological non-Abelian anyons from fractional quantum Hall states are a few of the promising approaches proposed in condensed matter physics to solve this problem.

Condensed matter physics also has important applications to biophysics, for example, the experimental technique of magnetic resonance imaging, which is widely used in medical diagnosis.

Atomic, Molecular and Optical Physics

Atomic, molecular, and optical physics (AMO) is the study of matter-matter and light-matter interactions; at the scale of one or a few atoms and energy scales around several electron volts.

The three areas are closely interrelated. AMO theory includes classical, semi-classical and quantum treatments. Typically, the theory and applications of emission, absorption, scattering of electromagnetic radiation (light) from excited atoms and molecules, analysis of spectroscopy, generation of lasers and masers, and the optical properties of matter in general, fall into these categories.

Atomic and Molecular Physics

Atomic physics is the subfield of AMO that studies atoms as an isolated system of electrons and an atomic nucleus, while molecular physics is the study of the physical properties of molecules. The term *atomic physics* is often associated with nuclear power and nuclear bombs, due to the synonymous use of *atomic* and *nuclear* in standard English. However, physicists distinguish between atomic physics — which deals with the atom as a system consisting of a nucleus and electrons — and nuclear physics, which considers atomic nuclei alone. The important experimental techniques are the various types of spectroscopy. Molecular physics, while closely related to atomic physics, also overlaps greatly with theoretical chemistry, physical chemistry and chemical physics.

Both subfields are primarily concerned with electronic structure and the dynamical processes by which these arrangements change. Generally this work involves using quantum mechanics. For molecular physics this approach is known as quantum chemistry. One important aspect of molecular physics is that the essential atomic orbital theory in the field of atomic physics expands to the molecular orbital theory. Molecular physics is concerned with atomic processes in molecules, but it is additionally concerned with effects due to the molecular structure. Additionally to the electronic excitation states which are known from atoms, molecules are able to rotate and to vibrate. These rotations and vibrations are quantized; there are discrete energy levels. The smallest energy differences exist between different rotational states, therefore pure rotational spectra are in the far infrared region (about 30 - 150 µm wavelength) of the electromagnetic spectrum. Vibrational spectra are in the near infrared (about 1 - 5 µm) and spectra resulting from electronic transitions are mostly in the visible and ultraviolet regions. From measuring rotational and vibrational spectra properties of molecules like the distance between the nuclei can be calculated.

As with many scientific fields, strict delineation can be highly contrived and atomic physics is often considered in the wider context of *atomic, molecular, and optical physics*. Physics research groups are usually so classified.

Optical Physics

Optical physics is the study of the generation of electromagnetic radiation, the properties of that radiation, and the interaction of that radiation with matter, especially its manipulation and control. It differs from general optics and optical engineering in that it is focused on the discovery and application of new phenomena. There is no strong distinction, however, between optical physics, applied optics, and optical engineering, since the devices of optical engineering and the applications of applied optics are necessary for basic research in optical physics, and that research leads to the development of new devices and applications. Often the same people are involved in both the basic research and the applied technology development.

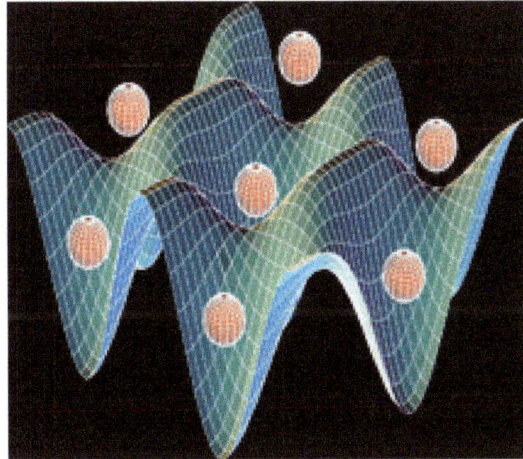

An optical lattice formed by laser interference.
Optical lattices are used to simulate interacting condensed matter systems.

Researchers in optical physics use and develop light sources that span the electromagnetic spectrum from microwaves to X-rays. The field includes the generation and detection of light, linear and nonlinear optical processes, and spectroscopy. Lasers and laser spectroscopy have transformed optical science. Major study in optical physics is also devoted to quantum optics and coherence, and to femtosecond optics. In optical physics, support is also provided in areas such as the nonlinear response of isolated atoms to intense, ultra-short electromagnetic fields, the atom-cavity interaction at high fields, and quantum properties of the electromagnetic field.

Other important areas of research include the development of novel optical techniques for nano-optical measurements, diffractive optics, low-coherence interferometry, optical coherence tomography, and near-field microscopy. Research in optical physics places an emphasis on ultrafast optical science and technology. The applications of optical physics create advancements in communications, medicine, manufacturing, and even entertainment.

History

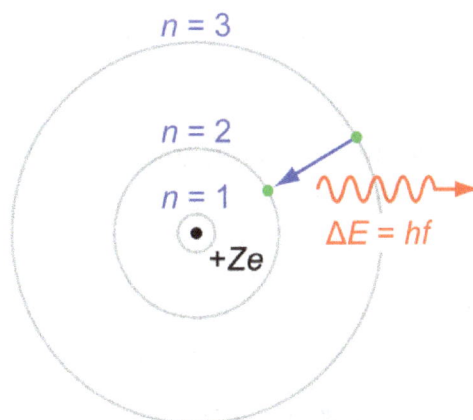

The Bohr model of the Hydrogen atom

One of the earliest steps towards *atomic physics* was the recognition that matter was composed of *atoms*, in modern terms the basic unit of a chemical element. This theory was developed by

John Dalton in the 18th century. At this stage, it wasn't clear what atoms were - although they could be described and classified by their observable properties in bulk; summarized by the developing periodic table, by John Newlands and Dmitri Mendeleyev around the mid to late 19th century.

Later, the connection between atomic physics *and* optical physics became apparent, by the discovery of spectral lines and attempts to describe the phenomenon - notably by Joseph von Fraunhofer, Fresnel, and others in the 19th century.

From that time to the 1920s, physicists were seeking to explain atomic spectra and blackbody radiation. One attempt to explain hydrogen spectral lines was the Bohr atom model.

Experiments including electromagnetic radiation and matter - such as the photoelectric effect, Compton effect, and spectra of sunlight the due to the unknown element of Helium, the limitation of the Bohr model to Hydrogen, and numerous other reasons, lead to an entirely new mathematical model of matter and light: quantum mechanics.

Classical Oscillator Model of Matter

Early models to explain the origin of the index of refraction treated an electron in an atomic system classically according to the model of Paul Drude and Hendrik Lorentz. The theory was developed to attempt to provide an origin for the wavelength-dependent refractive index n of a material. In this model, incident electromagnetic waves forced an electron bound to an atom to oscillate. The amplitude of the oscillation would then have a relationship to the frequency of the incident electromagnetic wave and the resonant frequencies of the oscillator. The superposition of these emitted waves from many oscillators would then lead to a wave which moved more slowly.

Early Quantum Model of Matter and Light

Max Planck derived a formula to describe the electromagnetic field inside a box when in thermal equilibrium in 1900. His model consisted of a superposition of standing waves. In one dimension, the box has length L, and only sinusodial waves of wavenumber

$$k = \frac{n\pi}{L}$$

can occur in the box, where n is a positive integer (mathematically denoted by $n \in_1$). The equation describing these standing waves is given by:

$$E = E_0 \sin\left(\frac{n\pi}{L}x\right)..$$

where E_0 is the magnitude of the electric field amplitude, and E is the magnitude of the electric field at position x. From this basic, Planck's law was derived.

In 1911, Ernest Rutherford concluded, based on alpha particle scattering, that an atom has a central pointlike proton. He also thought that an electron would be still attracted to the proton by Coulomb's law, which he had verified still held at small scales. As a result, he believed that electrons revolved around the proton. Niels Bohr, in 1913, combined the Rutherford model of the atom with the quantisation ideas of Planck. Only specific and well-defined orbits of the electron could exist, which also do not radiate light. In jumping orbit the electron would emit or absorb light corresponding to the difference in energy of the orbits. His prediction of the energy levels was then consistent with observation.

These results, based on a *discrete* set of specific standing waves, were inconsistent with the *continuous* classical oscillator model.

Work by Albert Einstein in 1905 on the photoelectric effect led to the association of a light wave of frequency v with a photon of energy hv. In 1917 Einstein created an extension to Bohrs model by the introduction of the three processes of stimulated emission, spontaneous emission and absorption (electromagnetic radiation).

Modern Treatments

The largest steps towards the modern treatment was the formulation of quantum mechanics with the matrix mechanics approach by Werner Heisenberg and the discovery of the Schrödinger equation by Erwin Schrödinger.

There are a variety of semi-classical treatments within AMO. Which aspects of the problem are treated quantum mechanically and which are treated classically is dependent on the specific problem at hand. The semi-classical approach is ubiquitous in computational work within AMO, largely due to the large decrease in computational cost and complexity associated with it.

For matter under the action of a laser, a fully quantum mechanical treatment of the atomic or molecular system is combined with the system being under the action of a classical electromagnetic field. Since the field is treated classically it can not deal with spontaneous emission. This semi-classical treatment is valid for most systems, particular those under the action of high intensity laser fields. The distinction between optical physics and quantum optics is the use of semi-classical and fully quantum treatments respectively.

Within collision dynamics and using the semi-classical treatment, the internal degrees of freedom may be treated quantum mechanically, whilst the relative motion of the quantum systems under consideration are treated classically. When considering medium to high speed collisions, the nuclei can be treated classically while the electron is treated quantum mechanically. In low speed collisions the approximation fails.

Classical Monte-Carlo methods for the dynamics of electrons can be described as semi-classical in that the initial conditions are calculated using a fully quantum treatment, but all further treatment is classical.

Isolated Atoms and Molecules

Atomic, Molecular and Optical physics frequently considers atoms and molecules in isolation. Atomic models will consist of a single nucleus that may be surrounded by one or more bound electrons, whilst molecular models are typically concerned with molecular hydrogen and its molecular hydrogen ion. It is not concerned with the formation of molecules (although much of the physics is identical) nor does it examine atoms in a solid state as condensed matter. It is concerned with processes such as ionization, above threshold ionization and excitation by photons or collisions with atomic particles.

While modelling atoms in isolation may not seem realistic, if one considers molecules in a gas or plasma then the time-scales for molecule-molecule interactions are huge in comparison to the atomic and molecular processes that we are concerned with. This means that the individual molecules can be treated as if each were in isolation for the vast majority of the time. By this consideration atomic and molecular physics provides the underlying theory in plasma physics and atmospheric physics even though both deal with huge numbers of molecules.

Electronic Configuration

Electrons form notional shells around the nucleus. These are naturally in a ground state but can be excited by the absorption of energy from light (photons), magnetic fields, or interaction with a colliding particle (typically other electrons).

Electrons that populate a shell are said to be in a bound state. The energy necessary to remove an electron from its shell (taking it to infinity) is called the binding energy. Any quantity of energy absorbed by the electron in excess of this amount is converted to kinetic energy according to the conservation of energy. The atom is said to have undergone the process of ionization.

In the event that the electron absorbs a quantity of energy less than the binding energy, it may transition to an excited state or to a virtual state. After a statistically sufficient quantity of time, an electron in an excited state will undergo a transition to a lower state via spontaneous emission. The change in energy between the two energy levels must be accounted for (conservation of energy). In a neutral atom, the system will emit a photon of the difference in energy. However, if the lower state is in an inner shell, a phenomenon known as the Auger effect may take place where the energy is transferred to another bound electrons causing it to go into the continuum. This allows one to multiply ionize an atom with a single photon.

There are strict selection rules as to the electronic configurations that can be reached by excitation by light—however there are no such rules for excitation by collision processes.

Atomic Physics

Atomic physics is the field of physics that studies atoms as an isolated system of electrons and an atomic nucleus. It is primarily concerned with the arrangement of electrons around the nucleus and the processes by which these arrangements change. This comprises ions, neutral atoms and, unless otherwise stated, it can be assumed that the term *atom* includes ions.

The term *atomic physics* can be associated with nuclear power and nuclear weapons, due to the synonymous use of *atomic* and *nuclear* in standard English. Physicists distinguish between atomic physics — which deals with the atom as a system consisting of a nucleus and electrons — and nuclear physics, which considers atomic nuclei alone.

As with many scientific fields, strict delineation can be highly contrived and atomic physics is often considered in the wider context of *atomic, molecular, and optical physics*. Physics research groups are usually so classified.

Isolated Atoms

Atomic physics primarily considers atoms in isolation. Atomic models will consist of a single nucleus that may be surrounded by one or more bound electrons. It is not concerned with the formation of molecules (although much of the physics is identical), nor does it examine atoms in a solid state as condensed matter. It is concerned with processes such as ionization and excitation by photons or collisions with atomic particles.

While modelling atoms in isolation may not seem realistic, if one considers atoms in a gas or plasma then the time-scales for atom-atom interactions are huge in comparison to the atomic processes that are generally considered. This means that the individual atoms can be treated as if each were in isolation, as the vast majority of the time they are. By this consideration atomic physics provides the underlying theory in plasma physics and atmospheric physics, even though both deal with very large numbers of atoms.

Electronic Configuration

Electrons form notional shells around the nucleus. These are normally in a ground state but can be excited by the absorption of energy from light (photons), magnetic fields, or interaction with a colliding particle (typically ions or other electrons).

Electrons that populate a shell are said to be in a bound state. The energy necessary to remove an electron from its shell (taking it to infinity) is called the binding energy. Any quantity of energy absorbed by the electron in excess of this amount is converted to kinetic energy according to the conservation of energy. The atom is said to have undergone the process of ionization.

If the electron absorbs a quantity of energy less than the binding energy, it will be transferred to an excited state. After a certain time, the electron in an excited state will "jump" (undergo a transition) to a lower state. In a neutral atom, the system will emit a photon of the difference in energy, since energy is conserved.

If an inner electron has absorbed more than the binding energy (so that the atom ionizes), then a more outer electron may undergo a transition to fill the inner orbital. In this case, a visible photon or a characteristic x-ray is emitted, or a phenomenon known as the Auger effect may take place, where the released energy is transferred to another bound electron, causing it to go into the continuum. The Auger effect allows one to multiply ionize an atom with a single photon.

There are rather strict selection rules as to the electronic configurations that can be reached by excitation by light — however there are no such rules for excitation by collision processes.

History and Developments

The majority of fields in physics can be divided between theoretical work and experimental work, and atomic physics is no exception. It is usually the case, but not always, that progress goes in alternate cycles from an experimental observation, through to a theoretical explanation followed by some predictions that may or may not be confirmed by experiment, and so on. Of course, the current state of technology at any given time can put limitations on what can be achieved experimentally and theoretically so it may take considerable time for theory to be refined.

One of the earliest steps towards atomic physics was the recognition that matter was composed of *atoms*. It forms a part of the texts written in 6th century BC to 2nd century BC such as those of Democritus or Vaisheshika Sutra written by Kanad. This theory was later developed in the modern sense of the basic unit of a chemical element by the British chemist and physicist John Dalton in the 18th century. At this stage, it wasn't clear what atoms were although they could be described and classified by their properties (in bulk). The invention of the periodic system of elements by Mendeleev was another great step forward.

The true beginning of atomic physics is marked by the discovery of spectral lines and attempts to describe the phenomenon, most notably by Joseph von Fraunhofer. The study of these lines led to the Bohr atom model and to the birth of quantum mechanics. In seeking to explain atomic spectra an entirely new mathematical model of matter was revealed. As far as atoms and their electron shells were concerned, not only did this yield a better overall description, i.e. the atomic orbital model, but it also provided a new theoretical basis for chemistry (quantum chemistry) and spectroscopy.

Since the Second World War, both theoretical and experimental fields have advanced at a rapid pace. This can be attributed to progress in computing technology, which has allowed larger and more sophisticated models of atomic structure and associated collision processes. Similar technological advances in accelerators, detectors, magnetic field generation and lasers have greatly assisted experimental work.

Significant Atomic Physicists

Pre quantum mechanics

- John Dalton
- Joseph von Fraunhofer
- Johannes Rydberg
- J. J. Thomson
- Ernest Rutherford (Mcgill University)

Post quantum mechanics

- Alexander Dalgarno
- David Bates
- Niels Bohr
- Max Born
- Clinton Joseph Davisson
- Enrico Fermi
- Charlotte Froese Fischer

- Vladimir Fock

- Douglas Hartree

- Ernest M. Henley

- Ratko Janev

- Harrie S. Massey

- Nevill Mott

- Mike Seaton

- John C. Slater

- George Paget Thomson

Molecular Physics

Molecular physics is the study of the physical properties of molecules, the chemical bonds between atoms as well as the molecular dynamics. Its most important experimental techniques are the various types of spectroscopy; scattering is also used. The field is closely related to atomic physics and overlaps greatly with theoretical chemistry, physical chemistry and chemical physics.

Additionally to the electronic excitation states which are known from atoms, molecules are able to rotate and to vibrate. These rotations and vibrations are quantized, there are discrete energy levels. The smallest energy differences exist between different rotational states, therefore pure rotational spectra are in the far infrared region (about 30 - 150 μm wavelength) of the electromagnetic spectrum. Vibrational spectra are in the near infrared (about 1 - 5 μm) and spectra resulting from electronic transitions are mostly in the visible and ultraviolet regions. From measuring rotational and vibrational spectra properties of molecules like the distance between the nuclei can be calculated.

One important aspect of molecular physics is that the essential atomic orbital theory in the field of atomic physics expands to the molecular orbital theory.

Optical Physics

Optical physics is a subfield of atomic, molecular, and optical physics. It is the study of the generation of electromagnetic radiation, the properties of that radiation, and the interaction of that radiation with matter, especially its manipulation and control. It differs from general optics and optical engineering in that it is focused on the discovery and application of new phenomena. There is no strong distinction, however, between optical physics, applied optics, and optical engineering, since the devices of optical engineering and the applications of applied optics are necessary for basic research in optical physics, and that research leads to the development of new devices and applications. Often the same people are involved in both the basic research and the applied technology development, for example the experimental demonstration of electromagnetically induced transparency by S. E. Harris and of slow light by Harris and Lene Vestergaard Hau.

Researchers in optical physics use and develop light sources that span the electromagnetic spectrum from microwaves to X-rays. The field includes the generation and detection of light, linear and nonlinear optical processes, and spectroscopy. Lasers and laser spectroscopy have transformed optical science. Major study in optical physics is also devoted to quantum optics and coherence, and to femtosecond optics. In optical physics, research is also encouraged in areas such as the nonlinear response of isolated atoms to intense, ultra-short electromagnetic fields, the atom-cavity interaction at high fields, and quantum properties of the electromagnetic field. Other important areas of research include the development of novel optical techniques for nano-optical measurements, diffractive optics, low-coherence interferometry, optical coherence tomography, and near-field microscopy. Research in optical physics places an emphasis on ultrafast optical science and technology. The applications of optical physics create advancements in communications, medicine, manufacturing, and even entertainment.

Electromagnetism

Electromagnetism is a branch of physics which involves the study of the electromagnetic force, a type of physical interaction that occurs between electrically charged particles. The electromagnetic force usually exhibits electromagnetic fields, such as electric fields, magnetic fields, and light. The electromagnetic force is one of the four fundamental interactions (commonly called forces) in nature. The other three fundamental interactions are the strong interaction, the weak interaction, and gravitation.

Lightning is an electrostatic discharge that travels between two charged regions.

Electromagnetic phenomena are defined in terms of the electromagnetic force, sometimes called the Lorentz force, which includes both electricity and magnetism as different manifestations of the same phenomenon.

The electromagnetic force plays a major role in determining the internal properties of most objects encountered in daily life. Ordinary matter takes its form as a result of intermolecular forces between individual atoms and molecules in matter, and are a manifestation of the electromagnetic force. Electrons are bound by the electromagnetic force to atomic nuclei, and their orbital shapes and their influence on nearby atoms with their electrons is described by quantum mechanics. The electromagnetic force governs the processes involved in chemistry, which arise from interactions between the electrons of neighboring atoms.

There are numerous mathematical descriptions of the electromagnetic field. In classical electrodynamics, electric fields are described as electric potential and electric current. In Faraday's law, magnetic fields are associated with electromagnetic induction and magnetism, and Maxwell's equations describe how electric and magnetic fields are generated and altered by each other and by charges and currents.

The theoretical implications of electromagnetism, in particular the establishment of the speed of light based on properties of the "medium" of propagation (permeability and permittivity), led to the development of special relativity by Albert Einstein in 1905.

Although electromagnetism is considered one of the four fundamental forces, at high energy the weak force and electromagnetic force are unified as a single electroweak force. In the history of the universe, during the quark epoch the unified force broke into the two separate forces as the universe cooled.

History of the Theory

Hans Christian Ørsted.

Originally, electricity and magnetism were thought of as two separate forces. This view changed, however, with the publication of James Clerk Maxwell's 1873 *A Treatise on Electricity and Magnetism* in which the interactions of positive and negative charges were shown to be mediated by one force. There are four main effects resulting from these interactions, all of which have been clearly demonstrated by experiments:

1. Electric charges attract or repel one another with a force inversely proportional to the square of the distance between them: unlike charges attract, like ones repel.

2. Magnetic poles (or states of polarization at individual points) attract or repel one another in a manner similar to positive and negative charges and always exist as pairs: every north pole is yoked to a south pole.

3. An electric current inside a wire creates a corresponding circumferential magnetic field outside the wire. Its direction (clockwise or counter-clockwise) depends on the direction of the current in the wire.

4. A current is induced in a loop of wire when it is moved toward or away from a magnetic field, or a magnet is moved towards or away from it; the direction of current depends on that of the movement.

André-Marie Ampère

While preparing for an evening lecture on 21 April 1820, Hans Christian Ørsted made a surprising observation. As he was setting up his materials, he noticed a compass needle deflected away from magnetic north when the electric current from the battery he was using was switched on and off. This deflection convinced him that magnetic fields radiate from all sides of a wire carrying an electric current, just as light and heat do, and that it confirmed a direct relationship between electricity and magnetism.

Michael Faraday

At the time of discovery, Ørsted did not suggest any satisfactory explanation of the phenomenon, nor did he try to represent the phenomenon in a mathematical framework. However, three months later he began more intensive investigations. Soon thereafter he published his findings, proving that an electric current produces a magnetic field as it flows through a wire. The CGS unit of magnetic induction (oersted) is named in honor of his contributions to the field of electromagnetism.

James Clerk Maxwell

His findings resulted in intensive research throughout the scientific community in electrodynamics. They influenced French physicist André-Marie Ampère's developments of a single mathematical form to represent the magnetic forces between current-carrying conductors. Ørsted's discovery also represented a major step toward a unified concept of energy.

This unification, which was observed by Michael Faraday, extended by James Clerk Maxwell, and partially reformulated by Oliver Heaviside and Heinrich Hertz, is one of the key accomplishments of 19th century mathematical physics. It had far-reaching consequences, one of which was the understanding of the nature of light. Unlike what was proposed by electromagnetic theory of that time, light and other electromagnetic waves are at present seen as taking the form of quantized, self-propagating oscillatory electromagnetic field disturbances called photons. Different frequencies of oscillation give rise to the different forms of electromagnetic radiation, from radio waves at the lowest frequencies, to visible light at intermediate frequencies, to gamma rays at the highest frequencies.

Ørsted was not the only person to examine the relationship between electricity and magnetism. In 1802, Gian Domenico Romagnosi, an Italian legal scholar, deflected a magnetic needle using electrostatic charges. Actually, no galvanic current existed in the setup and hence no electromagnetism was present. An account of the discovery was published in 1802 in an Italian newspaper, but it was largely overlooked by the contemporary scientific community.

Fundamental Forces

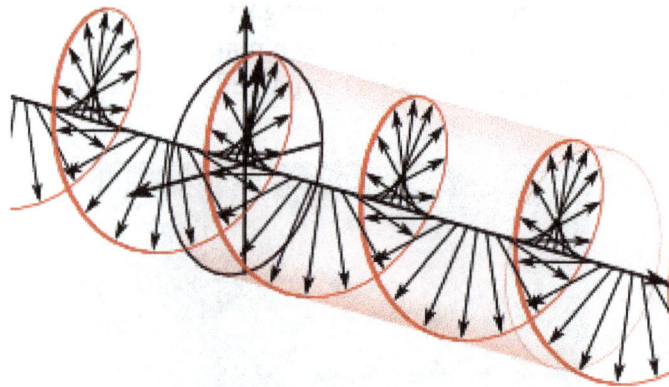

Representation of the electric field vector of a wave of circularly polarized electromagnetic radiation.

The electromagnetic force is one of the four known fundamental forces. The other fundamental forces are:

- the weak nuclear force, which binds to all known particles in the Standard Model, and causes certain forms of radioactive decay. (In particle physics though, the electroweak interaction is the unified description of two of the four known fundamental interactions of nature: electromagnetism and the weak interaction);

- the strong nuclear force, which binds quarks to form nucleons, and binds nucleons to form nuclei

- the gravitational force.

All other forces (e.g., friction, contact forces) are derived from these four fundamental forces (including momentum which is carried by the movement of particles).

The electromagnetic force is the one responsible for practically all the phenomena one encounters in daily life above the nuclear scale, with the exception of gravity. Roughly speaking, all the forces involved in interactions between atoms can be explained by the electromagnetic force acting between the electrically charged atomic nuclei and electrons of the atoms. Electromagnetic forces also explain how these particles carry momentum by their movement. This includes the forces we experience in "pushing" or "pulling" ordinary material objects, which result from the intermolecular forces that act between the individual molecules in our bodies and those in the objects. The electromagnetic force is also involved in all forms of chemical phenomena.

A necessary part of understanding the intra-atomic and intermolecular forces is the effective force generated by the momentum of the electrons' movement, such that as electrons move between interacting atoms they carry momentum with them. As a collection of electrons becomes more confined, their minimum momentum necessarily increases due to the Pauli exclusion principle. The behaviour of matter at the molecular scale including its density is determined by the balance between the electromagnetic force and the force generated by the exchange of momentum carried by the electrons themselves.

Classical Electrodynamics

In 1600, William Gilbert proposed, in his *De Magnete*, that electricity and magnetism, while both capable of causing attraction and repulsion of objects, were distinct effects. Mariners had noticed that lightning strikes had the ability to disturb a compass needle, but the link between lightning and electricity was not confirmed until Benjamin Franklin's proposed experiments in 1752. One of the first to discover and publish a link between man-made electric current and magnetism was Romagnosi, who in 1802 noticed that connecting a wire across a voltaic pile deflected a nearby compass needle. However, the effect did not become widely known until 1820, when Ørsted performed a similar experiment. Ørsted's work influenced Ampère to produce a theory of electromagnetism that set the subject on a mathematical foundation.

A theory of electromagnetism, known as classical electromagnetism, was developed by various physicists over the course of the 19th century, culminating in the work of James Clerk Maxwell, who unified the preceding developments into a single theory and discovered the electromagnetic nature of light. In classical electromagnetism, the behavior of the electromagnetic field is described by a set of equations known as Maxwell's equations, and the electromagnetic force is given by the Lorentz force law.

One of the peculiarities of classical electromagnetism is that it is difficult to reconcile with classical mechanics, but it is compatible with special relativity. According to Maxwell's equations, the speed of light in a vacuum is a universal constant that is dependent only on the electrical permittivity and magnetic permeability of free space. This violates Galilean invariance, a long-standing cornerstone of classical mechanics. One way to reconcile the two theories (electromagnetism and classical mechanics) is to assume the existence of a luminiferous aether through which the light propagates. However, subsequent experimental efforts failed to detect the presence of the aether. After important contributions of Hendrik Lorentz and Henri Poincaré, in 1905, Albert Einstein solved the problem with the introduction of special relativity, which replaced classical kinematics with a new theory of kinematics compatible with classical electromagnetism.

In addition, relativity theory implies that in moving frames of reference a magnetic field transforms to a field with a nonzero electric component and conversely, a moving electric field transforms to a nonzero magnetic component, thus firmly showing that the phenomena are two sides of the same coin. Hence the term "electromagnetism".

Quantum Mechanics
Early Quantum Theory

In a second paper published in 1905, Albert Einstein undermined the very foundations of classical electromagnetism. In his theory of the photoelectric effect, for which he won the Nobel prize in physics, he posited that light could exist in discrete particle-like quantities, which later came to be called photons. Einstein's theory of the photoelectric effect extended the insights that appeared in the solution of the ultraviolet catastrophe presented by Max Planck in 1900 and who coined the term "quanta". In his work, Planck showed that hot objects emit electromagnetic radiation in discrete packets ("quanta"), which leads to a finite total energy emitted as black body radiation. Both of these results were in direct contradiction with the classical view of light as a continuous wave.

Planck's and Einstein's theories were progenitors of quantum mechanics, which, when formulated in 1925, necessitated the invention of a quantum theory of electromagnetism.

Quantum Electrodynamics

Maxwell's equations have been superseded by quantum electrodynamics (QED). Richard Feynman called it "the jewel of physics"[Ch1] for its extremely accurate predictions of quantities like the Lamb shift, and measurement of the magnetic moment of the electron. The electromagnetic field is quantized by imagining that at every point in space and time is a quantum harmonic oscillator. The empty field (vacuum state) fluctuates randomly as a consequence of the uncertainty principle This theory, completed in the 1940s–1950s, is one of the most accurate theories known to physics in situations where perturbation theory can be applied. Like classical electromagnetism, QED is a linear U(1) gauge group.

Electroweak Interaction

The electroweak interaction is a unified field theory description of two of the four known fundamental interactions of nature: electromagnetism and the weak interaction. It is a SU(2) × U(1) gauge group. Although these two forces appear very different at everyday low energies, the theory models them as two different aspects of the same force. At energies greater than 100 GeV, called the unification energy, the two forces merge into a single electroweak force. Thus when the universe was hot enough (approximately 10^{15} K, a temperature that was exceeded until shortly after the Big Bang) the electromagnetic force and weak force were merged into the electroweak force. With the cooling of the universe, at the end of the electroweak epoch, the electroweak force separated into the electromagnetic force and the weak force. During the following quark epoch, it was still too hot for quarks to combine into hadrons and they moved about freely.

Chaotic and Emergent Phenomena

The mathematical models used in classical electromagnetism, quantum electrodynamics (QED) and the standard model all view the electromagnetic force as a linear set of equations. In these theories electromagnetism is a U(1) gauge theory, whose topological properties do not allow any complex nonlinear interaction of a field with and on itself. For example, in the QED vacuum the field fluctuates randomly as a consequence of the uncertainty principle but that these fluctuations cancel each over out with no overall observable effect. However, there are many observed nonlinear physical electromagnetic phenomena such as Aharonov–Bohm (AB) and Altshuler–Aronov–Spivak (AAS) effects, Berry, Aharonov– Anandan, Pancharatnam and Chiao–Wu phase rotation effects, Josephson effect, Quantum Hall effect, the de Haas–van Alphen effect, the Sagnac effect and many other physically observable phenomena which would indicate that the electromagnetic potential field has real physical meaning rather than being a mathematical artifact and therefore an all encompassing theory would not confine electromagnetism as a local force as is currently done, but as a SU(2) gauge theory or higher geometry. Higher symmetries allow for nonlinear, aperiodic behaviour which manifest as a variety of complex non-equilibrium phenomena that do not arise in the linearised U(1) theory, such as multiple stable states, symmetry breaking, chaos and emergence. In higher symmetry groups, the electromagnetic field is not a calm, randomly fluctuating, passive substance, but can at times can be viewed as a turbulent virtual plasma that can have complex vorticcs, cntanglcd statcs and a rich nonlinear structure.

What are called Maxwell's equation's today are in fact a simplified version of the original equations reformulated by Heaviside, FitzGerald, Lodge and Hertz. The original equations used Hamilton's more expressive quaternion notation, a kind of Clifford algebra, which fully subsumes the standard Maxwell vectorial equations largely used today. In the late 1880s there was a debate over the relative merits of vector analysis and quaternions. According to Heaviside the electromagnetic potential field was purely metaphysical, an arbitrary mathematical fiction, that needed to be *"murdered"*. It was concluded that there was no need for the greater physical insights provided by the quaternions if the theory was purely local in nature. Local vector analysis has become the dominant way of using Maxwell's equations ever since. However, this strictly vectorial approach as led to a restrictive topological understanding in some areas of electromagnetism, for example, a full understanding of the energy transfer dynamics in Tesla's oscilator-shuttle-circuit can only be achieved in quaternionic algebra or higher SU(2) symmetries. It has often been argued that quaternions are not compatible with special relativity, but multiple papers have shown ways of incorporating relativity

Plasma vortices give rise to solar flares on the sun

A good example of nonlinear electromagnetics is in high energy dense plasmas, where vortical phenomena occur which seemingly violate the second law of thermodynamics by increasing the energy gradient within the electromagnetic field and violate Maxwell's laws by creating ion currents which capture and concentrate their own and surrounding magnetic fields. In particular Lorentz force law, which elaborates Maxwell's equations is violated by these force free vortices. These apparent violations are due to the fact that the traditional conservation laws in classical and quantum electrodynamics (QED) only display linear U(1) symmetry (in particular, by the extended Noether theorem, conservation laws such as the laws of thermodynamics need not always apply to dissipative systems, which are expressed in gauges of higher symmetry). The second law of thermodynamics states that in a closed linear system entropy flow can be only be positive (or exactly zero at the end of a cycle). However, negative entropy (i.e. increased order, structure or self-organisation) can spontaneously appear in an open nonlinear thermodynamic system that is far from equilibrium, so long as this emergent order accelerates the overall flow of entropy in the total system.

Below its critical temperature, a superconductor becomes perfectly diamagnetic and excludes sufficiently weak magnetic fields from passing through it. This is called the Meissner effect which is described by the London equations.

Given the complex and adaptive behaviour that arises from nonlinear systems considerable attention in recent years has gone into studying a new class of phase transitions which occur at absolute zero temperature. These are quantum phase transitions which are driven by electomagnetic field fluctuations as a consequence of zero-point energy A good example of a spontaneous phase transition that are attributed to zero-point fluctuations can be found in superconductors. Superconductivity is one of the best known empirically quantified macroscopic electromagnetic phenomena whose basis is recognised to be quantum mechanical in origin. The behaviour of the electric and magnetic fields under superconductivity is governed by the London equations. However, it has been questioned in a series of journal articles whether the quantum mechanically canonised London equations can be given a purely classical derivation. Bostick for instance, has claimed to show that the London equations do indeed have a classical origin that applies to superconductors and to some collisionless plasmas as well. In particular it has been asserted that the Beltrami vortices in the plasma focus display the same paired flux-tube morphology as Type II superconductors. Others have also pointed out this connection, Fröhlich has shown that the hydrodynamic equations of compressible fluids, together with the London equations, lead to a macroscopic parameter (= electric charge density / mass density), without involving either quantum phase factors or Planck's constant. In essence, it has been asserted that Beltrami plasma vortex structures are able to at least simulate the morphology of Type I and Type II superconductors. This occurs because the *"organised"* dissipative energy of the vortex configuration comprising the ions and electrons far exceeds the *"disorganised"* dissipative random thermal energy. The transition from disorganised fluctuations to organised helical structures is a phase transition involving a change in the condensate's energy (i.e. the ground state or zero-point energy) but *without any associated rise in temperature*. This is an example of zero-point energy having multiple stable states and where the overall system structure is independent of a reductionist or deterministic view, that *"classical"* macroscopic order can also causally affect quantum phenomena.

Quantities and Units

Electromagnetic units are part of a system of electrical units based primarily upon the magnetic properties of electric currents, the fundamental SI unit being the ampere. The units are:

- ampere (electric current)

- coulomb (electric charge)

- farad (capacitance)

- henry (inductance)

- ohm (resistance)

- siemens (conductance)

- tesla (magnetic flux density)

- volt (electric potential)

- watt (power)

- weber (magnetic flux)

In the electromagnetic cgs system, electric current is a fundamental quantity defined via Ampère's law and takes the permeability as a dimensionless quantity (relative permeability) whose value in a vacuum is unity. As a consequence, the square of the speed of light appears explicitly in some of the equations interrelating quantities in this system.

SI electromagnetism units				
Symbol	**Name of Quantity**	**Derived Units**	**Unit**	**Base Units**
I	electric current	ampere (SI base unit)	A	A ($= W/V = C/s$)
Q	electric charge	coulomb	C	A·s
$U, \Delta V, \Delta\varphi; E$	potential difference; electromotive force	volt	V	kg·m²·s⁻³·A⁻¹ ($= J/C$)
$R; Z; X$	electric resistance; impedance; reactance	ohm	Ω	kg·m²·s⁻³·A⁻² ($= V/A$)
ρ	resistivity	ohm metre	Ω·m	kg·m³·s⁻³·A⁻²
P	electric power	watt	W	kg·m²·s⁻³ ($= V·A$)
C	capacitance	farad	F	kg⁻¹·m⁻²·s⁴·A² ($= C/V$)
E	electric field strength	volt per metre	V/m	kg·m·s⁻³·A⁻¹ ($= N/C$)
D	electric displacement field	coulomb per square metre	C/m²	A·s·m⁻²
ε	permittivity	farad per metre	F/m	kg⁻¹·m⁻³·s⁴·A²
χ_e	electric susceptibility	(dimensionless)	–	–
$G; Y; B$	conductance; admittance; susceptance	siemens	S	kg⁻¹·m⁻²·s³·A² ($= \Omega^{-1}$)
κ, γ, σ	conductivity	siemens per metre	S/m	kg⁻¹·m⁻³·s³·A²
B	magnetic flux density, magnetic induction	tesla	T	kg·s⁻²·A⁻¹ ($= Wb/m²$ $= N·A^{-1}·m^{-1}$)
	magnetic flux	weber	Wb	kg·m²·s⁻²·A⁻¹ ($= V·s$)
H	magnetic field strength	ampere per metre	A/m	A·m⁻¹
L, M	inductance	henry	H	kg·m²·s⁻²·A⁻² ($= Wb/A = V·s/A$)

μ	permeability	henry per metre	H/m	$kg{\cdot}m{\cdot}s^{-2}{\cdot}A^{-2}$
χ	magnetic susceptibility	(dimensionless)	–	–

Formulas for physical laws of electromagnetism (such as Maxwell's equations) need to be adjusted depending on what system of units one uses. This is because there is no one-to-one correspondence between electromagnetic units in SI and those in CGS, as is the case for mechanical units. Furthermore, within CGS, there are several plausible choices of electromagnetic units, leading to different unit "sub-systems", including Gaussian, "ESU", "EMU", and Heaviside–Lorentz. Among these choices, Gaussian units are the most common today, and in fact the phrase "CGS units" is often used to refer specifically to CGS-Gaussian units.

Quantum Mechanics

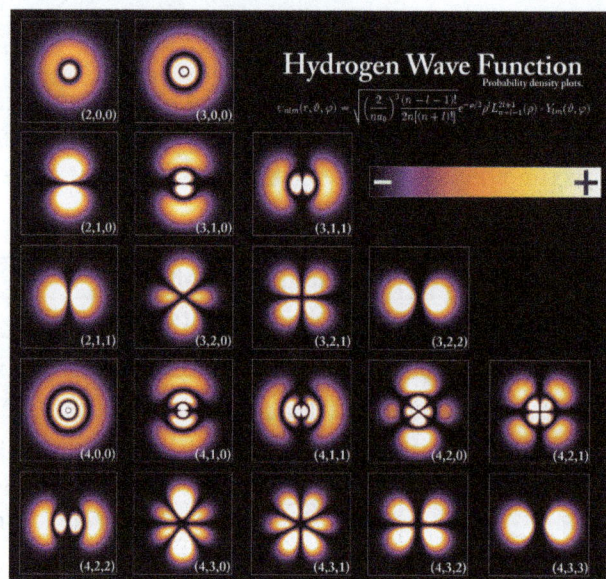

Solution to Schrödinger's equation for the hydrogen atom at different energy levels. The brighter areas represent a higher probability of finding an electron.

Quantum mechanics (QM; also known as quantum physics or quantum theory), including quantum field theory, is a fundamental branch of physics concerned with processes involving, for example, atoms and photons. Systems such as these which obey quantum mechanics can be in a quantum superposition of different states, unlike in classical physics.

Quantum mechanics gradually arose from Max Planck's solution in 1900 to the black-body radiation problem (reported 1859) and Albert Einstein's 1905 paper which offered a quantum-based theory to explain the photoelectric effect (reported 1887). Early quantum theory was profoundly reconceived in the mid-1920s.

The reconceived theory is formulated in various specially developed mathematical formalisms. In one of them, a mathematical function, the wave function, provides information about the probability amplitude of position, momentum, and other physical properties of a particle.

Important applications of quantum theory include superconducting magnets, light-emitting diodes, and the laser, the transistor and semiconductors such as the microprocessor, medical and research imaging such as magnetic resonance imaging and electron microscopy, and explanations for many biological and physical phenomena.

History

Scientific inquiry into the wave nature of light began in the 17th and 18th centuries, when scientists such as Robert Hooke, Christiaan Huygens and Leonhard Euler proposed a wave theory of light based on experimental observations. In 1803, Thomas Young, an English polymath, performed the famous double-slit experiment that he later described in a paper titled *On the nature of light and colours*. This experiment played a major role in the general acceptance of the wave theory of light.

In 1838, Michael Faraday discovered cathode rays. These studies were followed by the 1859 statement of the black-body radiation problem by Gustav Kirchhoff, the 1877 suggestion by Ludwig Boltzmann that the energy states of a physical system can be discrete, and the 1900 quantum hypothesis of Max Planck. Planck's hypothesis that energy is radiated and absorbed in discrete "quanta" (or energy packets) precisely matched the observed patterns of black-body radiation.

In 1896, Wilhelm Wien empirically determined a distribution law of black-body radiation, known as Wien's law in his honor. Ludwig Boltzmann independently arrived at this result by considerations of Maxwell's equations. However, it was valid only at high frequencies and underestimated the radiance at low frequencies. Later, Planck corrected this model using Boltzmann's statistical interpretation of thermodynamics and proposed what is now called Planck's law, which led to the development of quantum mechanics.

Following Max Planck's solution in 1900 to the black-body radiation problem (reported 1859), Albert Einstein offered a quantum-based theory to explain the photoelectric effect (1905, reported 1887). Around 1900-1910, the atomic theory and the corpuscular theory of light first came to be widely accepted as scientific fact; these latter theories can be viewed as quantum theories of matter and electromagnetic radiation, respectively.

Among the first to study quantum phenomena in nature were Arthur Compton, C. V. Raman, and Pieter Zeeman, each of whom has a quantum effect named after him. Robert Andrews Millikan studied the photoelectric effect experimentally, and Albert Einstein developed a theory for it. At the same time, Ernest Rutherford experimentally discovered the nuclear model of the atom, for which Niels Bohr developed his theory of the atomic structure, which was later confirmed by the experiments of Henry Moseley. In 1913, Peter Debye extended Niels Bohr's theory of atomic structure, introducing elliptical orbits, a concept also introduced by Arnold Sommerfeld. This phase is known as old quantum theory.

According to Planck, each energy element (E) is proportional to its frequency (v): $E = hv$

where h is Planck's constant.

Planck cautiously insisted that this was simply an aspect of the *processes* of absorption and emission of radiation and had nothing to do with the *physical reality* of the radiation itself. In fact, he considered his quantum hypothesis a mathematical trick to get the right answer rather than a sizable discovery. However, in 1905 Albert Einstein interpreted Planck's quantum hypothesis realistically and used it to explain the photoelectric effect, in which shining light on certain materials can eject electrons from the material. He won the 1921 Nobel Prize in Physics for this work.

Max Planck is considered the father of the quantum theory.

Einstein further developed this idea to show that an electromagnetic wave such as light could also be described as a particle (later called the photon), with a discrete quantum of energy that was dependent on its frequency.

The 1927 Solvay Conference in Brussels.

The foundations of quantum mechanics were established during the first half of the 20th century by Max Planck, Niels Bohr, Werner Heisenberg, Louis de Broglie, Arthur Compton, Albert Einstein, Erwin Schrödinger, Max Born, John von Neumann, Paul Dirac, Enrico Fermi, Wolfgang Pauli, Max von Laue, Freeman Dyson, David Hilbert, Wilhelm Wien, Satyendra Nath Bose, Arnold Sommerfeld, and others. The Copenhagen interpretation of Niels Bohr became widely accepted.

In the mid-1920s, developments in quantum mechanics led to its becoming the standard formulation for atomic physics. In the summer of 1925, Bohr and Heisenberg published results that closed the old quantum theory. Out of deference to their particle-like behavior in certain processes and measurements, light quanta came to be called photons (1926). From Einstein's simple postulation was born a flurry of debating, theorizing, and testing. Thus, the entire field of quantum physics emerged, leading to its wider acceptance at the Fifth Solvay Conference in 1927.

It was found that subatomic particles and electromagnetic waves are neither simply particle nor wave but have certain properties of each. This originated the concept of wave–particle duality.

By 1930, quantum mechanics had been further unified and formalized by the work of David Hilbert, Paul Dirac and John von Neumann with greater emphasis on measurement, the statistical nature of our knowledge of reality, and philosophical speculation about the 'observer'. It has since permeated many disciplines including quantum chemistry, quantum electronics, quantum optics, and quantum information science. Its speculative modern developments include string theory and quantum gravity theories. It also provides a useful framework for many features of the modern periodic table of elements, and describes the behaviors of atoms during chemical bonding and the flow of electrons in computer semiconductors, and therefore plays a crucial role in many modern technologies.

While quantum mechanics was constructed to describe the world of the very small, it is also needed to explain some macroscopic phenomena such as superconductors, and superfluids.

The word *quantum* derives from the Latin, meaning "how great" or "how much". In quantum mechanics, it refers to a discrete unit assigned to certain physical quantities such as the energy of an atom at rest. The discovery that particles are discrete packets of energy with wave-like properties led to the branch of physics dealing with atomic and subatomic systems which is today called quantum mechanics. It underlies the mathematical framework of many fields of physics and chemistry, including condensed matter physics, solid-state physics, atomic physics, molecular physics, computational physics, computational chemistry, quantum chemistry, particle physics, nuclear chemistry, and nuclear physics. Some fundamental aspects of the theory are still actively studied.

Quantum mechanics is essential to understanding the behavior of systems at atomic length scales and smaller. If the physical nature of an atom were solely described by classical mechanics, electrons would not *orbit* the nucleus, since orbiting electrons emit radiation (due to circular motion) and would eventually collide with the nucleus due to this loss of energy. This framework was unable to explain the stability of atoms. Instead, electrons remain in an uncertain, non-deterministic, *smeared*, probabilistic wave–particle orbital about the nucleus, defying the traditional assumptions of classical mechanics and electromagnetism.

Quantum mechanics was initially developed to provide a better explanation and description of the atom, especially the differences in the spectra of light emitted by different isotopes of the same chemical element, as well as subatomic particles. In short, the quantum-mechanical atomic model has succeeded spectacularly in the realm where classical mechanics and electromagnetism falter.

Broadly speaking, quantum mechanics incorporates four classes of phenomena for which classical physics cannot account:

- quantization of certain physical properties

- quantum entanglement

- principle of uncertainty

- wave–particle duality

Mathematical Formulations

In the mathematically rigorous formulation of quantum mechanics developed by Paul Dirac, David Hilbert, John von Neumann, and Hermann Weyl, the possible states of a quantum mechanical system are symbolized as unit vectors (called *state vectors*). Formally, these reside in a complex separable Hilbert space—variously called the *state space* or the *associated Hilbert space* of the system—that is well defined up to a complex number of norm 1 (the phase factor). In other words, the possible states are points in the projective space of a Hilbert space, usually called the complex projective space. The exact nature of this Hilbert space is dependent on the system—for example, the state space for position and momentum states is the space of square-integrable functions, while the state space for the spin of a single proton is just the product of two complex planes. Each observable is represented by a maximally Hermitian (precisely: by a self-adjoint) linear operator acting on the state space. Each eigenstate of an observable corresponds to an eigenvector of the operator, and the associated eigenvalue corresponds to the value of the observable in that eigenstate. If the operator's spectrum is discrete, the observable can attain only those discrete eigenvalues.

In the formalism of quantum mechanics, the state of a system at a given time is described by a complex wave function, also referred to as state vector in a complex vector space. This abstract mathematical object allows for the calculation of probabilities of outcomes of concrete experiments. For example, it allows one to compute the probability of finding an electron in a particular region around the nucleus at a particular time. Contrary to classical mechanics, one can never make simultaneous predictions of conjugate variables, such as position and momentum, to arbitrary precision. For instance, electrons may be considered (to a certain probability) to be located somewhere within a given region of space, but with their exact positions unknown. Contours of constant probability, often referred to as "clouds", may be drawn around the nucleus of an atom to conceptualize where the electron might be located with the most probability. Heisenberg's uncertainty principle quantifies the inability to precisely locate the particle given its conjugate momentum.

According to one interpretation, as the result of a measurement the wave function containing the probability information for a system collapses from a given initial state to a particular eigenstate. The possible results of a measurement are the eigenvalues of the operator representing the observable—which explains the choice of *Hermitian* operators, for which all the eigenvalues are real. The probability distribution of an observable in a given state can be found by computing the spectral decomposition of the corresponding operator. Heisenberg's uncertainty principle is represented by the statement that the operators corresponding to certain observables do not commute.

The probabilistic nature of quantum mechanics thus stems from the act of measurement. This is one of the most difficult aspects of quantum systems to understand. It was the central topic in the famous Bohr–Einstein debates, in which the two scientists attempted to clarify these fundamental principles by way of thought experiments. In the decades after the formulation of quantum mechanics, the question of what constitutes a "measurement" has been extensively studied. Newer interpretations of quantum mechanics have been formulated that do away with the concept of "wave function collapse". The basic idea is that when a quantum system interacts with a measuring apparatus, their respective wave functions become entangled, so that the original quantum system ceases to exist as an independent entity.

Generally, quantum mechanics does not assign definite values. Instead, it makes a prediction using a probability distribution; that is, it describes the probability of obtaining the possible outcomes from measuring an observable. Often these results are skewed by many causes, such as dense probability clouds. Probability clouds are approximate (but better than the Bohr model) whereby electron location is given by a probability function, the wave function eigenvalue, such that the probability is the squared modulus of the complex amplitude, or quantum state nuclear attraction. Naturally, these probabilities will depend on the quantum state at the "instant" of the measurement. Hence, uncertainty is involved in the value. There are, however, certain states that are associated with a definite value of a particular observable. These are known as eigenstates of the observable ("eigen" can be translated from German as meaning "inherent" or "characteristic").

In the everyday world, it is natural and intuitive to think of everything (every observable) as being in an eigenstate. Everything appears to have a definite position, a definite momentum, a definite energy, and a definite time of occurrence. However, quantum mechanics does not pinpoint the exact values of a particle's position and momentum (since they are conjugate pairs) or its energy and time (since they too are conjugate pairs); rather, it provides only a range of probabilities in which that particle might be given its momentum and momentum probability. Therefore, it is helpful to use different words to describe states having *uncertain* values and states having *definite* values (eigenstates). Usually, a system will not be in an eigenstate of the observable (particle) we are interested in. However, if one measures the observable, the wave function will instantaneously be an eigenstate (or "generalized" eigenstate) of that observable. This process is known as wave function collapse, a controversial and much-debated process that involves expanding the system under study to include the measurement device. If one knows the corresponding wave function at the instant before the measurement, one will be able to compute the probability of the wave function collapsing into each of the possible eigenstates. For example, the free particle in the previous example will usually have a wave function that is a wave packet centered around some mean position x_0 (neither an eigenstate of position nor of momentum). When one measures the position of the particle, it is impossible to predict with certainty the result. It is probable, but not certain, that it will be near x_0, where the amplitude of the wave function is large. After the measurement is performed, having obtained some result x, the wave function collapses into a position eigenstate centered at x.

The time evolution of a quantum state is described by the Schrödinger equation, in which the Hamiltonian (the operator corresponding to the total energy of the system) generates the time evolution. The time evolution of wave functions is deterministic in the sense that - given a wave function at an *initial* time - it makes a definite prediction of what the wave function will be at any *later* time.

During a measurement, on the other hand, the change of the initial wave function into another, later wave function is not deterministic, it is unpredictable (i.e., random). A time-evolution simulation can be seen here.

Wave functions change as time progresses. The Schrödinger equation describes how wave functions change in time, playing a role similar to Newton's second law in classical mechanics. The Schrödinger equation, applied to the aforementioned example of the free particle, predicts that the center of a wave packet will move through space at a constant velocity (like a classical particle with no forces acting on it). However, the wave packet will also spread out as time progresses, which means that the position becomes more uncertain with time. This also has the effect of turning a position eigenstate (which can be thought of as an infinitely sharp wave packet) into a broadened wave packet that no longer represents a (definite, certain) position eigenstate.

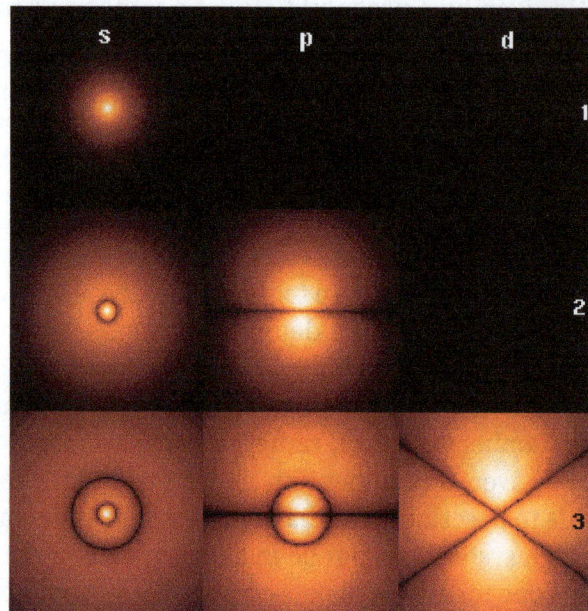

Probability densities corresponding to the wave functions of an electron in a hydrogen atom possessing definite energy levels (increasing from the top of the image to the bottom: $n = 1, 2, 3, ...$) and angular momenta (increasing across from left to right: s, p, d, ...). Brighter areas correspond to higher probability density in a position measurement. Such wave functions are directly comparable to Chladni's figures of acoustic modes of vibration in classical physics, and are modes of oscillation as well, possessing a sharp energy and, thus, a definite frequency. The angular momentum and energy are quantized, and take only discrete values like those shown (as is the case for resonant frequencies in acoustics)

Some wave functions produce probability distributions that are constant, or independent of time—such as when in a stationary state of constant energy, time vanishes in the absolute square of the wave function. Many systems that are treated dynamically in classical mechanics are described by such "static" wave functions. For example, a single electron in an unexcited atom is pictured classically as a particle moving in a circular trajectory around the atomic nucleus, whereas in quantum mechanics it is described by a static, spherically symmetric wave function surrounding the nucleus (Fig. 1) (note, however, that only the lowest angular momentum states, labeled s, are spherically symmetric).

The Schrödinger equation acts on the *entire* probability amplitude, not merely its absolute value. Whereas the absolute value of the probability amplitude encodes information about probabilities, its phase encodes information about the interference between quantum states. This gives rise to the "wave-like" behavior of quantum states. As it turns out, analytic solutions of the Schröding- er equation are available for only a very small number of relatively simple model Hamiltonians, of which the quantum harmonic oscillator, the particle in a box, the dihydrogen cation, and the hydrogen atom are the most important representatives. Even the helium atom—which contains just one more electron than does the hydrogen atom—has defied all attempts at a fully analytic treatment.

There exist several techniques for generating approximate solutions, however. In the important method known as perturbation theory, one uses the analytic result for a simple quantum mechani- cal model to generate a result for a more complicated model that is related to the simpler model by (for one example) the addition of a weak potential energy. Another method is the "semi-classical equation of motion" approach, which applies to systems for which quantum mechanics produces only weak (small) deviations from classical behavior. These deviations can then be computed based on the classical motion. This approach is particularly important in the field of quantum chaos.

Mathematically Equivalent Formulations of Quantum Mechanics

There are numerous mathematically equivalent formulations of quantum mechanics. One of the oldest and most commonly used formulations is the "transformation theory" proposed by Paul Dirac, which unifies and generalizes the two earliest formulations of quantum mechanics - matrix mechanics (invented by Werner Heisenberg) and wave mechanics (invented by Erwin Schrödinger).

Especially since Werner Heisenberg was awarded the Nobel Prize in Physics in 1932 for the cre- ation of quantum mechanics, the role of Max Born in the development of QM was overlooked until the 1954 Nobel award. The role is noted in a 2005 biography of Born, which recounts his role in the matrix formulation of quantum mechanics, and the use of probability amplitudes. Heisen- berg himself acknowledges having learned matrices from Born, as published in a 1940 *festschrift* honoring Max Planck. In the matrix formulation, the instantaneous state of a quantum system encodes the probabilities of its measurable properties, or "observables". Examples of observables include energy, position, momentum, and angular momentum. Observables can be either continu- ous (e.g., the position of a particle) or discrete (e.g., the energy of an electron bound to a hydrogen atom). An alternative formulation of quantum mechanics is Feynman's path integral formulation, in which a quantum-mechanical amplitude is considered as a sum over all possible classical and non-classical paths between the initial and final states. This is the quantum-mechanical counter- part of the action principle in classical mechanics.

Interactions with Other Scientific Theories

The rules of quantum mechanics are fundamental. They assert that the state space of a system is a Hilbert space and that observables of that system are Hermitian operators acting on that space— although they do not tell us which Hilbert space or which operators. These can be chosen appro- priately in order to obtain a quantitative description of a quantum system. An important guide for making these choices is the correspondence principle, which states that the predictions of quan- tum mechanics reduce to those of classical mechanics when a system moves to higher energies or, equivalently, larger quantum numbers, i.e. whereas a single particle exhibits a degree of random- ness, in systems incorporating millions of particles averaging takes over and, at the high energy limit, the statistical probability of random behaviour approaches zero. In other words, classical mechanics is simply a quantum mechanics of large systems. This "high energy" limit is known as the *classical* or *correspondence limit*. One can even start from an established classical model of a particular system, then attempt to guess the underlying quantum model that would give rise to the classical model in the correspondence limit.

When quantum mechanics was originally formulated, it was applied to models whose correspondence limit was non-relativistic classical mechanics. For instance, the well-known model of the quantum harmonic oscillator uses an explicitly non-relativistic expression for the kinetic energy of the oscillator, and is thus a quantum version of the classical harmonic oscillator.

Early attempts to merge quantum mechanics with special relativity involved the replacement of the Schrödinger equation with a covariant equation such as the Klein–Gordon equation or the Dirac equation. While these theories were successful in explaining many experimental results, they had certain unsatisfactory qualities stemming from their neglect of the relativistic creation and annihilation of particles. A fully relativistic quantum theory required the development of quantum field theory, which applies quantization to a field (rather than a fixed set of particles). The first complete quantum field theory, quantum electrodynamics, provides a fully quantum description of the electromagnetic interaction. The full apparatus of quantum field theory is often unnecessary for describing electrodynamic systems. A simpler approach, one that has been employed since the inception of quantum mechanics, is to treat charged particles as quantum mechanical objects being acted on by a classical electromagnetic field. For example, the elementary quantum model of the hydrogen atom describes the electric field of the hydrogen atom using a classical $-e^2 / (4\pi \epsilon_0 r)$ Coulomb potential. This "semi-classical" approach fails if quantum fluctuations in the electromagnetic field play an important role, such as in the emission of photons by charged particles.

Quantum field theories for the strong nuclear force and the weak nuclear force have also been developed. The quantum field theory of the strong nuclear force is called quantum chromodynamics, and describes the interactions of subnuclear particles such as quarks and gluons. The weak nuclear force and the electromagnetic force were unified, in their quantized forms, into a single quantum field theory (known as electroweak theory), by the physicists Abdus Salam, Sheldon Glashow and Steven Weinberg. These three men shared the Nobel Prize in Physics in 1979 for this work.

It has proven difficult to construct quantum models of gravity, the remaining fundamental force. Semi-classical approximations are workable, and have led to predictions such as Hawking radiation. However, the formulation of a complete theory of quantum gravity is hindered by apparent incompatibilities between general relativity (the most accurate theory of gravity currently known) and some of the fundamental assumptions of quantum theory. The resolution of these incompatibilities is an area of active research, and theories such as string theory are among the possible candidates for a future theory of quantum gravity.

Classical mechanics has also been extended into the complex domain, with complex classical mechanics exhibiting behaviors similar to quantum mechanics.

Quantum Mechanics and Classical Physics

Predictions of quantum mechanics have been verified experimentally to an extremely high degree of accuracy. According to the correspondence principle between classical and quantum mechanics, all objects obey the laws of quantum mechanics, and classical mechanics is just an approximation for large systems of objects (or a statistical quantum mechanics of a large collection of particles). The laws of classical mechanics thus follow from the laws of quantum mechanics as a statistical average at the limit of large systems or large quantum numbers. However, chaotic systems do not have good quantum numbers, and quantum chaos studies the relationship between classical and quantum descriptions in these systems.

Quantum coherence is an essential difference between classical and quantum theories as illustrated by the Einstein–Podolsky–Rosen (EPR) paradox — an attack on a certain philosophical interpretation of quantum mechanics by an appeal to local realism. Quantum interference involves adding together *probability amplitudes*, whereas classical "waves" infer that there is an adding together of *intensities*. For microscopic bodies, the extension of the system is much smaller than the coherence length, which gives rise to long-range entanglement and other nonlocal phenomena characteristic of quantum systems. Quantum coherence is not typically evident at macroscopic scales, though an exception to this rule may occur at extremely low temperatures (i.e. approaching absolute zero) at which quantum behavior may manifest itself macroscopically. This is in accordance with the following observations:

- Many macroscopic properties of a classical system are a direct consequence of the quantum behavior of its parts. For example, the stability of bulk matter (consisting of atoms and molecules which would quickly collapse under electric forces alone), the rigidity of solids, and the mechanical, thermal, chemical, optical and magnetic properties of matter are all results of the interaction of electric charges under the rules of quantum mechanics.

- While the seemingly "exotic" behavior of matter posited by quantum mechanics and relativity theory become more apparent when dealing with particles of extremely small size or velocities approaching the speed of light, the laws of classical, often considered "Newtonian", physics remain accurate in predicting the behavior of the vast majority of "large" objects (on the order of the size of large molecules or bigger) at velocities much smaller than the velocity of light.

Copenhagen Interpretation of Quantum Versus Classical Kinematics

A big difference between classical and quantum mechanics is that they use very different kinematic descriptions.

In Niels Bohr's mature view, quantum mechanical phenomena are required to be experiments, with complete descriptions of all the devices for the system, preparative, intermediary, and finally measuring. The descriptions are in macroscopic terms, expressed in ordinary language, supplemented with the concepts of classical mechanics. The initial condition and the final condition of the system are respectively described by values in a configuration space, for example a position space, or some equivalent space such as a momentum space. Quantum mechanics does not admit a completely precise description, in terms of both position and momentum, of an initial condition or "state" (in the classical sense of the word) that would support a precisely deterministic and causal prediction of a final condition. In this sense, advocated by Bohr in his mature writings, a quantum phenomenon is a process, a passage from initial to final condition, not an instantaneous "state" in the classical sense of that word. Thus there are two kinds of processes in quantum mechanics:

stationary and transitional. For a stationary process, the initial and final condition are the same. For a transition, they are different. Obviously by definition, if only the initial condition is given, the process is not determined. Given its initial condition, prediction of its final condition is possible, causally but only probabilistically, because the Schrödinger equation is deterministic for wave function evolution, but the wave function describes the system only probabilistically.

For many experiments, it is possible to think of the initial and final conditions of the system as being a particle. In some cases it appears that there are potentially several spatially distinct pathways or trajectories by which a particle might pass from initial to final condition. It is an important feature of the quantum kinematic description that it does not permit a unique definite statement of which of those pathways is actually followed. Only the initial and final conditions are definite, and, as stated in the foregoing paragraph, they are defined only as precisely as allowed by the configuration space description or its equivalent. In every case for which a quantum kinematic description is needed, there is always a compelling reason for this restriction of kinematic precision. An example of such a reason is that for a particle to be experimentally found in a definite position, it must be held motionless; for it to be experimentally found to have a definite momentum, it must have free motion; these two are logically incompatible.

Classical kinematics does not primarily demand experimental description of its phenomena. It allows completely precise description of an instantaneous state by a value in phase space, the Cartesian product of configuration and momentum spaces. This description simply assumes or imagines a state as a physically existing entity without concern about its experimental measurability. Such a description of an initial condition, together with Newton's laws of motion, allows a precise deterministic and causal prediction of a final condition, with a definite trajectory of passage. Hamiltonian dynamics can be used for this. Classical kinematics also allows the description of a process analogous to the initial and final condition description used by quantum mechanics. Lagrangian mechanics applies to this. For processes that need account to be taken of actions of a small number of Planck constants, classical kinematics is not adequate; quantum mechanics is needed.

General Relativity and Quantum Mechanics

Even with the defining postulates of both Einstein's theory of general relativity and quantum theory being indisputably supported by rigorous and repeated empirical evidence, and while they do not directly contradict each other theoretically (at least with regard to their primary claims), they have proven extremely difficult to incorporate into one consistent, cohesive model.

Gravity is negligible in many areas of particle physics, so that unification between general relativity and quantum mechanics is not an urgent issue in those particular applications. However, the lack of a correct theory of quantum gravity is an important issue in cosmology and the search by physicists for an elegant "Theory of Everything" (TOE). Consequently, resolving the inconsistencies between both theories has been a major goal of 20th and 21st century physics. Many prominent physicists, including Stephen Hawking, have labored for many years in the attempt to discover a theory underlying *everything*. This TOE would combine not only the different models of subatomic physics, but also derive the four fundamental forces of nature - the strong force, electromagnetism, the weak force, and gravity - from a single force or phenomenon. While Stephen Hawking was initially a believer in the Theory of Everything, after considering Gödel's Incompleteness Theorem, he has concluded that one is not obtainable, and has stated so publicly in his lecture "Gödel and the End of Physics" (2002).

Attempts at a Unified Field Theory

The quest to unify the fundamental forces through quantum mechanics is still ongoing. Quantum electrodynamics (or "quantum electromagnetism"), which is currently (in the perturbative regime at least) the most accurately tested physical theory in competition with general relativity, has been successfully merged with the weak nuclear force into the electroweak force and work is currently being done to merge the electroweak and strong force into the electrostrong force. Current predictions state that at around 10^{14} GeV the three aforementioned forces are fused into a single unified field. Beyond this "grand unification", it is speculated that it may be possible to merge gravity with the other three gauge symmetries, expected to occur at roughly 10^{19} GeV. However — and while special relativity is parsimoniously incorporated into quantum electrodynamics — the expanded general relativity, currently the best theory describing the gravitation force, has not been fully incorporated into quantum theory. One of those searching for a coherent TOE is Edward Witten, a theoretical physicist who formulated the M-theory, which is an attempt at describing the supersymmetrical based string theory. M-theory posits that our apparent 4-dimensional spacetime is, in reality, actually an 11-dimensional spacetime containing 10 spatial dimensions and 1 time dimension, although 7 of the spatial dimensions are - at lower energies - completely "compactified" (or infinitely curved) and not readily amenable to measurement or probing.

Another popular theory is Loop quantum gravity (LQG), a theory first proposed by Carlo Rovelli that describes the quantum properties of gravity. It is also a theory of quantum space and quantum time, because in general relativity the geometry of spacetime is a manifestation of gravity. LQG is an attempt to merge and adapt standard quantum mechanics and standard general relativity. The main output of the theory is a physical picture of space where space is granular. The granularity is a direct consequence of the quantization. It has the same nature of the granularity of the photons in the quantum theory of electromagnetism or the discrete levels of the energy of the atoms. But here it is space itself which is discrete. More precisely, space can be viewed as an extremely fine fabric or network "woven" of finite loops. These networks of loops are called spin networks. The evolution of a spin network over time is called a spin foam. The predicted size of this structure is the Planck length, which is approximately 1.616×10^{-35} m. According to theory, there is no meaning to length shorter than this (cf. Planck scale energy). Therefore, LQG predicts that not just matter, but also space itself, has an atomic structure.

Philosophical Implications

Since its inception, the many counter-intuitive aspects and results of quantum mechanics have provoked strong philosophical debates and many interpretations. Even fundamental issues, such as Max Born's basic rules concerning probability amplitudes and probability distributions, took decades to be appreciated by society and many leading scientists. Richard Feynman once said, "I think I can safely say that nobody understands quantum mechanics." According to Steven Weinberg, "There is now in my opinion no entirely satisfactory interpretation of quantum mechanics."

The Copenhagen interpretation — due largely to Niels Bohr and Werner Heisenberg — remains most widely accepted amongst physicists, some 75 years after its enunciation. According to this interpretation, the probabilistic nature of quantum mechanics is not a *temporary* feature which will eventually be replaced by a deterministic theory, but instead must be considered a *final* renunciation of the classical idea of "causality." It is also believed therein that any well-defined application of the quantum mechanical formalism must always make reference to the experimental arrangement, due to the conjugate nature of evidence obtained under different experimental situations.

Albert Einstein, himself one of the founders of quantum theory, did not accept some of the more philosophical or metaphysical interpretations of quantum mechanics, such as rejection of determinism and of causality. He is famously quoted as saying, in response to this aspect, "God does not play with dice". He rejected the concept that the state of a physical system depends on the experimental arrangement for its measurement. He held that a state of nature occurs in its own right, regardless of whether or how it might be observed. In that view, he is supported by the currently accepted definition of a quantum state, which remains invariant under arbitrary choice of configuration space for its representation, that is to say, manner of observation. He also held that underlying quantum mechanics there should be a theory that thoroughly and directly expresses the rule against action at a distance; in other words, he insisted on the principle of locality. He considered, but rejected on theoretical grounds, a particular proposal for hidden variables to obviate the indeterminism or acausality of quantum mechanical measurement. He considered that quantum mechanics was a currently valid but not a permanently definitive theory for quantum phenomena. He thought its future replacement would require profound conceptual advances, and would not come quickly or easily. The Bohr-Einstein debates provide a vibrant critique of the Copenhagen Interpretation from an epistemological point of view. In arguing for his views, he produced a series of objections, the most famous of which has become known as the Einstein–Podolsky–Rosen paradox.

John Bell showed that this "EPR" paradox led to experimentally testable differences between quantum mechanics and theories that rely on added hidden variables. Experiments have been performed confirming the accuracy of quantum mechanics, thereby demonstrating that quantum mechanics cannot be improved upon by addition of hidden variables. Alain Aspect's initial experiments in 1982, and many subsequent experiments since, have definitively verified quantum entanglement.

Entanglement, as demonstrated in Bell-type experiments, does not, however, violate causality, since no transfer of information happens. Quantum entanglement forms the basis of quantum cryptography, which is proposed for use in high-security commercial applications in banking and government.

The Everett many-worlds interpretation, formulated in 1956, holds that *all* the possibilities described by quantum theory *simultaneously* occur in a multiverse composed of mostly independent parallel universes. This is not accomplished by introducing some "new axiom" to quantum mechanics, but on the contrary, by *removing* the axiom of the collapse of the wave packet. *All* of the possible consistent states of the measured system and the measuring apparatus (including the observer) are present in a *real* physical - not just formally mathematical, as in other interpretations - quantum superposition. Such a superposition of consistent state combinations of different systems is called an entangled state. While the multiverse is deterministic, we perceive non-deterministic behavior governed by probabilities, because we can only observe the universe (i.e., the consistent state contribution to the aforementioned superposition) that we, as observers, inhabit. Everett's interpretation is perfectly consistent with John Bell's experiments and makes them intuitively understandable. However, according to the theory of quantum decoherence, these "parallel universes" will never be accessible to us. The inaccessibility can be understood as follows: once a measurement is done, the measured system becomes entangled with *both* the physicist who measured it *and* a huge number of other particles, some of which are photons flying away at the speed of light towards the other end of the universe. In order to prove that the wave function did not collapse, one would have to bring *all* these particles back and measure them again, together with the system that was originally measured. Not only is this completely impractical, but even if one *could* theoretically do this, it would have to destroy any evidence that the original measurement took place (including the physicist's memory). In light of these Bell tests, Cramer (1986) formulated his transactional interpretation. Relational quantum mechanics appeared in the late 1990s as the modern derivative of the Copenhagen Interpretation.

Applications

Quantum mechanics has had enormous success in explaining many of the features of our universe. Quantum mechanics is often the only tool available that can reveal the individual behaviors of the subatomic particles that make up all forms of matter (electrons, protons, neutrons, photons, and others). Quantum mechanics has strongly influenced string theories, candidates for a Theory of Everything.

Quantum mechanics is also critically important for understanding how individual atoms combine covalently to form molecules. The application of quantum mechanics to chemistry is known as quantum chemistry. Relativistic quantum mechanics can, in principle, mathematically describe most of chemistry. Quantum mechanics can also provide quantitative insight into ionic and covalent bonding processes by explicitly showing which molecules are energetically favorable to which others and the magnitudes of the energies involved. Furthermore, most of the calculations performed in modern computational chemistry rely on quantum mechanics.

In many aspects modern technology operates at a scale where quantum effects are significant.

Electronics

Many modern electronic devices are designed using quantum mechanics. Examples include the laser, the transistor (and thus the microchip), the electron microscope, and magnetic resonance imaging (MRI). The study of semiconductors led to the invention of the diode and the transistor, which are indispensable parts of modern electronics systems, computer and telecommunication devices. Another application is the light emitting diode which is a high-efficiency source of light.

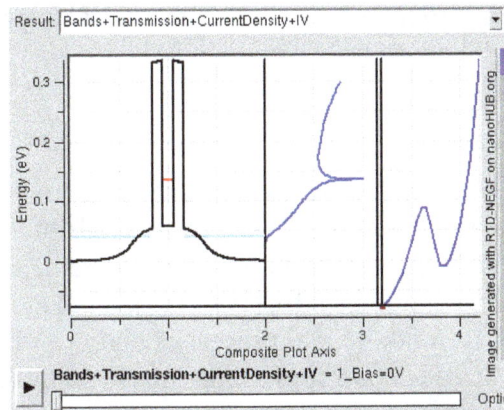

A working mechanism of a resonant tunneling diode device, based on the phenomenon of quantum tunneling through potential barriers. (Left: band diagram; Center: transmission coefficient; Right: current-voltage characteristics) As shown in the band diagram(left), although there are two barriers, electrons still tunnel through via the confined states between two barriers(center), conducting current.

Many electronic devices operate under effect of Quantum tunneling. It even exists in the simple light switch. The switch would not work if electrons could not quantum tunnel through the layer of oxidation on the metal contact surfaces. Flash memory chips found in USB drives use quantum tunneling to erase their memory cells. Some negative differential resistance devices also utilizes quantum tunneling effect, such as resonant tunneling diode. Unlike classical diodes, its current is carried by resonant tunneling through two potential barriers. Its negative resistance behavior can only be understood with quantum mechanics: As the confined state moves close to Fermi level, tunnel current increases. As it moves away, current decreases. Quantum mechanics is vital to understanding and designing such electronic devices.

Cryptography

Researchers are currently seeking robust methods of directly manipulating quantum states. Efforts are being made to more fully develop quantum cryptography, which will theoretically allow guaranteed secure transmission of information.

Quantum Computing

A more distant goal is the development of quantum computers, which are expected to perform certain computational tasks exponentially faster than classical computers. Instead of using classical bits, quantum computers use qubits, which can be in superpositions of states. Another active research topic is quantum teleportation, which deals with techniques to transmit quantum information over arbitrary distances.

Macroscale Quantum Effects

While quantum mechanics primarily applies to the smaller atomic regimes of matter and energy, some systems exhibit quantum mechanical effects on a large scale. Superfluidity, the frictionless flow of a liquid at temperatures near absolute zero, is one well-known example. So is the closely related phenomenon of superconductivity, the frictionless flow of an electron gas in a conducting material (an electric current) at sufficiently low temperatures. The fractional quantum hall effect is a topological ordered state which corresponds to patterns of long-range quantum entanglement. States with different topological orders (or different patterns of long range entanglements) cannot change into each other without a phase transition.

Quantum Theory

Quantum theory also provides accurate descriptions for many previously unexplained phenomena, such as black-body radiation and the stability of the orbitals of electrons in atoms. It has also given insight into the workings of many different biological systems, including smell receptors and protein structures. Recent work on photosynthesis has provided evidence that quantum correlations play an essential role in this fundamental process of plants and many other organisms. Even so, classical physics can often provide good approximations to results otherwise obtained by quantum physics, typically in circumstances with large numbers of particles or large quantum numbers. Since classical formulas are much simpler and easier to compute than quantum formulas, classical approximations are used and preferred when the system is large enough to render the effects of quantum mechanics insignificant.

Examples

Free Particle

For example, consider a free particle. In quantum mechanics, there is wave–particle duality, so the properties of the particle can be described as the properties of a wave. Therefore, its quantum state can be represented as a wave of arbitrary shape and extending over space as a wave function. The position and momentum of the particle are observables. The Uncertainty Principle states that both the position and the momentum cannot simultaneously be measured with complete precision. However, one *can* measure the position (alone) of a moving free particle, creating an eigenstate of position with a wave function that is very large (a Dirac delta) at a particular position x, and zero everywhere else. If one performs a position measurement on such a wave function, the resultant x will be obtained with 100% probability (i.e., with full certainty, or complete precision). This is called an eigenstate of position—or, stated in mathematical terms, a *generalized position eigenstate (eigendistribution)*. If the particle is in an eigenstate of position, then its momentum is completely unknown. On the other hand, if the particle is in an eigenstate of momentum, then its position is completely unknown. In an eigenstate of momentum having a plane wave form, it can be shown that the wavelength is equal to h/p, where h is Planck's constant and p is the momentum of the eigenstate.

Step Potential

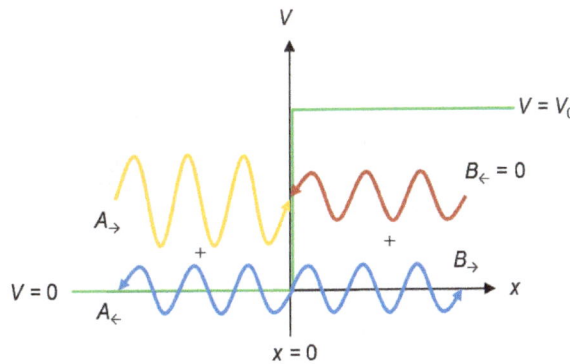

Scattering at a finite potential step of height V_0, shown in green. The amplitudes and direction of left- and right-moving waves are indicated. Yellow is the incident wave, blue are reflected and transmitted waves, red does not occur. $E > V_0$ for this figure.

The potential in this case is given by:

$$V(x) = \begin{cases} 0, & x < 0, \\ V_0, & x \geq 0. \end{cases}$$

The solutions are superpositions of left- and right-moving waves:

$$\psi_1(x) = \frac{1}{\sqrt{k_1}}\left(A_\rightarrow e^{ik_1 x} + A_\leftarrow e^{-ik_1 x}\right) \quad x < 0$$

$$\psi_2(x) = \frac{1}{\sqrt{k_2}}\left(B_\rightarrow e^{ik_2 x} + B_\leftarrow e^{-ik_2 x}\right) \quad x > 0$$

where the wave vectors are related to the energy via

$$k_1 = \sqrt{2mE/\hbar^2}\text{, and } k_2 = \sqrt{2m(E-V_0)/\hbar^2}$$

with coefficients A and B determined from the boundary conditions and by imposing a continuous derivative on the solution.

Each term of the solution can be interpreted as an incident, reflected, or transmitted component of the wave, allowing the calculation of transmission and reflection coefficients. Notably, in contrast to classical mechanics, incident particles with energies greater than the potential step are partially reflected.

Rectangular Potential Barrier

This is a model for the quantum tunneling effect which plays an important role in the performance of modern technologies such as flash memory and scanning tunneling microscopy. Quantum tunneling is central to physical phenomena involved in superlattices.

Particle in a Box

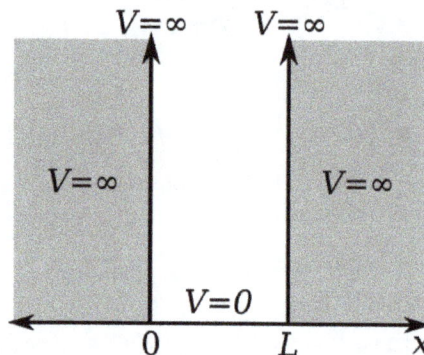

1-dimensional potential energy box (or infinite potential well)

The particle in a one-dimensional potential energy box is the most mathematically simple example where restraints lead to the quantization of energy levels. The box is defined as having zero potential energy everywhere *inside* a certain region, and infinite potential energy everywhere *outside* that region. For the one-dimensional case in the x direction, the time-independent Schrödinger equation may be written

$$-\frac{\hbar^2}{2m}\frac{d^2\psi}{dx^2} = E\psi.$$

With the differential operator defined by

$$\hat{p}_x = -i\hbar\frac{d}{dx}$$

the previous equation is evocative of the classic kinetic energy analogue,

$$\frac{1}{2m}\hat{p}_x^2 = E,$$

with state ψ in this case having energy E coincident with the kinetic energy of the particle.

The general solutions of the Schrödinger equation for the particle in a box are

$$\psi(x) = Ae^{ikx} + Be^{-ikx} \qquad E = \frac{\hbar^2 k^2}{2m}$$

or, from Euler's formula,

$$\psi(x) = C\sin kx + D\cos kx.$$

The infinite potential walls of the box determine the values of C, D, and k at $x = 0$ and $x = L$ where ψ must be zero. Thus, at $x = 0$,

$$\psi(0) = 0 = C\sin 0 + D\cos 0 = D$$

and $D = 0$. At $x = L$,

$$\psi(L) = 0 = C\sin kL.$$

in which C cannot be zero as this would conflict with the Born interpretation. Therefore, since $\sin(kL) = 0$, kL must be an integer multiple of π,

$$k = \frac{n\pi}{L} \qquad n = 1, 2, 3, \ldots.$$

The quantization of energy levels follows from this constraint on k, since

$$E = \frac{\hbar^2 \pi^2 n^2}{2mL^2} = \frac{n^2 h^2}{8mL^2}.$$

Finite Potential Well

A finite potential well is the generalization of the infinite potential well problem to potential wells having finite depth.

The finite potential well problem is mathematically more complicated than the infinite particle-in-a-box problem as the wave function is not pinned to zero at the walls of the well. Instead, the wave function must satisfy more complicated mathematical boundary conditions as it is nonzero in regions outside the well.

Harmonic Oscillator

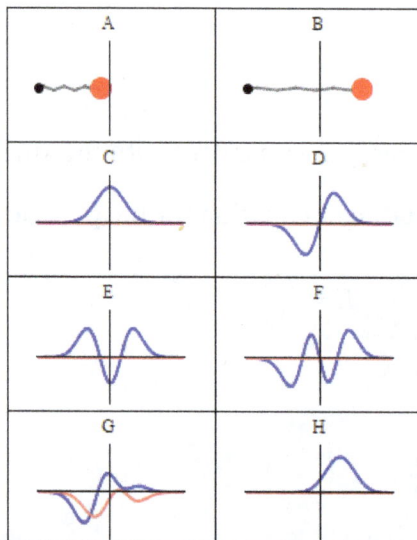

Some trajectories of a harmonic oscillator (i.e. a ball attached to a spring) in classical mechanics (A-B) and quantum mechanics (C-H). In quantum mechanics, the position of the ball is represented by a wave (called the wave function), with the real part shown in blue and the imaginary part shown in red. Some of the trajectories (such as C,D,E,and F) are standing waves (or "stationary states"). Each standing-wave frequency is proportional to a possible energy level of the oscillator. This "energy quantization" does not occur in classical physics, where the oscillator can have *any* energy.

As in the classical case, the potential for the quantum harmonic oscillator is given by

$$V(x) = \frac{1}{2}m\omega^2 x^2$$

This problem can either be treated by directly solving the Schrödinger equation, which is not trivial, or by using the more elegant "ladder method" first proposed by Paul Dirac. The eigenstates are given by

$$\psi_n(x) = \sqrt{\frac{1}{2^n n!}} \cdot \left(\frac{m\omega}{\pi\hbar}\right)^{1/4} \cdot e^{-\frac{m\omega x^2}{2\hbar}} \cdot H_n\left(\sqrt{\frac{m\omega}{\hbar}}x\right),$$

$$n = 0, 1, 2, \ldots$$

where H_n are the Hermite polynomials,

$$H_n(x) = (-1)^n e^{x^2} \frac{d^n}{dx^n}\left(e^{-x^2}\right)$$

and the corresponding energy levels are

$$E_n = \hbar\omega\left(n + \frac{1}{2}\right).$$

This is another example illustrating the quantization of energy for bound states.

Fluid Mechanics

Fluid mechanics is the branch of physics that studies the mechanics of fluids (liquids, gases, and plasmas) and the forces on them. Fluid mechanics has a wide range of applications, including for mechanical engineering, chemical engineering, geophysics, astrophysics, and biology. Fluid mechanics can be divided into fluid statics, the study of fluids at rest; and fluid dynamics, the study of the effect of forces on fluid motion. It is a branch of continuum mechanics, a subject which models matter without using the information that it is made out of atoms; that is, it models matter from a *macroscopic* viewpoint rather than from *microscopic*. Fluid mechanics, especially fluid dynamics, is an active field of research with many problems that are partly or wholly unsolved. Fluid mechanics can be mathematically complex, and can best be solved by numerical methods, typically using computers. A modern discipline, called computational fluid dynamics (CFD), is devoted to this approach to solving fluid mechanics problems. Particle image velocimetry, an experimental method for visualizing and analyzing fluid flow, also takes advantage of the highly visual nature of fluid flow.

Brief History

The study of fluid mechanics goes back at least to the days of ancient Greece, when Archimedes investigated fluid statics and buoyancy and formulated his famous law known now as the Archimedes' principle, which was published in his work *On Floating Bodies* – generally considered to be the first major work on fluid mechanics. Rapid advancement in fluid mechanics began with Leonardo da Vinci (observations and experiments), Evangelista Torricelli (invented the barometer), Isaac Newton (investigated viscosity) and Blaise Pascal (researched hydrostatics, formulated Pascal's law), and was continued by Daniel Bernoulli with the introduction of mathematical fluid dynamics in *Hydrodynamica* (1738).

Inviscid flow was further analyzed by various mathematicians (Leonhard Euler, Jean le Rond d'Alembert, Joseph Louis Lagrange, Pierre-Simon Laplace, Siméon Denis Poisson) and viscous flow was explored by a multitude of engineers including Jean Léonard Marie Poiseuille and Gotthilf Hagen. Further mathematical justification was provided by Claude-Louis Navier and George Gabriel Stokes in the Navier–Stokes equations, and boundary layers were investigated (Ludwig Prandtl, Theodore von Kármán), while various scientists such as Osborne Reynolds, Andrey Kolmogorov, and Geoffrey Ingram Taylor advanced the understanding of fluid viscosity and turbulence.

Main Branches

Fluid Statics

Fluid statics or hydrostatics is the branch of fluid mechanics that studies fluids at rest. It embraces the study of the conditions under which fluids are at rest in stable equilibrium; and is contrasted with fluid dynamics, the study of fluids in motion. Hydrostatics offers physical explanations for many phenomena of everyday life, such as why atmospheric pressure changes with altitude, why wood and oil float on water, and why the surface of water is always flat and horizontal whatever the shape of its container. Hydrostatics is fundamental to hydraulics, the engineering of equipment for storing, transporting and using fluids. It is also relevant to some aspect of geophysics and astrophysics (for example, in understanding plate tectonics and anomalies in the Earth's gravitational field), to meteorology, to medicine (in the context of blood pressure), and many other fields.

Fluid Dynamics

Fluid dynamics is a subdiscipline of fluid mechanics that deals with fluid flow—the science of liquids and gases in motion. Fluid dynamics offers a systematic structure—which underlies these practical disciplines—that embraces empirical and semi-empirical laws derived from flow measurement and used to solve practical problems. The solution to a fluid dynamics problem typically involves calculating various properties of the fluid, such as velocity, pressure, density, and temperature, as functions of space and time. It has several subdisciplines itself, including aerodynamics (the study of air and other gases in motion) and hydrodynamics (the study of liquids in motion). Fluid dynamics has a wide range of applications, including calculating forces and moments on aircraft, determining the mass flow rate of petroleum through pipelines, predicting evolving weather patterns, understanding nebulae in interstellar space and modeling explosions. Some fluid-dynamical principles are used in traffic engineering and crowd dynamics.

Relationship to Continuum Mechanics

Fluid mechanics is a subdiscipline of continuum mechanics, as illustrated in the following table.

Continuum mechanics The study of the physics of continuous materials	Solid mechanics The study of the physics of continuous materials with a defined rest shape.	Elasticity Describes materials that return to their rest shape after applied stresses are removed.	
		Plasticity Describes materials that permanently deform after a sufficient applied stress.	Rheology The study of materials with both solid and fluid characteristics.
	Fluid mechanics The study of the physics of continuous materials which deform when subjected to a force.	Non-Newtonian fluids do not undergo strain rates proportional to the applied shear stress.	
		Newtonian fluids undergo strain rates proportional to the applied shear stress.	

In a mechanical view, a fluid is a substance that does not support shear stress; that is why a fluid at rest has the shape of its containing vessel. A fluid at rest has no shear stress.

Assumptions

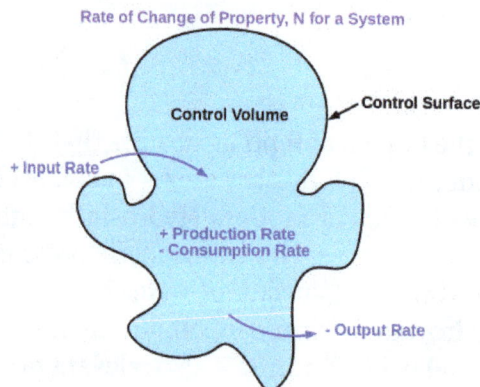

Balance for some integrated fluid quantity in a control volume enclosed by a control surface.

The assumptions inherent to a fluid mechanical treatment of a physical system can be expressed in terms of mathematical equations. Fundamentally, every fluid mechanical system is assumed to obey:

- Conservation of mass

- Conservation of energy

- Conservation of momentum

- The *continuum assumption*

For example, the assumption that mass is conserved means that for any fixed control volume (for example, a spherical volume) – enclosed by a control surface – the rate of change of the mass contained in that volume is equal to the rate at which mass is passing through the surface from *outside* to *inside*, minus the rate at which mass is passing from *inside* to *outside*. This can be expressed as an equation in integral form over the control volume.

The continuum assumption is an idealization of continuum mechanics under which fluids can be treated as continuous, even though, on a microscopic scale, they are composed of molecules. Under the continuum assumption, macroscopic (observed/measurable) properties such as density, pressure, temperature, and bulk velocity are taken to be well-defined at "infinitesimal" volume elements -- small in comparison to the characteristic length scale of the system, but large in comparison to molecular length scale. Fluid properties can vary continuously from one volume element to another and are average values of the molecular properties. The continuum hypothesis can lead to inaccurate results in applications like supersonic speed flows, or molecular flows on nano scale. Those problems for which the continuum hypothesis fails, can be solved using statistical mechanics. To determine whether or not the continuum hypothesis applies, the Knudsen number, defined as the ratio of the molecular mean free path to the characteristic length scale, is evaluated. Problems with Knudsen numbers below 0.1 can be evaluated using continuum hypothesis, but molecular approach (statistical mechanics) can be applied for all ranges of Knudsen numbers.

Navier–Stokes Equations

The Navier–Stokes equations (named after Claude-Louis Navier and George Gabriel Stokes) are differential equations that describe the force balance at a given point within a fluid. For an incompressible fluid with vector velocity field $\mathbf{u}$, the Navier–Stokes equations are

$$\frac{\partial \mathbf{u}}{\partial t} + (\mathbf{u} \cdot \nabla)\mathbf{u} = -\frac{1}{\rho}\nabla P + v\nabla^2 \mathbf{u},$$

where $\mathbf{u}$ is the flow velocity vector. Analogous to Newton's equations of motion, the Navier–Stokes equations describe changes in momentum (force) in response to pressure P and viscosity, parameterized, here, by the kinematic viscosity v.. Occasionally, body forces, such as the gravitational force or Lorentz force are added to the equations. Solutions of the Navier–Stokes equations for a given physical problem must be sought with the help of calculus. In practical terms only the simplest case can be solved exactly in this way. These cases generally involve non-turbulent, steady flow in which the Reynolds number is small. For more complex cases, especially those involving turbulence, such as global weather systems, aerodynamics, hydrodynamics and many more, solutions of the Navier–Stokes equations can currently only be found with the help of computers. This branch of science is called computational fluid dynamics.

Inviscid and Viscous Fluids

An inviscid fluid has no viscosity $v = 0$.. In practice, an inviscid flow is an idealization, one that facilitates mathematical treatment. In fact, purely inviscid flows are only known to be realized in the case of superfluidity. Otherwise, fluids are generally viscous, a property that is often most important within a boundary layer near a solid surface, where the flow must match onto the no-slip condition at the solid. In some cases, the mathematics of a fluid mechanical system can be treated by assuming that the fluid outside of boundary layers is inviscid, and then matching its solution onto that for a thin laminar boundary layer.

For fluid flow over a porous boundary, the fluid velocity can be discontinuous between the free fluid and the fluid in the porous media (this is related to the Beavers and Joseph condition). Further, it is useful at low subsonic speeds to assume that a gas is incompressible — that is, the density of the gas does not change even though the speed and static pressure change.

Newtonian Versus Non-newtonian fluids

A Newtonian fluid (named after Isaac Newton) is defined to be a fluid whose shear stress is linearly proportional to the velocity gradient in the direction perpendicular to the plane of shear. This definition means regardless of the forces acting on a fluid, it *continues to flow*. For example, water is a Newtonian fluid, because it continues to display fluid properties no matter how much it is stirred or mixed. A slightly less rigorous definition is that the drag of a small object being moved slowly through the fluid is proportional to the force applied to the object. (Compare friction). Important fluids, like water as well as most gases, behave – to good approximation – as a Newtonian fluid under normal conditions on Earth.

By contrast, stirring a non-Newtonian fluid can leave a "hole" behind. This will gradually fill up over time – this behaviour is seen in materials such as pudding, oobleck, or sand (although sand isn't strictly a fluid). Alternatively, stirring a non-Newtonian fluid can cause the viscosity to decrease, so the fluid appears "thinner" (this is seen in non-drip paints). There are many types of non-Newtonian fluids, as they are defined to be something that fails to obey a particular property – for example, most fluids with long molecular chains can react in a non-Newtonian manner.

Equations for a Newtonian Fluid

The constant of proportionality between the viscous stress tensor and the velocity gradient is known as the viscosity. A simple equation to describe incompressible Newtonian fluid behaviour is

$$\tau = -\mu \frac{dv}{dy}$$

where

τ is the shear stress exerted by the fluid ("drag")

μ is the fluid viscosity – a constant of proportionality

$\dfrac{dv}{dy}$ is the velocity gradient perpendicular to the direction of shear.

For a Newtonian fluid, the viscosity, by definition, depends only on temperature and pressure, not on the forces acting upon it. If the fluid is incompressible the equation governing the viscous stress (in Cartesian coordinates) is

$$\tau_{ij} = \mu \left(\frac{\partial v_i}{\partial x_j} + \frac{\partial v_j}{\partial x_i} \right)$$

where

τ_{ij} is the shear stress on the ith face of a fluid element in the jth direction

v_i is the velocity in the ith direction

x_j is the jth direction coordinate.

If the fluid is not incompressible the general form for the viscous stress in a Newtonian fluid is

$$\tau_{ij} = \mu \left(\frac{\partial v_i}{\partial x_j} + \frac{\partial v_j}{\partial x_i} - \frac{2}{3} \delta_{ij} \nabla \cdot \mathbf{v} \right) + \kappa \delta_{ij} \nabla \cdot \mathbf{v}$$

where κ is the second viscosity coefficient (or bulk viscosity). If a fluid does not obey this relation, it is termed a non-Newtonian fluid, of which there are several types. Non-Newtonian fluids can be either plastic, Bingham plastic, pseudoplastic, dilatant, thixotropic, rheopectic, viscoelastic.

In some applications another rough broad division among fluids is made: ideal and non-ideal fluids. An Ideal fluid is non-viscous and offers no resistance whatsoever to a shearing force. An ideal fluid really does not exist, but in some calculations, the assumption is justifiable. One example of this is the flow far from solid surfaces. In many cases the viscous effects are concentrated near the solid boundaries (such as in boundary layers) while in regions of the flow field far away from the boundaries the viscous effects can be neglected and the fluid there is treated as it were inviscid (ideal flow). When the viscosity is neglected, the term containing the viscous stress tensor τ in the Navier–Stokes equation vanishes. The equation reduced in this form is called the Euler equation.

Theoretical Physics

Theoretical physics is a branch of physics which employs mathematical models and abstractions of physical objects and systems to rationalize, explain and predict natural phenomena. This is in contrast to experimental physics, which uses experimental tools to probe these phenomena.

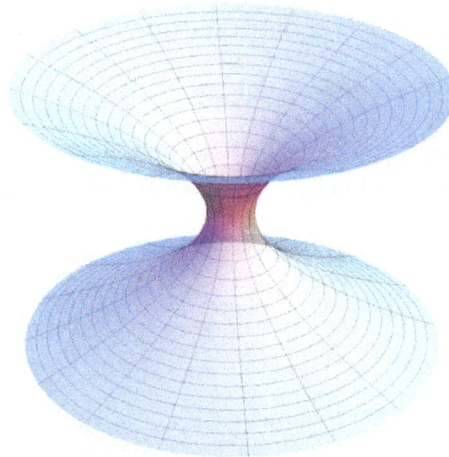

Visual representation of a Schwarzschild wormhole. Wormholes have never been observed, but they are predicted to exist through mathematical models and scientific theory.

The advancement of science depends in general on the interplay between experimental studies and theory. In some cases, theoretical physics adheres to standards of mathematical rigor while giving little weight to experiments and observations. For example, while developing special relativity, Albert Einstein was concerned with the Lorentz transformation which left Maxwell's equations invariant, but was apparently uninterested in the Michelson–Morley experiment on Earth's drift through a luminiferous ether. Conversely, Einstein was awarded the Nobel Prize for explaining the photoelectric effect, previously an experimental result lacking a theoretical formulation.

Overview

A physical theory is a model of physical events. It is judged by the extent to which its predictions agree with empirical observations. The quality of a physical theory is also judged on its ability to make new predictions which can be verified by new observations. A physical theory differs from a mathematical theorem in that while both are based on some form of axioms, judgment of mathematical applicability is not based on agreement with any experimental results. A physical theory similarly differs from a mathematical theory, in the sense that the word "theory" has a different meaning in mathematical terms.

Ric=Kg

" *The equations for an Einstein manifold, used in general relativity to describe the curvature of spacetime* "

A physical theory involves one or more relationships between various measurable quantities. Archimedes realized that a ship floats by displacing its mass of water, Pythagoras understood the relation between the length of a vibrating string and the musical tone it produces. Other examples include entropy as a measure of the uncertainty regarding the positions and motions of unseen particles and the quantum mechanical idea that (action and) energy are not continuously variable.

Theoretical physics consists of several different approaches. In this regard, theoretical particle physics forms a good example. For instance: "phenomenologists" might employ (semi-)

empirical formulas to agree with experimental results, often without deep physical under-standing. "Modelers" (also called "model-builders") often appear much like phenomenolo-gists, but try to model speculative theories that have certain desirable features (rather than on experimental data), or apply the techniques of mathematical modeling to physics problems. Some attempt to create approximate theories, called *effective theories*, because fully devel-oped theories may be regarded as unsolvable or too complicated. Other theorists may try to unify, formalise, reinterpret or generalise extant theories, or create completely new ones alto-gether. Sometimes the vision provided by pure mathematical systems can provide clues to how a physical system might be modeled; e.g., the notion, due to Riemann and others, that space itself might be curved. Theoretical problems that need computational investigation are often the concern of computational physics.

Theoretical advances may consist in setting aside old, incorrect paradigms (e.g., aether theory of light propagation, caloric theory of heat, burning consisting of evolving phlogiston, or astronom-ical bodies revolving around the Earth) or may be an alternative model that provides answers that are more accurate or that can be more widely applied. In the latter case, a correspondence principle will be required to recover the previously known result. Sometimes though, advances may proceed along different paths. For example, an essentially correct theory may need some conceptual or factual revisions; atomic theory, first postulated millennia ago (by several thinkers in Greece and India) and the two-fluid theory of electricity are two cases in this point. However, an exception to all the above is the wave–particle duality, a theory combining aspects of different, opposing models via the Bohr complementarity principle.

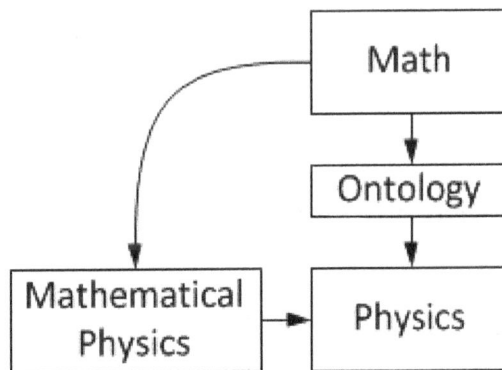

Relationship between mathematics and physics

Physical theories become accepted if they are able to make correct predictions and no (or few) incorrect ones. The theory should have, at least as a secondary objective, a certain economy and elegance (compare to mathematical beauty), a notion sometimes called "Occam's razor" after the 13th-century English philosopher William of Occam (or Ockham), in which the sim-pler of two theories that describe the same matter just as adequately is preferred (but concep-tual simplicity may mean mathematical complexity). They are also more likely to be accepted if they connect a wide range of phenomena. Testing the consequences of a theory is part of the scientific method.

Physical theories can be grouped into three categories: *mainstream theories*, *proposed theories* and *fringe theories*.

History

Theoretical physics began at least 2,300 years ago, under the Pre-socratic philosophy, and continued by Plato and Aristotle, whose views held sway for a millennium. During the rise of medieval universities, the only acknowledged intellectual disciplines were the seven liberal arts of the *Trivium* like grammar, logic, and rhetoric and of the *Quadrivium* like arithmetic, geometry, music and astronomy. During the Middle Ages and Renaissance, the concept of experimental science, the counterpoint to theory, began with scholars such as Ibn al-Haytham and Francis Bacon. As the Scientific Revolution gathered pace, the concepts of matter, energy, space, time and causality slowly began to acquire the form we know today, and other sciences spun off from the rubric of natural philosophy. Thus began the modern era of theory with the Copernican paradigm shift in astronomy, soon followed by Johannes Kepler's expressions for planetary orbits, which summarized the meticulous observations of Tycho Brahe; the works of these men (alongside Galileo's) can perhaps be considered to constitute the Scientific Revolution.

The great push toward the modern concept of explanation started with Galileo, one of the few physicists who was both a consummate theoretician and a great experimentalist. The analytic geometry and mechanics of Descartes were incorporated into the calculus and mechanics of Isaac Newton, another theoretician/experimentalist of the highest order, writing Principia Mathematica. In it contained a grand synthesis of the work of Copernicus, Galileo and Kepler; as well as Newton's theories of mechanics and gravitation, which held sway as worldviews until the early 20th century. Simultaneously, progress was also made in optics (in particular colour theory and the ancient science of geometrical optics), courtesy of Newton, Descartes and the Dutchmen Snell and Huygens. In the 18th and 19th centuries Joseph-Louis Lagrange, Leonhard Euler and William Rowan Hamilton would extend the theory of classical mechanics considerably. They picked up the interactive intertwining of mathematics and physics begun two millennia earlier by Pythagoras.

Among the great conceptual achievements of the 19th and 20th centuries were the consolidation of the idea of energy (as well as its global conservation) by the inclusion of heat, electricity and magnetism, and then light. The laws of thermodynamics, and most importantly the introduction of the singular concept of entropy began to provide a macroscopic explanation for the properties of matter. Statistical mechanics (followed by statistical physics) emerged as an offshoot of thermodynamics late in the 19th century. Another important event in the 19th century was the discovery of electromagnetic theory, unifying the previously separate phenomena of electricity, magnetism and light.

The pillars of modern physics, and perhaps the most revolutionary theories in the history of physics, have been relativity theory and quantum mechanics. Newtonian mechanics was subsumed under special relativity and Newton's gravity was given a kinematic explanation by general relativity. Quantum mechanics led to an understanding of blackbody radiation (which indeed, was an original motivation for the theory) and of anomalies in the specific heats of solids — and finally to an understanding of the internal structures of atoms and molecules. Quantum mechanics soon gave way to the formulation of quantum field theory (QFT), begun in the late 1920s. In the aftermath of World War 2, more progress brought much renewed interest in QFT, which had since the early efforts, stagnated. The same period also saw fresh attacks on the problems of superconductivity and phase transitions, as well as the first applications of QFT in the area of theoretical condensed matter. The 1960s and 70s saw the formulation of the Standard model of particle physics using

QFT and progress in condensed matter physics (theoretical foundation of superconductivity and critical phenomena, among others), in parallel to the applications of relativity to problems in astronomy and cosmology respectively.

All of these achievements depended on the theoretical physics as a moving force both to suggest experiments and to consolidate results — often by ingenious application of existing mathematics, or, as in the case of Descartes and Newton (with Leibniz), by inventing new mathematics. Fourier's studies of heat conduction led to a new branch of mathematics: infinite, orthogonal series.

Modern theoretical physics attempts to unify theories and explain phenomena in further attempts to understand the Universe, from the cosmological to the elementary particle scale. Where experimentation cannot be done, theoretical physics still tries to advance through the use of mathematical models.

Mainstream Theories

Mainstream theories (sometimes referred to as *central theories*) are the body of knowledge of both factual and scientific views and possess a usual scientific quality of the tests of repeatability, consistency with existing well-established science and experimentation. There do exist mainstream theories that are generally accepted theories based solely upon their effects explaining a wide variety of data, although the detection, explanation, and possible composition are subjects of debate.

Examples

- Black hole thermodynamics
- Classical mechanics
- Condensed matter physics (including solid state physics and the electronic structure of materials)
- Conservation of energy
- Dark Energy
- Dark matter
- Dynamics
- Electromagnetism
- Field theory
- Fluid dynamics
- General relativity
- Particle physics
- Physical cosmology
- Quantum chromodynamics
- Quantum computers

- Quantum electrochemistry
- Quantum electrodynamics
- Quantum field theory
- Quantum information theory
- Quantum mechanics
- Quantum Gravity
- Solid mechanics
- Special relativity
- Standard Model
- Statistical mechanics
- Thermodynamics

Proposed Theories

The proposed theories of physics are usually relatively new theories which deal with the study of physics which include scientific approaches, means for determining the validity of models and new types of reasoning used to arrive at the theory. However, some proposed theories include theories that have been around for decades and have eluded methods of discovery and testing. Proposed theories can include fringe theories in the process of becoming established (and, sometimes, gaining wider acceptance). Proposed theories usually have not been tested.

Examples

- Causal Sets
- Dark energy or Einstein's Cosmological Constant
- Einstein–Rosen Bridge
- Emergence
- Grand unification theory
- Loop quantum gravity
- M-theory
- Scale relativity
- String theory
- Supersymmetry
- Theory of everything
- Unparticle physics

Fringe Theories

Fringe theories include any new area of scientific endeavor in the process of becoming established and some proposed theories. It can include speculative sciences. This includes physics fields and physical theories presented in accordance with known evidence, and a body of associated predictions have been made according to that theory.

Some fringe theories go on to become a widely accepted part of physics. Other fringe theories end up being disproven. Some fringe theories are a form of protoscience and others are a form of pseudoscience. The falsification of the original theory sometimes leads to reformulation of the theory.

Examples

- Electrogravitics
- Tesla's dynamic theory of gravity
- Luminiferous aether
- Orgone

Thought Experiments Vs Real Experiments

"Thought" experiments are situations created in one's mind, asking a question akin to "suppose you are in this situation, assuming such is true, what would follow?". They are usually created to investigate phenomena that are not readily experienced in every-day situations. Famous examples of such thought experiments are Schrödinger's cat, the EPR thought experiment, simple illustrations of time dilation, and so on. These usually lead to real experiments designed to verify that the conclusion (and therefore the assumptions) of the thought experiments are correct. The EPR thought experiment led to the Bell inequalities, which were then tested to various degrees of rigor, leading to the acceptance of the current formulation of quantum mechanics and probabilism as a working hypothesis.

References

- Braibant, S.; Giacomelli, G.; Spurio, M. (2009). Particles and Fundamental Interactions: An Introduction to Particle Physics. Springer. pp. 313–314. ISBN 978-94-007-2463-1.
- Lloyd, G.E.R. (1968). Aristotle: The Growth and Structure of His Thought. Cambridge: Cambridge University Press. pp. 134–5. ISBN 0-521-09456-9.
- Galilei, Galileo (1989), Van Helden, Albert, ed., Sidereus Nuncius or The Sidereal Messenger, Chicago: University of Chicago Press, pp. 21, 47, ISBN 0-226-27903-0
- Westfall, Richard S. (1980), Never at Rest: A Biography of Isaac Newton, Cambridge: Cambridge University Press, pp. 731–732, ISBN 0-521-27435-4
- Burtt, Edwin Arthur (2003) [First published 1924], The Metaphysical Foundations of Modern Science (second revised ed.), Mineola, NY: Dover Publications, pp. 30, 41, 241–2, ISBN 9780486425511
- Eddington, A. S. (1988) [1926], Internal Constitution of the Stars, New York: Cambridge University Press, ISBN 0-521-33708-9
- Taylor, Philip L. (2002). A Quantum Approach to Condensed Matter Physics. Cambridge University Press. ISBN 0-521-77103-X.

- Hoddeson, Lillian (1992). Out of the Crystal Maze: Chapters from The History of Solid State Physics. Oxford University Press. ISBN 978-0-19-505329-6.

- Kragh, Helge (2002). Quantum Generations: A History of Physics in the Twentieth Century (Reprint ed.). Princeton University Press. ISBN 978-0-691-09552-3.

- Amusia, M.; Popov, K.; Shaginyan, V.; Stephanovich, V. (2014). "Theory of Heavy-Fermion Compounds - Theory of Strongly Correlated Fermi-Systems". Springer. ISBN 978-3-319-10825-4.

- David J Thouless (12 March 1998). Topological Quantum Numbers in Nonrelativistic Physics. World Scientific. ISBN 978-981-4498-03-6.

- Malcolm F. Collins Professor of Physics McMaster University. Magnetic Critical Scattering. Oxford University Press, USA. ISBN 978-0-19-536440-8.

- Richardson, Robert C. (1988). Experimental Techniques in Condensed Matter Physics at Low Temperatures. Addison-Wesley. ISBN 0-201-15002-6.

- Chaikin, P. M.; Lubensky, T. C. (1995). Principles of condensed matter physics. Cambridge University Press. ISBN 0-521-43224-3.

- Cowan, Robert D. (1981). The Theory of Atomic Structure and Spectra. University of California Press. ISBN 0-520-03821-5.

- J. R. Hook; H. E. Hall (2010). Solid State Physics (2nd ed.). Manchester Physics Series, John Wiley & Sons. ISBN 978-0-471-92804-1.

- Y. B. Band (2010). Light and Matter: Electromagnetism, Optics, Spectroscopy and Lasers. John Wiley & Sons. ISBN 978-0-471-89931-0.

- R. Eisberg; R. Resnick (1985). Quantum Physics of Atoms, Molecules, Solids, Nuclei, and Particles (2nd ed.). John Wiley & Sons. ISBN 978-0-471-87373-0.

- P.W. Atkins (1977). Molecular Quantum Mechanics Parts I and II: An Introduction to QUANTUM CHEMISTRY (Volume 1). Oxford University Press. ISBN 0-19-855129-0.

- Ravaioli, Fawwaz T. Ulaby, Eric Michielssen, Umberto (2010). Fundamentals of applied electromagnetics (6th ed.). Boston: Prentice Hall. p. 13. ISBN 978-0-13-213931-1.

Physics: Theories, Principles and Laws

Classical physics studies the widely applicable theories of physics. Hamiltonian mechanics, Archimedes' principle, physical law, conservation law, inverse-square law, Stokes' law etc. have been explained in this section. The chapter serves as a source to understand the major theories and principles of physics.

Classical Physics

Classical physics refers to theories of physics that predate modern, more complete, or more widely applicable theories. If a currently accepted theory is considered to be "modern," and its introduction represented a major paradigm shift, then the previous theories, or new theories based on the older paradigm, will often be referred to as belonging to the realm of "classical" physics.

As such, the definition of a classical theory depends on context. Classical physical concepts are often used when modern theories are unnecessarily complex for a particular situation.

Overview

Classical theory has at least two distinct meanings in physics. In the context of quantum mechanics, classical theory refers to theories of physics that do not use the quantisation paradigm, which includes classical mechanics and relativity. Likewise, classical field theories, such as general relativity and classical electromagnetism, are those that do not use quantum mechanics. In the context of general and special relativity, classical theories are those that obey Galilean relativity.

Among the branches of theory included in classical physics are:

- Classical mechanics
    - o Newton's laws of motion
    - o Classical Lagrangian and Hamiltonian formalisms
- Classical electrodynamics (Maxwell's Equations)
- Classical thermodynamics
- Special relativity and general relativity
- Classical chaos theory and nonlinear dynamics

Comparison with Modern Physics

In contrast to classical physics, "modern physics" is a slightly looser term which may refer to just quantum physics or to 20th and 21st century physics in general. Modern physics includes quantum theory and relativity, when applicable.

A physical system can be described by classical physics when it satisfies conditions such that the laws of classical physics are approximately valid. In practice, physical objects ranging from those larger than atoms and molecules, to objects in the macroscopic and astronomical realm, can be well-described (understood) with classical mechanics. Beginning at the atomic level and lower, the laws of classical physics break down and generally do not provide a correct description of nature. Electromagnetic fields and forces can be described well by classical electrodynamics at length scales and field strengths large enough that quantum mechanical effects are negligible. Unlike quantum physics, classical physics is generally characterized by the principle of complete determinism, although deterministic interpretations of quantum mechanics do exist.

From the point of view of classical physics as being non-relativistic physics, the predictions of general and special relativity are significantly different than those of classical theories, particularly concerning the passage of time, the geometry of space, the motion of bodies in free fall, and the propagation of light. Traditionally, light was reconciled with classical mechanics by assuming the existence of a stationary medium through which light propagated, the luminiferous aether, which was later shown not to exist.

Mathematically, classical physics equations are those in which Planck's constant does not appear. According to the correspondence principle and Ehrenfest's theorem, as a system becomes larger or more massive the classical dynamics tends to emerge, with some exceptions, such as superfluidity. This is why we can usually ignore quantum mechanics when dealing with everyday objects and the classical description will suffice. However, one of the most vigorous on-going fields of research in physics is classical-quantum correspondence. This field of research is concerned with the discovery of how the laws of quantum physics give rise to classical physics found at the limit of the large scales of the classical level.

Computer Modeling and Manual Calculation, Modern and Classic Comparison

Today a computer performs millions of arithmetic operations in seconds to solve a classical differential equation, while Newton (one of the fathers of the differential calculus) would take hours to solve the same equation by manual calculation, even if he were the discoverer of that particular equation.

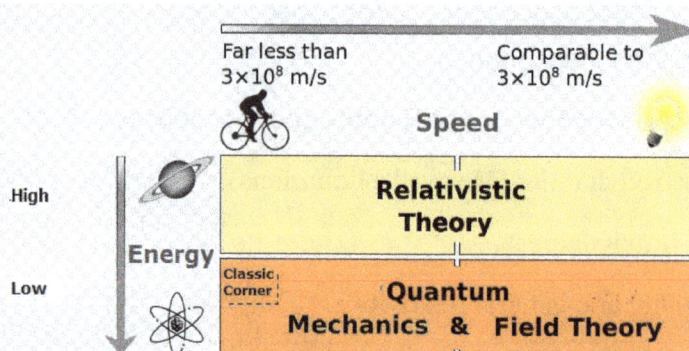

A computer model would use quantum theory and relativistic theory only

Computer modeling is essential for quantum and relativistic physics. Classic physics is considered the limit of quantum mechanics for large number of particles. On the other hand, classic mechanics is derived from relativistic mechanics. For example, in many formulations from special relativity, a correction factor $(v/c)^2$ appears, where v is the velocity of the object and c is the speed of light. For velocities much smaller than that of light, one can neglect the terms with c^2 and higher that appear. These formulas then reduce to the standard definitions of Newtonian kinetic energy and momentum. This is as it should be, for special relativity must agree with Newtonian mechanics at low velocities. Computer modeling has to be as real as possible. Classical physics would introduce an error as in the superfluidity case. In order to produce reliable models of the world, we can not use classic physics. It is true that quantum theories consume time and computer resources, and the equations of classical physics could be resorted to provide a quick solution, but such a solution would lack reliability.

Computer modeling would use only the energy criteria to determine which theory to use: relativity or quantum theory, when attempting to describe the behavior of an object. A physicist would use a classical model to provide an approximation before more exacting models are applied and those calculations proceed.

In a computer model, there is no need to use the speed of the object if classical physics is excluded. Low energy objects would be handled by quantum theory and high energy objects by relativity theory.

Hamiltonian Mechanics

Hamiltonian mechanics is a theory developed as a reformulation of classical mechanics and predicts the same outcomes as non-Hamiltonian classical mechanics. It uses a different mathematical formalism, providing a more abstract understanding of the theory. Historically, it was an important reformulation of classical mechanics, which later contributed to the formulation of statistical mechanics and quantum mechanics.

Hamiltonian mechanics was first formulated by William Rowan Hamilton in 1833, starting from Lagrangian mechanics, a previous reformulation of classical mechanics introduced by Joseph Louis Lagrange in 1788.

Overview

In Hamiltonian mechanics, a classical physical system is described by a set of canonical coordinates $r = (q, p)$, where each component of the coordinate q_i, p_i is indexed to the frame of reference of the system.

The time evolution of the system is uniquely defined by Hamilton's equations:

$$d\frac{d\mathbf{p}}{dt} = -\frac{\partial \mathcal{H}}{\partial \mathbf{q}} \frac{d\mathbf{q}}{dt} = +\frac{\partial \mathcal{H}}{\partial \mathbf{p}}$$

where $\mathcal{H} = \mathcal{H}(q, p, t)$ is the Hamiltonian, which often corresponds to the total energy of the system. For a closed system, it is the sum of the kinetic and potential energy in the system.

In Newtonian mechanics, the time evolution is obtained by computing the total force being exerted on each particle of the system, and from Newton's second law, the time-evolutions of both position and velocity are computed. In contrast, in Hamiltonian mechanics, the time evolution is obtained by computing the Hamiltonian of the system in the generalized coordinates and inserting it in the Hamilton's equations. This approach is equivalent to the one used in Lagrangian mechanics. In fact, as is shown below, the Hamiltonian is the Legendre transform of the Lagrangian when holding q and t fixed and denoting p as the dual variable, and thus both approaches give the same equations for the same generalized momentum. The main motivation to use Hamiltonian mechanics instead of Lagrangian mechanics comes from the symplectic structure of Hamiltonian systems.

While Hamiltonian mechanics can be used to describe simple systems such as a bouncing ball, a pendulum or an oscillating spring in which energy changes from kinetic to potential and back again over time, its strength is shown in more complex dynamic systems, such as planetary orbits in celestial mechanics. The more degrees of freedom the system has, the more complicated its time evolution is and, in most cases, it becomes chaotic.

Basic Physical Interpretation

A simple interpretation of the Hamiltonian mechanics comes from its application on a one-dimensional system consisting of one particle of mass m. The Hamiltonian represents the total energy of the system, which is the sum of kinetic and potential energy, traditionally denoted T and V, respectively. Here q is the space coordinate and p is the momentum, mv. Then

$$\mathcal{H} = T + V, \quad T = \frac{p^2}{2m}, \quad V = V(q).$$

Note that T is a function of p alone, while V is a function of q alone (i.e., T and V are scleronomic).

In this example, the time-derivative of the momentum p equals the *Newtonian force*, and so the first Hamilton equation means that the force equals the negative gradient of potential energy. The time-derivative of q is the velocity, and so the second Hamilton equation means that the particle's velocity equals the derivative of its kinetic energy with respect to its momentum.

Calculating a Hamiltonian from a Lagrangian

Given a Lagrangian in terms of the generalized coordinates q_i and generalized velocities and time,

1. The momenta are calculated by differentiating the Lagrangian with respect to the (generalized) velocities: $p_i(q_i, \dot{q}_i, t) = \frac{\partial \mathcal{L}}{\partial \dot{q}_i}$.

2. The velocities $\dot{q}_i$ are expressed in terms of the momenta p_i by inverting the expressions in the previous step.

3. The Hamiltonian is calculated using the usual definition of $\mathcal{H}$ as the Legendre transformation of $\mathcal{L}$:

$$\mathcal{H} = \sum_i \dot{q}_i \frac{\partial \mathcal{L}}{\partial \dot{q}_i} - \mathcal{L} = \sum_i \dot{q}_i p_i - \mathcal{L}.$$

Then the velocities are substituted for through the above results.

Deriving Hamilton's Equations

Hamilton's equations can be derived by looking at how the total differential of the Lagrangian depends on time, generalized positions q_i and generalized velocities $\dot{q}_i$

$$d\mathcal{L} = \sum_i \left(\frac{\partial \mathcal{L}}{\partial q_i} dq_i + \frac{\partial \mathcal{L}}{\partial \dot{q}_i} d\dot{q}_i \right) + \frac{\partial \mathcal{L}}{\partial t} dt.$$

The generalized momenta were defined as

$$p_i = \frac{\partial \mathcal{L}}{\partial \dot{q}_i}.$$

If this is substituted into the total differential of the Lagrangian, one gets

$$d\mathcal{L} = \sum_i \left(\frac{\partial \mathcal{L}}{\partial q_i} dq_i + p_i d\dot{q}_i \right) + \frac{\partial \mathcal{L}}{\partial t} dt.$$

This can be re-written as

$$d\mathcal{L} = \sum_i \left(\frac{\partial \mathcal{L}}{\partial q_i} dq_i + d(p_i \dot{q}_i) - \dot{q}_i dp_i \right) + \frac{\partial \mathcal{L}}{\partial t} dt.$$

which after re-arranging leads to

$$d\left(\sum_i p_i \dot{q}_i - \mathcal{L} \right) = \sum_i \left(-\frac{\partial \mathcal{L}}{\partial q_i} dq_i + \dot{q}_i dp_i \right) - \frac{\partial \mathcal{L}}{\partial t} dt.$$

The term on the left-hand side is just the Hamiltonian that defined before, therefore

$$d\mathcal{H} = \sum_i \left(-\frac{\partial \mathcal{L}}{\partial q_i} dq_i + \dot{q}_i dp_i \right) - \frac{\partial \mathcal{L}}{\partial t} dt.$$

It is also possible to calculate the total differential of the Hamiltonian $\mathcal{H}$ with respect to time directly, similar to what was carried on with the Lagrangian $\mathcal{L}$ above, yielding:

$$d\mathcal{H} = \sum_i \left(\frac{\partial \mathcal{H}}{\partial q_i} dq_i + \frac{\partial \mathcal{H}}{\partial p_i} dp_i \right) + \frac{\partial \mathcal{H}}{\partial t} dt.$$

It follows from the previous two independent equations that their right-hand sides are equal with each other. The result is

$$\sum_i \left(-\frac{\partial \mathcal{L}}{\partial q_i} dq_i + \dot{q}_i dp_i \right) - \frac{\partial \mathcal{L}}{\partial t} dt = \sum_i \left(\frac{\partial \mathcal{H}}{\partial q_i} dq_i + \frac{\partial \mathcal{H}}{\partial p_i} dp_i \right) + \frac{\partial \mathcal{H}}{\partial t} dt.$$

Since this calculation was done off-shell, one can associate corresponding terms from both sides of this equation to yield:

$$\frac{\partial \mathcal{H}}{\partial q_i} = -\frac{\partial \mathcal{L}}{\partial q_i}, \quad \frac{\partial \mathcal{H}}{\partial p_i} = \dot{q}_i, \quad \frac{\partial \mathcal{H}}{\partial t} = -\frac{\partial \mathcal{L}}{\partial t}.$$

On-shell, Lagrange's equations indicate that

$$\frac{d}{dt} \frac{\partial \mathcal{L}}{\partial \dot{q}_i} - \frac{\partial \mathcal{L}}{\partial q_i} = 0.$$

A re-arrangement of this yields

$$\frac{\partial \mathcal{L}}{\partial q_i} = \dot{p}_i.$$

Thus Hamilton's equations hold on-shell:

$$\frac{\partial \mathcal{H}}{\partial q_j} = -\dot{p}_j, \quad \frac{\partial \mathcal{H}}{\partial p_j} = \dot{q}_j, \quad \frac{\partial \mathcal{H}}{\partial t} = -\frac{\partial \mathcal{L}}{\partial t}.$$

As a Reformulation of Lagrangian Mechanics

Starting with Lagrangian mechanics, the equations of motion are based on generalized coordinates

$$\{q_j \mid j = 1, \dots, N\}$$

and matching generalized velocities

$$\{\dot{q}_j \mid j = 1, \dots, N\}.$$

We write the Lagrangian as

$$\mathcal{L}(q_j, \dot{q}_j, t)$$

with the subscripted variables understood to represent all N variables of that type. Hamiltonian mechanics aims to replace the generalized velocity variables with generalized momentum variables, also known as *conjugate momenta*. By doing so, it is possible to handle certain systems, such as aspects of quantum mechanics, that would otherwise be even more complicated.

For each generalized velocity, there is one corresponding conjugate momentum, defined as:

$$p_j = \frac{\partial \mathcal{L}}{\partial \dot{q}_j}.$$

In Cartesian coordinates, the generalized momenta are precisely the physical linear momenta. In circular polar coordinates, the generalized momentum corresponding to the angular velocity is the physical angular momentum. For an arbitrary choice of generalized coordinates, it may not be possible to obtain an intuitive interpretation of the conjugate momenta.

One thing which is not too obvious in this coordinate dependent formulation is that different generalized coordinates are really nothing more than different coordinate patches on the same symplectic manifold.

The *Hamiltonian* is the Legendre transform of the Lagrangian:

$$\mathcal{H}\left(q_j, p_j, t\right) = \left(\sum_i \dot{q}_i p_i\right) - \mathcal{L}(q_j, \dot{q}_j, t).$$

If the transformation equations defining the generalized coordinates are independent of t, and the Lagrangian is a sum of products of functions (in the generalized coordinates) which are homogeneous of order 0, 1 or 2, then it can be shown that H is equal to the total energy $E = T + V$.

Each side in the definition of $\mathcal{H}$ produces a differential:

$$d\mathcal{H} = \sum_i \left[\left(\frac{\partial \mathcal{H}}{\partial q_i}\right) dq_i + \left(\frac{\partial \mathcal{H}}{\partial p_i}\right) dp_i\right] + \left(\frac{\partial \mathcal{H}}{\partial t}\right) dt$$

$$= \sum_i \left[\dot{q}_i \, dp_i + p_i \, d\dot{q}_i - \left(\frac{\partial \mathcal{L}}{\partial q_i}\right) dq_i - \left(\frac{\partial \mathcal{L}}{\partial \dot{q}_i}\right) d\dot{q}_i\right] - \left(\frac{\partial \mathcal{L}}{\partial t}\right) dt.$$

Substituting the previous definition of the conjugate momenta into this equation and matching coefficients, we obtain the equations of motion of Hamiltonian mechanics, known as the canonical equations of Hamilton:

$$\frac{\partial \mathcal{H}}{\partial q_j} = -\dot{p}_j, \qquad \frac{\partial \mathcal{H}}{\partial p_j} = \dot{q}_j, \qquad \frac{\partial \mathcal{H}}{\partial t} = -\frac{\partial \mathcal{L}}{\partial t}.$$

Hamilton's equations consist of 2n first-order differential equations, while Lagrange's equations consist of n second-order equations. However, Hamilton's equations usually don't reduce the difficulty of finding explicit solutions. They still offer some advantages, since important theoretical results can be derived because coordinates and momenta are independent variables with nearly symmetric roles.

Hamilton's equations have another advantage over Lagrange's equations: if a system has a symmetry, such that a coordinate does not occur in the Hamiltonian, the corresponding momentum is conserved, and that coordinate can be ignored in the other equations of the set. Effectively, this reduces the problem from n coordinates to (n-1) coordinates. In the Lagrangian framework, of course the result that the corresponding momentum is conserved still follows immediately, but all the generalized velocities still occur in the Lagrangian - we still have to solve a system of equations in n coordinates.

The Lagrangian and Hamiltonian approaches provide the groundwork for deeper results in the theory of classical mechanics, and for formulations of quantum mechanics.

Geometry of Hamiltonian Systems

A Hamiltonian system may be understood as a fiber bundle E over time R, with the fibers E_t, $t \in R$, being the position space. The Lagrangian is thus a function on the jet bundle J over E; taking the fiberwise Legendre transform of the Lagrangian produces a function on the dual bundle over time whose fiber at t is the cotangent space T^*E_t, which comes equipped with a natural symplectic form, and this latter function is the Hamiltonian.

Generalization to Quantum Mechanics Through Poisson Bracket

Hamilton's equations above work well for classical mechanics, but not for quantum mechanics, since the differential equations discussed assume that one can specify the exact position and momentum of the particle simultaneously at any point in time. However, the equations can be further generalized to then be extended to apply to quantum mechanics as well as to classical mechanics, through the deformation of the Poisson algebra over p and q to the algebra of Moyal brackets.

Specifically, the more general form of the Hamilton's equation reads

$$\frac{df}{dt} = \{f, \mathcal{H}\} + \frac{\partial f}{\partial t}$$

where f is some function of p and q, and H is the Hamiltonian. To find out the rules for evaluating a Poisson bracket without resorting to differential equations, a Poisson bracket is the name for the Lie bracket in a Poisson algebra. These Poisson brackets can then be extended to Moyal brackets comporting to an inequivalent Lie algebra, as proven by H. Groenewold, and thereby describe quantum mechanical diffusion in phase space. This more algebraic approach not only permits ultimately extending probability distributions in phase space to Wigner quasi-probability distributions, but, at the mere Poisson bracket classical setting, also provides more power in helping analyze the relevant conserved quantities in a system.

Mathematical Formalism

Any smooth real-valued function H on a symplectic manifold can be used to define a Hamiltonian system. The function H is known as the Hamiltonian or the energy function. The symplectic manifold is then called the phase space. The Hamiltonian induces a special vector field on the symplectic manifold, known as the Hamiltonian vector field.

The Hamiltonian vector field (a special type of symplectic vector field) induces a Hamiltonian flow on the manifold. This is a one-parameter family of transformations of the manifold (the parameter of the curves is commonly called the time); in other words an isotopy of symplectomorphisms, starting with the identity. By Liouville's theorem, each symplectomorphism preserves the volume form on the phase space. The collection of symplectomorphisms induced by the Hamiltonian flow is commonly called the Hamiltonian mechanics of the Hamiltonian system.

The symplectic structure induces a Poisson bracket. The Poisson bracket gives the space of functions on the manifold the structure of a Lie algebra.

Given a function f

$$\frac{\mathrm{d}}{\mathrm{d}t}f = \frac{\partial}{\partial t}f + \{f, \mathcal{H}\}.$$

If we have a probability distribution, ρ, then (since the phase space velocity ($\dot{p}_i, \dot{q}_i$) has zero divergence, and probability is conserved) its convective derivative can be shown to be zero and so

$$\frac{\partial}{\partial t}\rho = -\{\rho, \mathcal{H}\}.$$

This is called Liouville's theorem. Every smooth function G over the symplectic manifold generates a one-parameter family of symplectomorphisms and if $\{G, H\} = 0$, then G is conserved and the symplectomorphisms are symmetry transformations.

A Hamiltonian may have multiple conserved quantities G_i. If the symplectic manifold has dimension $2n$ and there are n functionally independent conserved quantities G_i which are in involution (i.e., $\{G_i, G_j\} = 0$), then the Hamiltonian is Liouville integrable. The Liouville-Arnold theorem says that locally, any Liouville integrable Hamiltonian can be transformed via a symplectomorphism into a new Hamiltonian with the conserved quantities G_i as coordinates; the new coordinates are called *action-angle coordinates*. The transformed Hamiltonian depends only on the G_i, and hence the equations of motion have the simple form

$$\dot{G}_i = 0, \qquad \dot{\varphi}_i = F(G),$$

for some function F (Arnol'd et al., 1988). There is an entire field focusing on small deviations from integrable systems governed by the KAM theorem.

The integrability of Hamiltonian vector fields is an open question. In general, Hamiltonian systems are chaotic; concepts of measure, completeness, integrability and stability are poorly defined. At this time, the study of dynamical systems is primarily qualitative, and not a quantitative science.

Riemannian Manifolds

An important special case consists of those Hamiltonians that are quadratic forms, that is, Hamiltonians that can be written as

$$\mathcal{H}(q, p) = \frac{1}{2}\langle p, p\rangle_q$$

where $\langle\cdot, \cdot\rangle_q$ is a smoothly varying inner product on the fibers $T_q^*Q,$, the cotangent space to the point q in the configuration space, sometimes called a cometric. This Hamiltonian consists entirely of the kinetic term.

If one considers a Riemannian manifold or a pseudo-Riemannian manifold, the Riemannian metric induces a linear isomorphism between the tangent and cotangent bundles. Using this isomorphism, one can define a cometric. (In coordinates, the matrix defining the cometric is the inverse of the matrix defining the metric.) The solutions to the Hamilton–Jacobi equations for this Hamiltonian are then the same as the geodesics on the manifold. In particular, the Hamiltonian flow in this case is the same thing as the geodesic flow.

Sub-riemannian Manifolds

When the cometric is degenerate, then it is not invertible. In this case, one does not have a Riemannian manifold, as one does not have a metric. However, the Hamiltonian still exists. In the case where the cometric is degenerate at every point q of the configuration space manifold Q, so that the rank of the cometric is less than the dimension of the manifold Q, one has a sub-Riemannian manifold.

The Hamiltonian in this case is known as a sub-Riemannian Hamiltonian. Every such Hamiltonian uniquely determines the cometric, and vice versa. This implies that every sub-Riemannian manifold is uniquely determined by its sub-Riemannian Hamiltonian, and that the converse is true: every sub-Riemannian manifold has a unique sub-Riemannian Hamiltonian. The existence of sub-Riemannian geodesics is given by the Chow–Rashevskii theorem.

The continuous, real-valued Heisenberg group provides a simple example of a sub-Riemannian manifold. For the Heisenberg group, the Hamiltonian is given by

$$\mathcal{H}(x,y,z,p_x,p_y,p_z) = \frac{1}{2}\left(p_x^2 + p_y^2\right).$$

p_z is not involved in the Hamiltonian.

Poisson Algebras

Hamiltonian systems can be generalized in various ways. Instead of simply looking at the algebra of smooth functions over a symplectic manifold, Hamiltonian mechanics can be formulated on general commutative unital real Poisson algebras. A state is a continuous linear functional on the Poisson algebra (equipped with some suitable topology) such that for any element A of the algebra, A^2 maps to a nonnegative real number.

Charged Particle in an Electromagnetic Field

A good illustration of Hamiltonian mechanics is given by the Hamiltonian of a charged particle in an electromagnetic field. In Cartesian coordinates (i.e. $q_i = x_i$), the Lagrangian of a non-relativistic classical particle in an electromagnetic field is (in SI Units):

$$\mathcal{L} = \sum_i \tfrac{1}{2}m\dot{x}_i^2 + \sum_i e\dot{x}_i A_i - e\phi,$$

where e is the electric charge of the particle (not necessarily the elementary charge), ϕ is the electric scalar potential, and the A_i are the components of the magnetic vector potential (these may be modified through a gauge transformation). This is called minimal coupling.

The generalized momenta are given by:

$$p_i = \frac{\partial \mathcal{L}}{\partial \dot{x}_i} = m\dot{x}_i + eA_i.$$

Rearranging, the velocities are expressed in terms of the momenta:

$$\dot{x}_i - \frac{p_i - eA_i}{m}.$$

If we substitute the definition of the momenta, and the definitions of the velocities in terms of the momenta, into the definition of the Hamiltonian given above, and then simplify and rearrange, we get:

$$\mathcal{H} = \left\{ \sum_i \dot{x}_i p_i \right\} - \mathcal{L} = \sum_i \frac{(p_i - eA_i)^2}{2m} + e\phi.$$

This equation is used frequently in quantum mechanics.

Relativistic Charged Particle in an Electromagnetic Field

The Lagrangian for a relativistic charged particle is given by:

$$\mathcal{L}(t) = -mc^2 \sqrt{1 - \frac{\dot{\vec{x}}(t)^2}{c^2}} - e\phi(\vec{x}(t),t) + e\dot{\vec{x}}(t) \cdot \vec{A}(\vec{x}(t),t).$$

Thus the particle's canonical (total) momentum is

$$\vec{P}(t) = \frac{\partial \mathcal{L}(t)}{\partial \dot{\vec{x}}(t)} = \frac{m\dot{\vec{x}}(t)}{\sqrt{1 - \frac{\dot{\vec{x}}(t)^2}{c^2}}} + e\vec{A}(\vec{x}(t),t),$$

that is, the sum of the kinetic momentum and the potential momentum.

Solving for the velocity, we get

$$\dot{\vec{x}}(t) = \frac{\vec{P}(t) - e\vec{A}(\vec{x}(t),t)}{\sqrt{m^2 + \frac{1}{c^2}\left(\vec{P}(t) - e\vec{A}(\vec{x}(t),t)\right)^2}}.$$

So the Hamiltonian is

$$\mathcal{H}(t) = \dot{\vec{x}}(t) \cdot \vec{P}(t) - \mathcal{L}(t) = c\sqrt{m^2c^2 + \left(\vec{P}(t) - e\vec{A}(\vec{x}(t),t)\right)^2} + e\phi(\vec{x}(t),t).$$

From this we get the force equation (equivalent to the Euler–Lagrange equation)

$$\dot{\vec{P}} = -\frac{\partial \mathcal{H}}{\partial \vec{x}} = e(\vec{\nabla}\vec{A}) \cdot \dot{\vec{x}} - e\vec{\nabla}\phi$$

from which one can derive

$$\frac{d}{dt}\left(\frac{m\dot{\vec{x}}}{\sqrt{1 - \frac{\dot{x}^2}{c^2}}}\right) = e\vec{E} + e\dot{\vec{x}} \times \vec{B}.$$

An equivalent expression for the Hamiltonian as function of the relativistic (kinetic) momentum, $\vec{p} = \gamma m \dot{\vec{x}}(t),$ is

$$\mathcal{H}(t) = \dot{\vec{x}}(t) \cdot \vec{p}(t) + \frac{mc^2}{\gamma} + e\phi(\vec{x}(t),t) = \gamma mc^2 + e\phi(\vec{x}(t),t) = E + V.$$

This has the advantage that $\vec{p}$ can be measured experimentally whereas $\vec{P}$ cannot. Notice that the Hamiltonian (total energy) can be viewed as the sum of the relativistic energy (kinetic+rest), plus the potential energy,

Archimedes' Principle

Archimedes' principle indicates that the upward buoyant force that is exerted on a body immersed in a fluid, whether fully or partially submerged, is equal to the weight of the fluid that the body displaces and it acts in the upward direction at the centre of mass of the displaced fluid. Archimedes' principle is a law of physics fundamental to fluid mechanics. It was formulated by Archimedes of Syracuse.

Explanation

In *On Floating Bodies*, Archimedes suggested that (c. 250 BC):

Any object, wholly or partially immersed in a fluid, is buoyed up by a force equal to the weight of the fluid displaced by the object.

Practically, the Archimedes' principle allows the buoyancy of an object partially or fully immersed in a liquid to be calculated. The downward force on the object is simply its weight. The upward, or buoyant force on the object is that stated by Archimedes' principle, above. Thus the net upward force on the object is the difference between the buoyant force and its weight. If this net force is positive, the object rises; if negative, the object sinks; and if zero, the object is neutrally buoyant - that is, it remains in place without either rising or sinking. In simple words, Archimedes' principle states that when a body is partially or completely immersed in a fluid, it experiences an apparent loss in weight which is equal to the weight of the fluid displaced by the immersed part of the body.

Formula

Consider a cube immersed in a fluid, with its sides parallel to the direction of gravity. The fluid will exert a normal force on each face, and therefore only the forces on the top and bottom faces will contribute to buoyancy. The pressure difference between the bottom and the top face is directly proportional to the height (difference in depth). Multiplying the pressure difference by the area of a face gives the net force on the cube – the buoyancy, or the weight of the fluid displaced. By extending this reasoning to irregular shapes, we can see that, whatever the shape of the submerged body, the buoyant force is equal to the weight of the fluid displaced. apparent loss in wt of water= wt of object in air-wt of object in water

The weight of the displaced fluid is directly proportional to the volume of the displaced fluid (if the surrounding fluid is of uniform density). The weight of the object in the fluid is reduced, because of the force acting on it, which is called upthrust. In simple terms, the principle states that the buoyant force on an object is equal to the weight of the fluid displaced by the object, or the density of the fluid multiplied by the submerged volume times the gravitational constant, g. Thus, among completely submerged objects with equal masses, objects with greater volume have greater buoyancy.

Suppose a rock's weight is measured as 10 newtons when suspended by a string in a vacuum with gravity acting on it. Suppose that when the rock is lowered into water, it displaces water of weight 3 newtons. The force it then exerts on the string from which it hangs would be 10 newtons minus the 3 newtons of buoyant force: 10 − 3 = 7 newtons. Buoyancy reduces the apparent weight of objects that have sunk completely to the sea floor. It is generally easier to lift an object up through the water than it is to pull it out of the water.

For a fully submerged object, Archimedes' principle can be reformulated as follows:

apparent immersed weight = weight of object − weight of displaced fluid

then inserted into the quotient of weights, which has been expanded by the mutual volume

$$\frac{\text{density of object}}{\text{density of fluid}} = \frac{\text{weight}}{\text{weight of displaced fluid}}$$

yields the formula below. The density of the immersed object relative to the density of the fluid can easily be calculated without measuring any volumes is

$$\frac{\text{density of object}}{\text{density of fluid}} = \frac{\text{weight}}{\text{weight} - \text{apparent immersed weight}}.$$

(This formula is used for example in describing the measuring principle of a dasymeter and of hydrostatic weighing.)

Example: If you drop wood into water, buoyancy will keep it afloat.

Example: A helium balloon in a moving car. When increasing speed or driving in a curve, the air moves in the opposite direction to the car's acceleration. However, due to buoyancy, the balloon is pushed "out of the way" by the air, and will actually drift in the same direction as the car's acceleration.

When an object is immersed in a liquid, the liquid exerts an upward force, which is known as the buoyant force, that is proportional to the weight of the displaced liquid. The sum force acting on the object, then, is equal to the difference between the weight of the object ('down' force) and the weight of displaced liquid ('up' force). Equilibrium, or neutral buoyancy, is achieved when these two weights (and thus forces) are equal.

Refinements

Archimedes' principle does not consider the surface tension (capillarity) acting on the body. Moreover, Archimedes' principle has been found to break down in complex fluids.

Principle of Flotation

Archimedes' principle shows buoyant force and displacement of fluid. However, the concept of Archimedes' principle can be applied when considering why objects float. Proposition 5 of Archimedes' treatise *On Floating Bodies* states that:

Any floating object displaces its own weight of fluid.

—Archimedes of Syracuse

In other words, for an object floating on a liquid surface (like a boat) or floating submerged in a fluid (like a submarine in water or dirigible in air) the weight of the displaced liquid equals the weight of the object. Thus, only in the special case of floating does the buoyant force acting on an object equal the objects weight. Consider a 1-ton block of solid iron. As iron is nearly eight times denser than water, it displaces only 1/8 ton of water when submerged, which is not enough to keep it afloat. Suppose the same iron block is reshaped into a bowl. It still weighs 1 ton, but when it is put in water, it displaces a greater volume of water than when it was a block. The deeper the iron bowl is immersed, the more water it displaces, and the greater the buoyant force acting on it. When the buoyant force equals 1 ton, it will sink no farther.

When any boat displaces a weight of water equal to its own weight, it floats. This is often called the "principle of flotation": A floating object displaces a weight of fluid equal to its own weight. Every ship, submarine, and dirigible must be designed to displace a weight of fluid at least equal to its own weight. A 10,000-ton ship must be built wide enough to displace 10,000 tons of water before it sinks too deep in the water. The same is true for vessels in air: a dirigible that weighs 100 tons needs to displace 100 tons of air. If it displaces more, it rises; if it displaces less, it falls. If the dirigible displaces exactly its weight, it hovers at a constant altitude.

It is important to realize that, while they are related to it, the principle of flotation and the concept that a submerged object displaces a volume of fluid equal to its own volume are *not* Archimedes' principle. Archimedes' principle, as stated above, equates the *buoyant force* to the weight of the fluid displaced.

One common point of confusion regarding Archimedes' principle is the meaning of displaced volume. Common demonstrations involve measuring the rise in water level when an object floats on the surface in order to calculate the displaced water.This measurement approach fails with a buoyant submerged object because the rise in the water level is directly related to the volume of the object and not the mass (except if the effective density of the object equals exactly the fluid density). Instead, in the case of submerged buoyant objects, the whole volume of fluid directly above the sample should be considered as the displaced volume. Another common point of confusion regarding Archimedes' principle is that it only applies to submerged objects that are buoyant, not sunk objects. In the case of a sunk object the mass of displaced fluid is less than the mass of the object and the difference is associated with the object's potential energy.

Bernoulli's Principle

In fluid dynamics, Bernoulli's principle states that an increase in the speed of a fluid occurs simultaneously with a decrease in pressure or a decrease in the fluid's potential energy. The principle is named after Daniel Bernoulli who published it in his book *Hydrodynamica* in 1738.

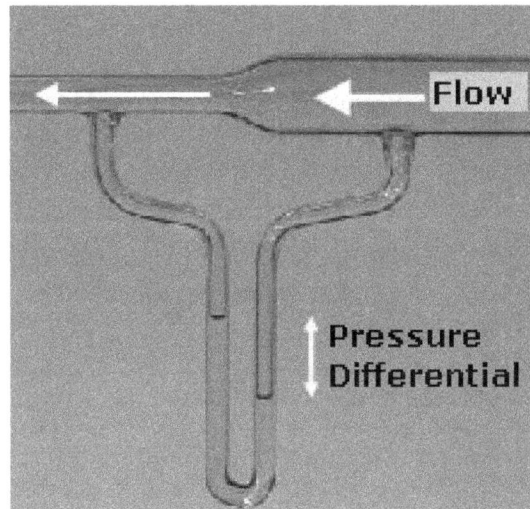

A flow of water into a venturi meter. The kinetic energy increases at the expense of the fluid pressure, as shown by the difference in height of the two columns of water.

Bernoulli's principle can be applied to various types of fluid flow, resulting in various forms of Bernoulli's equation; there are different forms of Bernoulli's equation for different types of flow. The simple form of Bernoulli's equation is valid for incompressible flows (e.g. most liquid flows and gases moving at low Mach number). More advanced forms may be applied to compressible flows at higher Mach numbers.

Bernoulli's principle can be derived from the principle of conservation of energy. This states that, in a steady flow, the sum of all forms of energy in a fluid along a streamline is the same at all points on that streamline. This requires that the sum of kinetic energy, potential energy and internal energy remains constant. Thus an increase in the speed of the fluid – implying an increase in both its dynamic pressure and kinetic energy – occurs with a simultaneous decrease in (the sum of) its static pressure, potential energy and internal energy. If the fluid is flowing out of a reservoir, the sum of all forms of energy is the same on all streamlines because in a reservoir the energy per unit volume (the sum of pressure and gravitational potential $\rho\, g\, h$) is the same everywhere.

Bernoulli's principle can also be derived directly from Newton's 2nd law. If a small volume of fluid is flowing horizontally from a region of high pressure to a region of low pressure, then there is more pressure behind than in front. This gives a net force on the volume, accelerating it along the streamline.

Fluid particles are subject only to pressure and their own weight. If a fluid is flowing horizontally and along a section of a streamline, where the speed increases it can only be because the fluid on that section has moved from a region of higher pressure to a region of lower pressure; and if its speed decreases, it can only be because it has moved from a region of lower pressure to a region of higher pressure. Consequently, within a fluid flowing horizontally, the highest speed occurs where the pressure is lowest, and the lowest speed occurs where the pressure is highest.

Incompressible Flow Equation

In most flows of liquids, and of gases at low Mach number, the density of a fluid parcel can be considered to be constant, regardless of pressure variations in the flow. Therefore, the fluid can be considered to be incompressible and these flows are called incompressible flows. Bernoulli performed his experiments on liquids, so his equation in its original form is valid only for incompressible flow. A common form of Bernoulli's equation, valid at any arbitrary point along a streamline, is:

$$\frac{v^2}{2} + gz + \frac{p}{\rho} = \text{constant} \tag{A}$$

where:

> v is the fluid flow speed at a point on a streamline,

> g is the acceleration due to gravity,

> z is the elevation of the point above a reference plane, with the positive z-direction pointing upward – so in the direction opposite to the gravitational acceleration,

> p is the pressure at the chosen point, and

> ρ is the density of the fluid at all points in the fluid.

The constant on the right-hand side of the equation depends only on the streamline chosen, whereas v, z and p depend on the particular point on that streamline.

The following assumptions must be met for this Bernoulli equation to apply:

- the flow must be steady, i.e. the fluid velocity at a point cannot change with time,

- the flow must be incompressible – even though pressure varies, the density must remain constant along a streamline;

- friction by viscous forces has to be negligible.

For conservative force fields (not limited to the gravitational field), Bernoulli's equation can be generalized as:

$$\frac{v^2}{2} + \Psi + \frac{p}{\rho} = \text{constant}$$

where Ψ is the force potential at the point considered on the streamline. E.g. for the Earth's gravity $\Psi = gz$.

By multiplying with the fluid density ρ, equation (A) can be rewritten as:

$$\tfrac{1}{2}\rho v^2 + \rho g z + p = \text{constant}$$

or:

$$q + \rho g h = p_0 + \rho g z = \text{constant}$$

where:

$q=\frac{1}{2}\rho v^2$ is dynamic pressure,

$h=z+\dfrac{p}{\rho g}$ is the piezometric head or hydraulic head (the sum of the elevation z and the pressure head) and

$p_0=p+q$ is the total pressure (the sum of the static pressure p and dynamic pressure q).

The constant in the Bernoulli equation can be normalised. A common approach is in terms of total head or energy head H:

$$H=z+\frac{p}{\rho g}+\frac{v^2}{2g}=h+\frac{v^2}{2g},$$

The above equations suggest there is a flow speed at which pressure is zero, and at even higher speeds the pressure is negative. Most often, gases and liquids are not capable of negative absolute pressure, or even zero pressure, so clearly Bernoulli's equation ceases to be valid before zero pressure is reached. In liquids – when the pressure becomes too low – cavitation occurs. The above equations use a linear relationship between flow speed squared and pressure. At higher flow speeds in gases, or for sound waves in liquid, the changes in mass density become significant so that the assumption of constant density is invalid.

Simplified Form

In many applications of Bernoulli's equation, the change in the ρ g z term along the streamline is so small compared with the other terms that it can be ignored. For example, in the case of aircraft in flight, the change in height z along a streamline is so small the ρ g z term can be omitted. This allows the above equation to be presented in the following simplified form:

$p+q = p_0$

where p_0 is called 'total pressure', and q is 'dynamic pressure'. Many authors refer to the pressure p as static pressure to distinguish it from total pressure p_0 and dynamic pressure q. In *Aerodynamics*, L.J. Clancy writes: "To distinguish it from the total and dynamic pressures, the actual pressure of the fluid, which is associated not with its motion but with its state, is often referred to as the static pressure, but where the term pressure alone is used it refers to this static pressure."

The simplified form of Bernoulli's equation can be summarized in the following memorable word equation:

static pressure + dynamic pressure = total pressure

Every point in a steadily flowing fluid, regardless of the fluid speed at that point, has its own unique static pressure p and dynamic pressure q. Their sum $p + q$ is defined to be the total pressure p_0. The significance of Bernoulli's principle can now be summarized as *total pressure is constant along a streamline*.

If the fluid flow is irrotational, the total pressure on every streamline is the same and Bernoulli's principle can be summarized as *total pressure is constant everywhere in the fluid flow*. It is reasonable to assume that irrotational flow exists in any situation where a large body of fluid is flowing past a solid body. Examples are aircraft in flight, and ships moving in open bodies of water. However, it is important to remember that Bernoulli's principle does not apply in the boundary layer or in fluid flow through long pipes.

If the fluid flow at some point along a stream line is brought to rest, this point is called a stagnation point, and at this point the total pressure is equal to the stagnation pressure.

Applicability of Incompressible Flow Equation to Flow of Gases

Bernoulli's equation is sometimes valid for the flow of gases: provided that there is no transfer of kinetic or potential energy from the gas flow to the compression or expansion of the gas. If both the gas pressure and volume change simultaneously, then work will be done on or by the gas. In this case, Bernoulli's equation – in its incompressible flow form – cannot be assumed to be valid. However, if the gas process is entirely isobaric, or isochoric, then no work is done on or by the gas, (so the simple energy balance is not upset). According to the gas law, an isobaric or isochoric process is ordinarily the only way to ensure constant density in a gas. Also the gas density will be proportional to the ratio of pressure and absolute temperature, however this ratio will vary upon compression or expansion, no matter what non-zero quantity of heat is added or removed. The only exception is if the net heat transfer is zero, as in a complete thermodynamic cycle, or in an individual isentropic (frictionless adiabatic) process, and even then this reversible process must be reversed, to restore the gas to the original pressure and specific volume, and thus density. Only then is the original, unmodified Bernoulli equation applicable. In this case the equation can be used if the flow speed of the gas is sufficiently below the speed of sound, such that the variation in density of the gas (due to this effect) along each streamline can be ignored. Adiabatic flow at less than Mach 0.3 is generally considered to be slow enough.

Unsteady Potential Flow

The Bernoulli equation for unsteady potential flow is used in the theory of ocean surface waves and acoustics.

For an irrotational flow, the flow velocity can be described as the gradient $\nabla\varphi$ of a velocity potential φ. In that case, and for a constant density ρ, the momentum equations of the Euler equations can be integrated to:

$$\frac{\partial\varphi}{\partial t}+\frac{1}{2}v^2+\frac{p}{\rho}+gz = f(t),$$

which is a Bernoulli equation valid also for unsteady—or time dependent—flows. Here $\partial\varphi/\partial t$ denotes the partial derivative of the velocity potential φ with respect to time t, and $v = |\nabla\varphi|$ is the flow speed. The function $f(t)$ depends only on time and not on position in the fluid. As a result, the Bernoulli equation at some moment t does not only apply along a certain streamline, but in the whole fluid domain. This is also true for the special case of a steady irrotational flow, in which case f is a constant.

Further $f(t)$ can be made equal to zero by incorporating it into the velocity potential using the transformation

$$\Phi = \varphi - \int_{t_0}^{t} f(\tau)d\tau,$$

resulting in $\quad \dfrac{\partial \Phi}{\partial t} + \tfrac{1}{2}v^2 + \dfrac{p}{\rho} + gz = 0$

Note that the relation of the potential to the flow velocity is unaffected by this transformation: $\nabla\Phi = \nabla\varphi$.

The Bernoulli equation for unsteady potential flow also appears to play a central role in Luke's variational principle, a variational description of free-surface flows using the Lagrangian.

Compressible Flow Equation

Bernoulli developed his principle from his observations on liquids, and his equation is applicable only to incompressible fluids, and compressible fluids up to approximately Mach number 0.3. It is possible to use the fundamental principles of physics to develop similar equations applicable to compressible fluids. There are numerous equations, each tailored for a particular application, but all are analogous to Bernoulli's equation and all rely on nothing more than the fundamental principles of physics such as Newton's laws of motion or the first law of thermodynamics.

Compressible Flow in Fluid Dynamics

For a compressible fluid, with a barotropic equation of state, and under the action of conservative forces,

$$\frac{v^2}{2} + \int_{p_1}^{p} \frac{d\tilde{p}}{\rho(\tilde{p})} + \Psi = \text{constant} \quad \text{(along a streamline)}$$

where:

   p is the pressure

   ρ is the density

   v is the flow speed

   Ψ is the potential associated with the conservative force field, often the gravitational potential

In engineering situations, elevations are generally small compared to the size of the Earth, and the time scales of fluid flow are small enough to consider the equation of state as adiabatic. In this case, the above equation becomes

$$\frac{v^2}{2} + gz + \left(\frac{\gamma}{\gamma - 1}\right)\frac{p}{\rho} = \text{constant} \quad \text{(along a streamline)}$$

where, in addition to the terms listed above:

γ is the ratio of the specific heats of the fluid

g is the acceleration due to gravity

z is the elevation of the point above a reference plane

In many applications of compressible flow, changes in elevation are negligible compared to the other terms, so the term gz can be omitted. A very useful form of the equation is then:

$$\frac{v^2}{2} + \left(\frac{\gamma}{\gamma-1}\right)\frac{p}{\rho} = \left(\frac{\gamma}{\gamma-1}\right)\frac{p_0}{\rho_0}$$

where:

p_0 is the total pressure

ρ_0 is the total density

Compressible Flow in Thermodynamics

The most general form of the equation, suitable for use in thermodynamics in case of (quasi) steady flow, is:

$$\frac{v^2}{2} + \Psi + w = \text{constant}.$$

Here w is the enthalpy per unit mass, which is also often written as h.

Note that $w = \epsilon + \dfrac{p}{\rho}$ where ε is the thermodynamic energy per unit mass, also known as the specific internal energy. So, for constant internal energy ε the equation reduces to the incompressible-flow form.

The constant on the right hand side is often called the Bernoulli constant and denoted b. For steady inviscid adiabatic flow with no additional sources or sinks of energy, b is constant along any given streamline. More generally, when b may vary along streamlines, it still proves a useful parameter, related to the "head" of the fluid.

When the change in Ψ can be ignored, a very useful form of this equation is:

$$\frac{v^2}{2} + w = w_0$$

where w_0 is total enthalpy. For a calorically perfect gas such as an ideal gas, the enthalpy is directly proportional to the temperature, and this leads to the concept of the total (or stagnation) temperature.

When shock waves are present, in a reference frame in which the shock is stationary and the flow is steady, many of the parameters in the Bernoulli equation suffer abrupt changes in passing through the shock. The Bernoulli parameter itself, however, remains unaffected. An exception to this rule is radiative shocks, which violate the assumptions leading to the Bernoulli equation, namely the lack of additional sinks or sources of energy.

Derivations of Bernoulli Equation

Bernoulli Equation for Incompressible Fluids

The Bernoulli equation for incompressible fluids can be derived by either integrating Newton's second law of motion or by applying the law of conservation of energy between two sections along a streamline, ignoring viscosity, compressibility, and thermal effects.

Derivation through Integrating Newton's Second Law of Motion

The simplest derivation is to first ignore gravity and consider constrictions and expansions in pipes that are otherwise straight, as seen in Venturi effect. Let the x axis be directed down the axis of the pipe. Define a parcel of fluid moving through a pipe with cross-sectional area A, the length of the parcel is dx, and the volume of the parcel $A\,dx$. If mass density is ρ, the mass of the parcel is density multiplied by its volume $m = \rho\,A\,dx$. The change in pressure over distance dx is dp and flow velocity $v = dx\,/\,dt$. Apply Newton's second law of motion (force = mass × acceleration) and recognizing that the effective force on the parcel of fluid is $-A\,dp$. If the pressure decreases along the length of the pipe, dp is negative but the force resulting in flow is positive along the x axis.

$$m\frac{dv}{dt} = F$$

$$\rho A\,dx\frac{dv}{dt} = -A\,dp$$

$$\rho\frac{dv}{dt} = -\frac{dp}{dx}$$

In steady flow the velocity field is constant with respect to time, $v = v(x) = v(x(t))$, so v itself is not directly a function of time t. It is only when the parcel moves through x that the cross sectional area changes: v depends on t only through the cross-sectional position $x(t)$.

$$\frac{dv}{dt} = \frac{dv}{dx}\frac{dx}{dt} = \frac{dv}{dx}v = \frac{d}{dx}\left(\frac{v^2}{2}\right).$$

With density ρ constant, the equation of motion can be written as

$$\frac{d}{dx}\left(\rho\frac{v^2}{2} + p\right) = 0$$

by integrating with respect to x

$$\frac{v^2}{2} + \frac{p}{\rho} = C$$

where C is a constant, sometimes referred to as the Bernoulli constant. It is not a universal constant, but rather a constant of a particular fluid system. The deduction is: where the speed is large, pressure is low and vice versa. In the above derivation, no external work-energy principle is invoked. Rather, Bernoulli's principle was derived by a simple manipulation of Newton's second law.

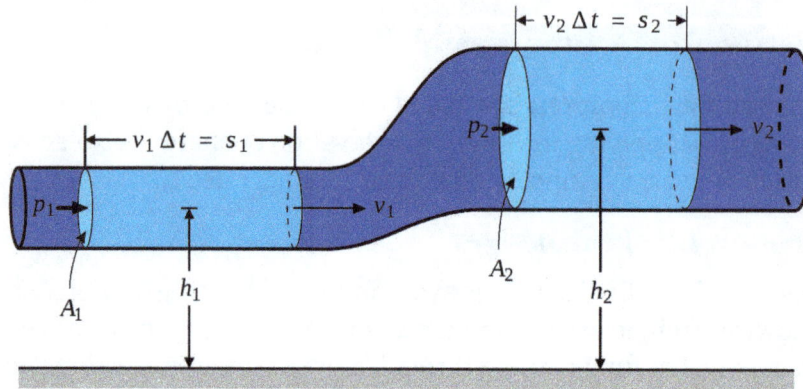

A streamtube of fluid moving to the right. Indicated are pressure, elevation, flow speed, distance (s), and cross-sectional area. Note that in this figure elevation is denoted as h, contrary to the text where it is given by z.

Derivation by using Conservation of Energy

Another way to derive Bernoulli's principle for an incompressible flow is by applying conservation of energy. In the form of the work-energy theorem, stating that

the change in the kinetic energy E_{kin} *of the system equals the net work* W *done on the system;*

$$W = \Delta E_{kin}.$$

Therefore,

the work done by the forces in the fluid = increase in kinetic energy.

The system consists of the volume of fluid, initially between the cross-sections A_1 and A_2. In the time interval Δt fluid elements initially at the inflow cross-section A_1 move over a distance $s_1 = v_1 \Delta t$, while at the outflow cross-section the fluid moves away from cross-section A_2 over a distance $s_2 = v_2 \Delta t$. The displaced fluid volumes at the inflow and outflow are respectively $A_1 s_1$ and $A_2 s_2$. The associated displaced fluid masses are – when ρ is the fluid's mass density – equal to density times volume, so $\rho A_1 s_1$ and $\rho A_2 s_2$. By mass conservation, these two masses displaced in the time interval Δt have to be equal, and this displaced mass is denoted by Δm:

$$\rho A_1 s_1 = \rho A_1 v_1 \Delta t = \Delta m,$$

$$\rho A_2 s_2 = \rho A_2 v_2 \Delta t = \Delta m.$$

The work done by the forces consists of two parts:

- The *work done by the pressure* acting on the areas A_1 and A_2

$$W_{1,pressure} = F_{1,pressure}s_1 - F_{2,pressure}s_2 = p_1A_1s_1 - p_2A_2s_2 = \Delta m\frac{p_1}{\rho} - \Delta m\frac{p_2}{\rho}.$$

- The *work done by gravity*: the gravitational potential energy in the volume $A_1\,s_1$ is lost, and at the outflow in the volume $A_2\,s_2$ is gained. So, the change in gravitational potential energy $\Delta E_{pot,gravity}$ in the time interval Δt is

$$\Delta E_{pot,gravity} = \Delta m\,gz_2 - \Delta m\,gz_1.$$

Now, the work by the force of gravity is opposite to the change in potential energy, $W_{gravity} = -\Delta E_{pot,gravity}$: while the force of gravity is in the negative z-direction, the work— gravity force times change in elevation—will be negative for a positive elevation change $\Delta z = z_2 - z_1$, while the corresponding potential energy change is positive. So:

$$W_{gravity} = -\Delta E_{pot,gravity} = \Delta m\,gz_1 - \Delta m\,gz_2.$$

And the total work done in this time interval Δt is

$$W = W_{pressure} + W_{gravity}.$$

The *increase in kinetic energy* is

$$\Delta E_{kin} = \tfrac{1}{2}\Delta m v_2^2 - \tfrac{1}{2}\Delta m v_1^2.$$

Putting these together, the work-kinetic energy theorem $W = \Delta E_{kin}$ gives:

$$\Delta m\frac{p_1}{\rho} - \Delta m\frac{p_2}{\rho} + \Delta m\,gz_1 - \Delta m\,gz_2 = \frac{1}{2}\Delta m v_2^2 - \frac{1}{2}\Delta m v_1^2$$

or

$$\frac{1}{2}\Delta m v_1^2 + \Delta m\,gz_1 + \Delta m\frac{p_1}{\rho} = \frac{1}{2}\Delta m v_2^2 + \Delta m\,gz_2 + \Delta m\frac{p_2}{\rho}.$$

After dividing by the mass $\Delta m = \rho\,A_1\,v_1\,\Delta t = \rho\,A_2\,v_2\,\Delta t$ the result is:

$$\frac{1}{2}v_1^2 + gz_1 + \frac{p_1}{\rho} = \frac{1}{2}v_2^2 + gz_2 + \frac{p_2}{\rho}$$

or, as stated in the first paragraph:

$$\frac{v^2}{2} + gz + \frac{p}{\rho} = C \quad \text{(Eqn. 1), Which is also Equation (A)}$$

Further division by g produces the following equation. Note that each term can be described in the length dimension (such as meters). This is the head equation derived from Bernoulli's principle:

$$\frac{v^2}{2g} + z + \frac{p}{\rho g} = C \quad \text{(Eqn. 2a)}$$

The middle term, z, represents the potential energy of the fluid due to its elevation with respect to a reference plane. Now, z is called the elevation head and given the designation $z_{elevation}$.

A free falling mass from an elevation $z > 0$ (in a vacuum) will reach a speed

$$v = \sqrt{2gz},$$

when arriving at elevation $z = 0$. Or when we rearrange it as a *head*:

$$h_v = \frac{v^2}{2g}$$

The term $v^2 / (2g)$ is called the *velocity head*, expressed as a length measurement. It represents the internal energy of the fluid due to its motion. The hydrostatic pressure p is defined as

$$p = p_0 - \rho gz,$$

with p_0 some reference pressure, or when we rearrange it as a *head*:

$$\psi = \frac{p}{\rho g}.$$

The term $p / (\rho g)$ is also called the *pressure head*, expressed as a length measurement. It represents the internal energy of the fluid due to the pressure exerted on the container. When we combine the head due to the flow speed and the head due to static pressure with the elevation above a reference plane, we obtain a simple relationship useful for incompressible fluids using the velocity head, elevation head, and pressure head.

$$h_v + z_{elevation} + \psi = C \quad \text{(Eqn. 2b)}$$

If we were to multiply Eqn. 1 by the density of the fluid, we would get an equation with three pressure terms:

$$\frac{\rho v^2}{2} + \rho gz + p = C \quad \text{(Eqn. 3)}$$

We note that the pressure of the system is constant in this form of the Bernoulli Equation. If the static pressure of the system (the far right term) increases, and if the pressure due to elevation (the middle term) is constant, then we know that the dynamic pressure (the left term) must have decreased. In other words, if the speed of a fluid decreases and it is not due to an elevation difference, we know it must be due to an increase in the static pressure that is resisting the flow.

All three equations are merely simplified versions of an energy balance on a system.

Bernoulli Equation for Compressible Fluids

The derivation for compressible fluids is similar. Again, the derivation depends upon (1) conservation of mass, and (2) conservation of energy. Conservation of mass implies that in the above figure, in the interval of time Δt, the amount of mass passing through the boundary defined by the area A_1 is equal to the amount of mass passing outwards through the boundary defined by the area A_2:

$$0 = \Delta M_1 - \Delta M_2 = \rho_1 A_1 v_1 \Delta t - \rho_2 A_2 v_2 \Delta t .$$

Conservation of energy is applied in a similar manner: It is assumed that the change in energy of the volume of the streamtube bounded by A_1 and A_2 is due entirely to energy entering or leaving through one or the other of these two boundaries. Clearly, in a more complicated situation such as a fluid flow coupled with radiation, such conditions are not met. Nevertheless, assuming this to be the case and assuming the flow is steady so that the net change in the energy is zero,

$$0 = \Delta E_1 - \Delta E_2$$

where ΔE_1 and ΔE_2 are the energy entering through A_1 and leaving through A_2, respectively. The energy entering through A_1 is the sum of the kinetic energy entering, the energy entering in the form of potential gravitational energy of the fluid, the fluid thermodynamic internal energy per unit of mass (ε_1) entering, and the energy entering in the form of mechanical $p\,dV$ work:

$$\Delta E_1 = \left[\frac{1}{2} \rho_1 v_1^2 + \Psi_1 \rho_1 + \epsilon_1 \rho_1 + p_1 \right] A_1 v_1 \Delta t$$

where $\Psi = gz$ is a force potential due to the Earth's gravity, g is acceleration due to gravity, and z is elevation above a reference plane. A similar expression for may easily be constructed. So now setting $0 = \Delta E_1 - \Delta E_2$:

$$0 = \left[\frac{1}{2} \rho_1 v_1^2 + \Psi_1 \rho_1 + \epsilon_1 \rho_1 + p_1 \right] A_1 v_1 \Delta t - \left[\frac{1}{2} \rho_2 v_2^2 + \Psi_2 \rho_2 + \epsilon_2 \rho_2 + p_2 \right] A_2 v_2 \Delta t$$

which can be rewritten as:

$$0 = \left[\frac{1}{2} v_1^2 + \Psi_1 + \epsilon_1 + \frac{p_1}{\rho_1} \right] \rho_1 A_1 v_1 \Delta t - \left[\frac{1}{2} v_2^2 + \Psi_2 + \epsilon_2 + \frac{p_2}{\rho_2} \right] \rho_2 A_2 v_2 \Delta t$$

Now, using the previously-obtained result from conservation of mass, this may be simplified to obtain

$$\frac{1}{2} v^2 + \Psi + \epsilon + \frac{p}{\rho} = \text{constant} \equiv b$$

which is the Bernoulli equation for compressible flow.

An equivalent expression can be written in terms of fluid enthalphy (h):

$$\frac{1}{2} v^2 + \Psi + h = \text{constant} \equiv b$$

Applications

Condensation visible over the upper surface of an Airbus A340 wing caused by
the fall in temperature accompanying the fall in pressure.

In modern everyday life there are many observations that can be successfully explained by application of Bernoulli's principle, even though no real fluid is entirely inviscid and a small viscosity often has a large effect on the flow.

- Bernoulli's principle can be used to calculate the lift force on an airfoil, if the behaviour of the fluid flow in the vicinity of the foil is known. For example, if the air flowing past the top surface of an aircraft wing is moving faster than the air flowing past the bottom surface, then Bernoulli's principle implies that the pressure on the surfaces of the wing will be lower above than below. This pressure difference results in an upwards lifting force. Whenever the distribution of speed past the top and bottom surfaces of a wing is known, the lift forces can be calculated (to a good approximation) using Bernoulli's equations – established by Bernoulli over a century before the first man-made wings were used for the purpose of flight. Bernoulli's principle does not explain why the air flows faster past the top of the wing and slower past the underside.

- The carburettor used in many reciprocating engines contains a venturi to create a region of low pressure to draw fuel into the carburettor and mix it thoroughly with the incoming air. The low pressure in the throat of a venturi can be explained by Bernoulli's principle; in the narrow throat, the air is moving at its fastest speed and therefore it is at its lowest pressure.

- An injector on a steam locomotive (or static boiler).

- The pitot tube and static port on an aircraft are used to determine the airspeed of the aircraft. These two devices are connected to the airspeed indicator, which determines the dynamic pressure of the airflow past the aircraft. Dynamic pressure is the difference between stagnation pressure and static pressure. Bernoulli's principle is used to calibrate the airspeed indicator so that it displays the indicated airspeed appropriate to the dynamic pressure.

- The flow speed of a fluid can be measured using a device such as a Venturi meter or an orifice plate, which can be placed into a pipeline to reduce the diameter of the flow. For a horizontal device, the continuity equation shows that for an incompressible fluid, the reduction in diameter will cause an increase in the fluid flow speed. Subsequently Bernoulli's principle then shows that there must be a decrease in the pressure in the reduced diameter region. This phenomenon is known as the Venturi effect.

- The maximum possible drain rate for a tank with a hole or tap at the base can be calculated directly from Bernoulli's equation, and is found to be proportional to the square root of the height of the fluid in the tank. This is Torricelli's law, showing that Torricelli's law is compatible with Bernoulli's principle. Viscosity lowers this drain rate. This is reflected in the discharge coefficient, which is a function of the Reynolds number and the shape of the orifice.

- The Bernoulli grip relies on this principle to create a non-contact adhesive force between a surface and the gripper.

Misunderstandings About the Generation of Lift

Many explanations for the generation of lift (on airfoils, propeller blades, etc.) can be found; some of these explanations can be misleading, and some are false, in particular the idea that air particles flowing above and below a cambered wing should reach simultaneously the trailing edge. This has been a source of heated discussion over the years. In particular, there has been debate about whether lift is best explained by Bernoulli's principle or Newton's laws of motion. Modern writings agree that both Bernoulli's principle and Newton's laws are relevant and either can be used to correctly describe lift.

Several of these explanations use the Bernoulli principle to connect the flow kinematics to the flow-induced pressures. In cases of incorrect (or partially correct) explanations relying on the Bernoulli principle, the errors generally occur in the assumptions on the flow kinematics and how these are produced. It is not the Bernoulli principle itself that is questioned because this principle is well established (the airflow above the wing *is* faster, the question is *why* it is faster).

Misapplications of Bernoulli's Principle in Common Classroom Demonstrations

There are several common classroom demonstrations that are sometimes incorrectly explained using Bernoulli's principle. One involves holding a piece of paper horizontally so that it droops downward and then blowing over the top of it. As the demonstrator blows over the paper, the paper rises. It is then asserted that this is because "faster moving air has lower pressure".

One problem with this explanation can be seen by blowing along the bottom of the paper: were the deflection due simply to faster moving air one would expect the paper to deflect downward, but the paper deflects upward regardless of whether the faster moving air is on the top or the bottom. Another problem is that when the air leaves the demonstrator's mouth it has the *same* pressure as the surrounding air; the air does not have lower pressure just because it is moving; in the demonstration, the static pressure of the air leaving the demonstrator's mouth is *equal* to the pressure of the surrounding air. A third problem is that it is false to make a connection between the flow on the two sides of the paper using Bernoulli's equation since the air above and below are *different* flow fields and Bernoulli's principle only applies within a flow field.

As the wording of the principle can change its implications, stating the principle correctly is important. What Bernoulli's principle actually says is that within a flow of constant energy, when fluid flows through a region of lower pressure it speeds up and vice versa. Thus, Bernoulli's principle concerns itself with *changes* in speed and *changes* in pressure *within* a flow field. It cannot be used to compare different flow fields.

A correct explanation of why the paper rises would observe that the plume follows the curve of the paper and that a curved streamline will develop a pressure gradient perpendicular to the direction of flow, with the lower pressure on the inside of the curve. Bernoulli's principle predicts that the decrease in pressure is associated with an increase in speed, i.e. that as the air passes over the paper it speeds up and moves faster than it was moving when it left the demonstrator's mouth. But this is not apparent from the demonstration.

Other common classroom demonstrations, such as blowing between two suspended spheres, inflating a large bag, or suspending a ball in an airstream are sometimes explained in a similarly misleading manner by saying "faster moving air has lower pressure".

Physical Law

A physical law or scientific law is a theoretical statement "inferred from particular facts, applicable to a defined group or class of phenomena, and expressible by the statement that a particular phenomenon always occurs if certain conditions be present." Physical laws are typically conclusions based on repeated scientific experiments and observations over many years and which have become accepted universally within the scientific community. The production of a summary description of our environment in the form of such laws is a fundamental aim of science. These terms are not used the same way by all authors.

The distinction between natural law in the political-legal sense and law of nature or physical law in the scientific sense is a modern one, both concepts being equally derived from *physis*, the Greek word (translated into Latin as *natura*) for *nature*.

Description

Several general properties of physical laws have been identified. Physical laws are:

- True, at least within their regime of validity. By definition, there have never been repeatable contradicting observations.
- Universal. They appear to apply everywhere in the universe.
- Simple. They are typically expressed in terms of a single mathematical equation.
- Absolute. Nothing in the universe appears to affect them.
- Stable. Unchanged since first discovered (although they may have been shown to be ap-

proximations of more accurate laws

- Omnipotent. Everything in the universe apparently must comply with them (according to observations).

- Generally conservative of quantity.

- Often expressions of existing homogeneities (symmetries) of space and time.

- Typically theoretically reversible in time (if non-quantum), although time itself is irreversible.

Examples

Some of the more famous laws of nature are found in Isaac Newton's theories of (now) classical mechanics, presented in his *Philosophiae Naturalis Principia Mathematica*, and in Albert Einstein's theory of relativity. Other examples of laws of nature include Boyle's law of gases, conservation laws, the four laws of thermodynamics, etc.

Laws as Definitions

Many scientific laws are couched in mathematical terms (e.g. Newton's Second law $F = \frac{dp}{dt}$, or the uncertainty principle, or the principle of least action, or causality). While these scientific laws explain what our senses perceive, they are still empirical, and so are not "mathematical" laws. (Mathematical laws can be proved purely by mathematics and not by scientific experiment.)

Laws Being Consequences of Mathematical Symmetries

Other laws reflect mathematical symmetries found in Nature (say, Pauli exclusion principle reflects identity of electrons, conservation laws reflect homogeneity of space, time, Lorentz transformations reflect rotational symmetry of space–time). Laws are constantly being checked experimentally to higher and higher degrees of precision. This is one of the main goals of science. Just because laws have never been observed to be violated does not preclude testing them at increased accuracy or in new kinds of conditions to confirm whether they continue to hold, or whether they break, and what can be discovered in the process. It is always possible for laws to be invalidated or proven to have limitations, by repeatable experimental evidence, should any be observed.

Well-established laws have indeed been invalidated in some special cases, but the new formulations created to explain the discrepancies can be said to generalize upon, rather than overthrow, the originals. That is, the invalidated laws have been found to be only close approximations, to which other terms or factors must be added to cover previously unaccounted-for conditions, e.g. very large or very small scales of time or space, enormous speeds or masses, etc. Thus, rather than unchanging knowledge, physical laws are better viewed as a series of improving and more precise generalizations.

Laws as Approximations

Some laws are only approximations of other more general laws, and are good approximations with a restricted domain of applicability. For example, Newtonian dynamics (which is based on Galilean transformations) is the low speed limit of special relativity (since the Galilean transformation is the low-speed approximation to the Lorentz transformation). Similarly, the Newtonian gravitation law is a low-mass approximation of general relativity, and Coulomb's law is an approximation to Quantum Electrodynamics at large distances (compared to the range of weak interactions). In such cases it is common to use the simpler, approximate versions of the laws, instead of the more accurate general laws.

Physical Laws Derived from Symmetry Principles

Many fundamental physical laws are mathematical consequences of various symmetries of space, time, or other aspects of nature. Specifically, Noether's theorem connects some conservation laws to certain symmetries. For example, conservation of energy is a consequence of the shift symmetry of time (no moment of time is different from any other), while conservation of momentum is a consequence of the symmetry (homogeneity) of space (no place in space is special, or different than any other). The indistinguishability of all particles of each fundamental type (say, electrons, or photons) results in the Dirac and Bose quantum statistics which in turn result in the Pauli exclusion principle for fermions and in Bose–Einstein condensation for bosons. The rotational symmetry between time and space coordinate axes (when one is taken as imaginary, another as real) results in Lorentz transformations which in turn result in special relativity theory. Symmetry between inertial and gravitational mass results in general relativity.

The inverse square law of interactions mediated by massless bosons is the mathematical consequence of the 3-dimensionality of space.

One strategy in the search for the most fundamental laws of nature is to search for the most general mathematical symmetry group that can be applied to the fundamental interactions.

History

Compared to pre-modern accounts of causality, laws of nature fill the role played by divine causality on the one hand, and accounts such as Plato's theory of forms on the other.

The observation that there are underlying regularities in nature dates from prehistoric times, since the recognition of cause-and-effect relationships is an implicit recognition that there are laws of nature. The recognition of such regularities as independent scientific laws *per se*, though, was limited by their entanglement in animism, and by the attribution of many effects that do not have readily obvious causes—such as meteorological, astronomical and biological phenomena—to the actions of various gods, spirits, supernatural beings, etc. Observation and speculation about nature were intimately bound up with metaphysics and morality.

In Europe, systematic theorizing about nature (*physis*) began with the early Greek philosophers and scientists and continued into the Hellenistic and Roman imperial periods, during which times the intellectual influence of Roman law increasingly became paramount.

The formula "law of nature" first appears as "a live metaphor" favored by Latin poets Lucretius, Virgil, Ovid, Manilius, in time gaining a firm theoretical presence in the prose treatises of Seneca and Pliny. Why this Roman origin? According to [historian and classicist Daryn] Lehoux's persuasive narrative, the idea was made possible by the pivotal role of codified law and forensic argument in Roman life and culture.

For the Romans . . . the place par excellence where ethics, law, nature, religion and politics overlap is the law court. When we read Seneca's *Natural Questions*, and watch again and again just how he applies standards of evidence, witness evaluation, argument and proof, we can recognize that we are reading one of the great Roman rhetoricians of the age, thoroughly immersed in forensic method. And not Seneca alone. Legal models of scientific judgment turn up all over the place, and for example prove equally integral to Ptolemy's approach to verification, where the mind is assigned the role of magistrate, the senses that of disclosure of evidence, and dialectical reason that of the law itself.

The precise formulation of what are now recognized as modern and valid statements of the laws of nature dates from the 17th century in Europe, with the beginning of accurate experimentation and development of advanced forms of mathematics. During this period, natural philosophers such as Isaac Newton were influenced by a religious view which held that God had instituted absolute, universal and immutable physical laws. The modern scientific method which took shape at this time (with Francis Bacon and Galileo) aimed at total separation of science from theology, with minimal speculation about metaphysics and ethics. Natural law in the political sense, conceived as universal (i.e., divorced from sectarian religion and accidents of place), was also elaborated in this period (by Grotius, Spinoza, and Hobbes, to name a few).

Other Fields

Some mathematical theorems and axioms are referred to as laws because they provide logical foundation to empirical laws.

Examples of other observed phenomena sometimes described as laws include the Titius-Bode law of planetary positions, Zipf's law of linguistics, Moore's law of technological growth. Many of these laws fall within the scope of uncomfortable science. Other laws are pragmatic and observational, such as the law of unintended consequences. By analogy, principles in other fields of study are sometimes loosely referred to as "laws". These include Occam's razor as a principle of philosophy and the Pareto principle of economics.

Conservation Law

In physics, a conservation law states that a particular measurable property of an isolated physical system does not change as the system evolves over time. Exact conservation laws include conservation of energy, conservation of linear momentum, conservation of angular momentum, and conservation of electric charge. There are also many approximate conservation laws, which apply to such quantities as mass, parity, lepton number, baryon number, strangeness, hypercharge, etc. These quantities are conserved in certain classes of physics processes, but not in all.

A local conservation law is usually expressed mathematically as a continuity equation, a partial differential equation which gives a relation between the amount of the quantity and the "transport" of that quantity. It states that the amount of the conserved quantity at a point or within a volume can only change by the amount of the quantity which flows in or out of the volume.

From Noether's theorem, each conservation law is associated with a symmetry in the underlying physics.

Conservation Laws as Fundamental Laws of Nature

Conservation laws are fundamental to our understanding of the physical world, in that they describe which processes can or cannot occur in nature. For example, the conservation law of energy states that the total quantity of energy in an isolated system does not change, though it may change form. In general, the total quantity of the property governed by that law remains unchanged during physical processes. With respect to classical physics, conservation laws include conservation of energy, mass (or matter), linear momentum, angular momentum, and electric charge. With respect to particle physics, particles cannot be created or destroyed except in pairs, where one is ordinary and the other is an antiparticle. With respect to symmetries and invariance principles, three special conservation laws have been described, associated with inversion or reversal of space, time, and charge.

Conservation laws are considered to be fundamental laws of nature, with broad application in physics, as well as in other fields such as chemistry, biology, geology, and engineering.

Most conservation laws are exact, or absolute, in the sense that they apply to all possible processes. Some conservation laws are partial, in that they hold for some processes but not for others.

One particularly important result concerning conservation laws is Noether's theorem, which states that there is a one-to-one correspondence between each one of them and a differentiable symmetry in the system. For example, the conservation of energy follows from the time-invariance of physical systems, and the fact that physical systems behave the same regardless of how they are oriented in space gives rise to the conservation of angular momentum.

Exact Laws

A partial listing of physical conservation equations due to symmetry that are said to be exact laws, or more precisely *have never been proven to be violated:*

Conservation Law	Respective Noether symmetry invariance			Number of dimensions	
Conservation of mass-energy	Time invariance	Lorentz invariance symmetry	1		translation about time axis
Conservation of linear momentum	Translation symmetry		3		translation about x,y,z position
Conservation of angular momentum	Rotation invariance		3		rotation about x,y,z axes
CPT symmetry (combining charge, parity and time conjugation)	Lorentz invariance		1+1+1		(charge inversion q→-q) + (position inversion r→-r) + (time inversion t→-t)
Conservation of electric charge	Gauge invariance		1→4		scalar field (1D) in 4D spacetime (x,y,z + time evolution)

Conservation of color charge	SU(3) Gauge invariance	3	r,g,b
Conservation of weak isospin	SU(2)$_L$ Gauge invariance	1	weak charge
Conservation of probability	Probability invariance	1→4	total probability always=1 in whole x,y,z space, during time evolution

Approximate Laws

There are also approximate conservation laws. These are approximately true in particular situations, such as low speeds, short time scales, or certain interactions.

- Conservation of rest mass

- Conservation of baryon number

- Conservation of lepton number (In the Standard Model)

- Conservation of flavor (violated by the weak interaction)

- Conservation of parity

- Invariance under charge conjugation

- Invariance under time reversal

- CP symmetry, the combination of charge and parity conjugation (equivalent to time reversal if CPT holds)

Differential Forms

In continuum mechanics, the most general form of an exact conservation law is given by a continuity equation. For example, conservation of electric charge q is

$$\frac{\partial \rho}{\partial t} = -\nabla \cdot \mathbf{j}$$

where $\nabla \cdot$ is the divergence operator, ρ is the density of q (amount per unit volume), j is the flux of q (amount crossing a unit area in unit time), and t is time.

If we assume that the motion u of the charge is a continuous function of position and time, then

$$\mathbf{j} = \rho \mathbf{u}$$

$$\frac{\partial \rho}{\partial t} = -\nabla \cdot (\rho \mathbf{u}).$$

In one space dimension this can be put into the form of a homogeneous first-order quasilinear hyperbolic equation:

$$y_t + A(y)y_x = 0$$

where the dependent variable y is called the *density* of a *conserved quantity*, and $A(y)$ is called the *current jacobian*, and the subscript notation for partial derivatives has been employed. The more general inhomogeneous case:

$$y_t + A(y)y_x = s$$

is not a conservation equation but the general kind of balance equation describing a dissipative system. The dependent variable y is called a *nonconserved quantity*, and the inhomogeneous term $s(y,x,t)$ is the-*source*, or dissipation. For example, balance equations of this kind are the momentum and energy Navier-Stokes equations, or the entropy balance for a general isolated system.

In the one-dimensional space a conservation equation is a first-order quasilinear hyperbolic equation that can be put into the *advection* form:

$$y_t + a(y)y_x = 0$$

where the dependent variable $y(x,t)$ is called the density of the *conserved* (scalar) quantity (c.q.(d.) = conserved quantity (density)), and $a(y)$ is called the current coefficient, usually corresponding to the partial derivative in the conserved quantity of a current density (c.d.) of the conserved quantity $j(y)$:

$$a(y) = j_y(y)$$

In this case since the chain rule applies:

$$j_x = j_y(y)y_x = a(y)y_x$$

the conservation equation can be put into the current density form:

$$y_t + j_x(y) = 0$$

In a space with more than one dimension the former definition can be extended to an equation that can be put into the form:

$$y_t + \mathbf{a}(y) \cdot \nabla y = 0$$

where the *conserved quantity* is $y(r,t)$, $\cdot$ denotes the scalar product, ∇ is the nabla operator, here indicating a gradient, and $a(y)$ is a vector of current coefficients, analogously corresponding to the divergence of a vector c.d. associated to the c.q. j(y):

$$y_t + \nabla \cdot \mathbf{j}(y) = 0$$

This is the case for the continuity equation:

$$\rho_t + \nabla \cdot (\rho \mathbf{u}) = 0$$

Here the conserved quantity is the mass, with density $\rho(r,t)$ and current density ρu, identical to the momentum density, while $u(r,t)$ is the flow velocity.

In the general case a conservation equation can be also a system of this kind of equations (a vector equation) in the form:

$$\mathbf{y}_t + \mathbf{A}(\mathbf{y}) \cdot \nabla \mathbf{y} = \mathbf{0}$$

where y is called the *conserved* (vector) quantity, ∇ y is its gradient, 0 is the zero vector, and A(y) is called the Jacobian of the current density. In fact as in the former scalar case, also in the vector case A(y) usually corresponding to the Jacobian of a current density matrix J(y):

$$\mathbf{A}(\mathbf{y}) = \mathbf{J}_y(\mathbf{y})$$

and the conservation equation can be put into the form:

$$\mathbf{y}_t + \nabla \cdot \mathbf{J}(\mathbf{y}) = \mathbf{0}$$

For example, this the case for Euler equations (fluid dynamics). In the simple incompressible case they are:

$$\nabla \cdot \mathbf{u} = 0$$

$$\frac{\partial \mathbf{u}}{\partial t} + \mathbf{u} \cdot \nabla \mathbf{u} + \nabla s = \mathbf{0},$$

where:

- u is the flow velocity vector, with components in a N-dimensional space $u_1, u_2 \ldots u_N$,

- s is the specific pressure (pressure per unit density) giving the source term,

It can be shown that the conserved (vector) quantity and the c.d. matrix for these equations are respectively:

$$\mathbf{y} = \begin{pmatrix} 1 \\ \mathbf{u} \end{pmatrix}; \qquad \mathbf{J} = \begin{pmatrix} \mathbf{u} \\ \mathbf{u} \otimes \mathbf{u} + s\mathbf{I} \end{pmatrix};$$

where $\otimes$ denotes the outer product.

Integral and Weak Forms

Conservation equations can be also expressed in integral form: the advantage of the latter is substantially that it requires less smoothness of the solution, which paves the way to weak form, extending the class of admissible solutions to include discontinuous solutions. By integrating in any space-time domain the current density form in 1-D space:

$$y_t + j_x(y) = 0$$

and by using Green's theorem, the integral form is:

$$\int_{-\infty}^{\infty} y dx + \int_{0}^{\infty} j(y) dt = 0$$

In a similar fashion, for the scalar multidimensional space, the integral form is:

$$\oint [y d^N r + j(y) dt] = 0$$

where the line integration is performed along the boundary of the domain, in an anticlock-wise manner.

Moreover, by defining a test function $\varphi(r,t)$ continuously differentiable both in time and space with compact support, the weak form can be obtained pivoting on the initial condition. In 1-D space it is:

$$\int_0^\infty \int_{-\infty}^\infty \phi_t y + \phi_x j(y) dx dt = -\int_{-\infty}^\infty \phi(x,0) y(x,0) dx$$

Note that in the weak form all the partial derivatives of the density and current density have been passed on to the test function, which with the former hypothesis is sufficiently smooth to admit these derivatives.

Inverse-square Law

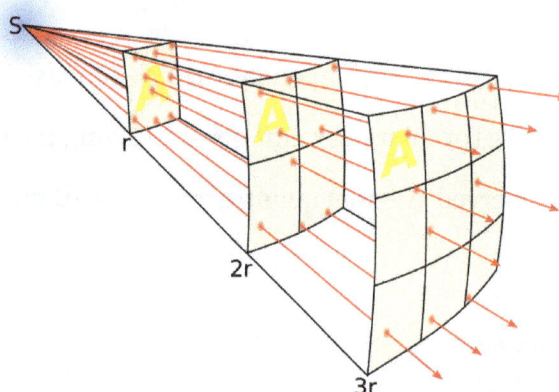

S represents the light source, while r represents the measured points. The lines represent the flux emanating from the source. The total number of flux lines depends on the strength of the source and is constant with increasing distance. A greater density of flux lines (lines per unit area) means a stronger field. The density of flux lines is inversely proportional to the square of the distance from the source because the surface area of a sphere increases with the square of the radius. Thus the strength of the field is inversely proportional to the square of the distance from the source.

In physics, an inverse-square law is any physical law stating that a specified physical quantity or intensity is inversely proportional to the square of the distance from the source of that physical quantity. The fundamental cause for this can be understood as geometric dilution corresponding to point-source radiation into three-dimensional space. Mathematically formulated:

$$\text{intensity} \propto \frac{1}{\text{distance}^2}$$

It can also be mathematically expressed as:

$$\frac{\text{intensity}_1}{\text{intensity}_2} = \frac{\text{distance}_2{}^2}{\text{distance}_1{}^2}$$

or as the formulation of a constant quantity:

$$\text{Intensity}_1 \times \text{distance}_1{}^2 = \text{Intensity}_2 \times \text{distance}_2{}^2$$

The divergence of a vector field which is the resultant of radial inverse-square law fields with respect to one or more sources is everywhere proportional to the strength of the local sources, and hence zero outside sources. Newton's law of universal gravitation follows an inverse-square law, as do the effects of electric, magnetic, light, sound, and radiation phenomena.

Radar energy expands during both the signal transmission and also on the reflected return, so the inverse square for both paths means that the radar will receive energy according to $1/4r^2$ power.

In order to prevent dilution of energy while propagating a signal, certain methods can be used such as a waveguide, which acts like a canal does for water, or how a gun barrel restricts hot gas expansion to one dimension in order to prevent loss of energy transfer to a bullet.

Justification

The inverse-square law generally applies when some force, energy, or other conserved quantity is evenly radiated outward from a point source in three-dimensional space. Since the surface area of a sphere (which is $4\pi r^2$) is proportional to the square of the radius, as the emitted radiation gets farther from the source, it is spread out over an area that is increasing in proportion to the square of the distance from the source. Hence, the intensity of radiation passing through any unit area (directly facing the point source) is inversely proportional to the square of the distance from the point source. Gauss's law is similarly applicable, and can be used with any physical quantity that acts in accord to the inverse-square relationship.

Occurrences

Gravitation

Gravitation is the attraction of two objects with mass. Newton's law states:

> The gravitational attraction force between two point masses is directly proportional to the product of their masses and inversely proportional to the square of their separation distance. The force is always attractive and acts along the line joining them.

If the distribution of matter in each body is spherically symmetric, then the objects can be treated as point masses without approximation, as shown in the shell theorem. Otherwise, if we want to calculate the attraction between massive bodies, we need to add all the point-point attraction forces vectorially and the net attraction might not be exact inverse square. However, if the separation between the massive bodies is much larger compared to their sizes, then to a good approximation, it is reasonable to treat the masses as point mass while calculating the gravitational force.

As the law of gravitation, this law was suggested in 1645 by Ismael Bullialdus. But Bullialdus did not accept Kepler's second and third laws, nor did he appreciate Christiaan Huygens's solution for circular motion (motion in a straight line pulled aside by the central force). Indeed, Bullialdus maintained the sun's force was attractive at aphelion and repulsive at perihelion. Robert Hooke and Giovanni Alfonso Borelli both expounded gravitation in 1666 as an attractive force (Hooke's lecture "On gravity" at the Royal Society, London, on 21 March; Borelli's "Theory of the Planets", published later in 1666). Hooke's 1670 Gresham lecture explained that gravitation applied to "all celestiall bodys" and added the principles that the gravitating power decreases with distance and that in the absence of any such power bodies move in straight lines. By 1679, Hooke thought gravitation had inverse square dependence and communicated this in a letter to Isaac Newton. Hooke remained bitter about Newton claiming the invention of this principle, even though Newton's *Principia* acknowledged that Hooke, along with Wren and Halley, had separately appreciated the inverse square law in the solar system, as well as giving some credit to Bullialdus.

Electrostatics

The force of attraction or repulsion between two electrically charged particles, in addition to being directly proportional to the product of the electric charges, is inversely proportional to the square of the distance between them; this is known as Coulomb's law. The deviation of the exponent from 2 is less than one part in 10^{15}.

Light and Other Electromagnetic Radiation

The intensity (or illuminance or irradiance) of light or other linear waves radiating from a point source (energy per unit of area perpendicular to the source) is inversely proportional to the square of the distance from the source; so an object (of the same size) twice as far away, receives only one-quarter the energy (in the same time period).

More generally, the irradiance, *i.e.*, the intensity (or power per unit area in the direction of propagation), of a spherical wavefront varies inversely with the square of the distance from the source (assuming there are no losses caused by absorption or scattering).

For example, the intensity of radiation from the Sun is 9126 watts per square meter at the distance of Mercury (0.387 AU); but only 1367 watts per square meter at the distance of Earth (1 AU)—an approximate threefold increase in distance results in an approximate ninefold decrease in intensity of radiation.

For non-isotropic radiators such as parabolic antennas, headlights, and lasers, the effective origin is located far behind the beam aperture. If you are close to the origin, you don't have to go far to double the radius, so the signal drops quickly. When you are far from the origin and still have a strong signal, like with a laser, you have to travel very far to double the radius and reduce the signal. This means you have a stronger signal or have antenna gain in the direction of the narrow beam relative to a wide beam in all directions of an isotropic antenna.

In photography and stage lighting, the inverse-square law is used to determine the "fall off" or the difference in illumination on a subject as it moves closer to or further from the light source. For quick approximations, it is enough to remember that doubling the distance reduces illumination to one quarter; or similarly, to halve the illumination increase the distance by a factor of 1.4 (the square root of 2), and to double illumination, reduce the distance to 0.7 (square root of 1/2). When the illuminant is not a point source, the inverse square rule is often still a useful approximation; when the size of the light source is less than one-fifth of the distance to the subject, the calculation error is less than 1%.

The fractional reduction in electromagnetic fluence (Φ) for indirectly ionizing radiation with increasing distance from a point source can be calculated using the inverse-square law. Since emissions from a point source have radial directions, they intercept at a perpendicular incidence. The area of such a shell is $4\pi r^2$ where r is the radial distance from the center. The law is particularly important in diagnostic radiography and radiotherapy treatment planning, though this proportionality does not hold in practical situations unless source dimensions are much smaller than the distance. As stated in fourier theory of heat "as the point source is magnification by distances , its radiation is dilute proportional to the sin of the angle, of the increasing circumference arc from the point of origin"

Example

Let the total power radiated from a point source, for example, an omnidirectional isotropic radiator, be P. At large distances from the source (compared to the size of the source), this power is distributed over larger and larger spherical surfaces as the distance from the source increases. Since the surface area of a sphere of radius r is $A = 4\pi r^2$, then intensity I (power per unit area) of radiation at distance r is

$$I = \frac{P}{A} = \frac{P}{4\pi r^2}.$$

The energy or intensity decreases (divided by 4) as the distance r is doubled; measured in dB it would decrease by 6.02 dB per doubling of distance.

Sound in a Gas

In acoustics, the sound pressure of a spherical wavefront radiating from a point source decreases by 50% as the distance r is doubled; measured in dB, the decrease is still 6.02 dB, since dB represents an intensity ratio. The pressure ratio (as opposed to power ratio) is not inverse-square, but is inverse-proportional (inverse distance law):

$$p \propto \frac{1}{r}$$

The same is true for the component of particle velocity v that is in-phase with the instantaneous sound pressure p:

$$v \propto \frac{1}{r}$$

In the near field is a quadrature component of the particle velocity that is 90° out of phase with the sound pressure and does not contribute to the time-averaged energy or the intensity of the sound. The sound intensity is the product of the RMS sound pressure and the in-phase component of the RMS particle velocity, both of which are inverse-proportional. Accordingly, the intensity follows an inverse-square behaviour:

$$I = pv \propto \frac{1}{r^2}.$$

Field Theory Interpretation

For an irrotational vector field in three-dimensional space, the inverse-square law corresponds to the property that the divergence is zero outside the source. This can be generalized to higher dimensions. Generally, for an irrotational vector field in n-dimensional Euclidean space, the intensity "I" of the vector field falls off with the distance "r" following the inverse $(n-1)^{th}$ power law

$$I \propto \frac{1}{r^{n-1}},$$

given that the space outside the source is divergence free.

History

John Dumbleton of the 14th-century Oxford Calculators, was one of the first to express functional relationships in graphical form. He gave a proof of the mean speed theorem stating that "the latitude of a uniformly difform movement corresponds to the degree of the midpoint" and used this method to study the quantitative decrease in intensity of illumination in his *Summa logicæ et philosophiæ naturalis* (ca. 1349), stating that it was not linearly proportional to the distance, but was unable to expose the Inverse-square law.

In proposition 9 of Book 1 in his book *Ad Vitellionem paralipomena, quibus astronomiae pars optica traditur* (1604), the astronomer Johannes Kepler argued that the spreading of light from a point source obeys an inverse square law:

Original: *Sicut se habent spharicae superificies, quibus origo lucis pro centro est, amplior ad angustiorem: ita se habet fortitudo seu densitas lucis radiorum in angustiori, ad illamin in laxiori sphaerica, hoc est, conversim. Nam per 6. 7. tantundem lucis est in angustiori sphaerica superficie, quantum in fusiore, tanto ergo illie stipatior & densior quam hic.*

Translation: Just as [the ratio of] spherical surfaces, for which the source of light is the center, [is] from the wider to the narrower, so the density or fortitude of the rays of light in the narrower [space], towards the more spacious spherical surfaces, that is, inversely. For according to [propositions] 6 & 7, there is as much light in the narrower spherical surface, as in the wider, thus it is as much more compressed and dense here than there.

In 1645 in his book *Astronomia Philolaica ...*, the French astronomer Ismaël Bullialdus (1605 – 1694) refuted Johannes Kepler's suggestion that "gravity" weakens as the inverse of the distance; instead, Bullialdus argued, "gravity" weakens as the inverse square of the distance:

Original: *Virtus autem illa, qua Sol prehendit seu harpagat planetas, corporalis quae ipsi pro manibus est, lineis rectis in omnem mundi amplitudinem emissa quasi species solis cum illius corpore rotatur: cum ergo sit corporalis imminuitur, & extenuatur in maiori spatio & intervallo, ratio autem huius imminutionis eadem est, ac luminus, in ratione nempe dupla intervallorum, sed eversa.*

Translation: As for the power by which the Sun seizes or holds the planets, and which, being corporeal, functions in the manner of hands, it is emitted in straight lines throughout the whole extent of the world, and like the species of the Sun, it turns with the body of the Sun; now, seeing that it is corporeal, it becomes weaker and attenuated at a greater distance or interval, and the ratio of its decrease in strength is the same as in the case of light, namely, the duplicate proportion, but inversely, of the distances [that is, $1/d^2$].

In England, the Anglican bishop Seth Ward (1617 – 1689) publicized the ideas of Bullialdus in his critique *In Ismaelis Bullialdi astronomiae philolaicae fundamenta inquisitio brevis*

(1653) and publicized the planetary astronomy of Kepler in his book *Astronomia geometrica* (1656).

In 1663–1664, the English scientist Robert Hooke was writing his book *Micrographia* (1666) in which he discussed, among other things, the relation between the height of the atmosphere and the barometric pressure at the surface. Since the atmosphere surrounds the earth, which itself is a sphere, the volume of atmosphere bearing on any unit area of the earth's surface is a truncated cone (which extends from the earth's center to the vacuum of space; obviously only the section of the cone from the earth's surface to space bears on the earth's surface). Although the volume of a cone is proportional to the cube of its height, Hooke argued that the air's pressure at the earth's surface is instead proportional to the height of the atmosphere because gravity diminishes with altitude. Although Hooke did not explicitly state so, the relation that he proposed would be true only if gravity decreases as the inverse square of the distance from the earth's center.

Stokes' Law

In 1851, George Gabriel Stokes derived an expression, now known as Stokes' law, for the frictional force – also called drag force – exerted on spherical objects with very small Reynolds numbers (i.e. very small particles) in a viscous fluid. Stokes' law is derived by solving the Stokes flow limit for small Reynolds numbers of the Navier–Stokes equations.

Statement of the Law

The force of viscosity on a small sphere moving through a viscous fluid is given by:

where:

- F_d is the frictional force – known as Stokes' drag – acting on the interface between the fluid and the particle

- η is the dynamic viscosity (Some authors use the symbol μ)

- r is the radius of the spherical object

- v is the flow velocity relative to the object.

In SI units, F_d is given in Newtons, μ in Pa·s, R in meters, and V in m/s.

Stokes' law makes the following assumptions for the behavior of a particle in a fluid:

- Laminar Flow

- Spherical particles

- Homogeneous (uniform in composition) material

- Smooth surfaces

- Particles do not interfere with each other.

Note that for molecules Stokes' law is used to define their Stokes radius.

The CGS unit of kinematic viscosity was named "stokes" after his work.

Applications

Stokes' law is the basis of the falling-sphere viscometer, in which the fluid is stationary in a vertical glass tube. A sphere of known size and density is allowed to descend through the liquid. If correctly selected, it reaches terminal velocity, which can be measured by the time it takes to pass two marks on the tube. Electronic sensing can be used for opaque fluids. Knowing the terminal velocity, the size and density of the sphere, and the density of the liquid, Stokes' law can be used to calculate the viscosity of the fluid. A series of steel ball bearings of different diameters are normally used in the classic experiment to improve the accuracy of the calculation. The school experiment uses glycerine or golden syrup as the fluid, and the technique is used industrially to check the viscosity of fluids used in processes. Several school experiments often involve varying the temperature and/or concentration of the substances used in order to demonstrate the effects this has on the viscosity. Industrial methods include many different oils, and polymer liquids such as solutions.

The importance of Stokes' law is illustrated by the fact that it played a critical role in the research leading to at least 3 Nobel Prizes.

Stokes' law is important for understanding the swimming of microorganisms and sperm; also, the sedimentation of small particles and organisms in water, under the force of gravity.

In air, the same theory can be used to explain why small water droplets (or ice crystals) can remain suspended in air (as clouds) until they grow to a critical size and start falling as rain (or snow and hail). Similar use of the equation can be made in the settlement of fine particles in water or other fluids.

Terminal Velocity of Sphere Falling in a Fluid

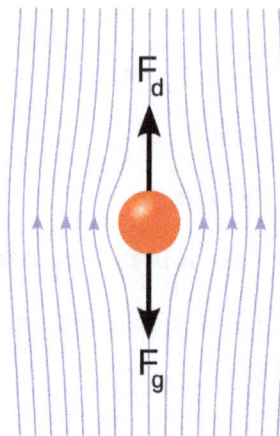

Creeping flow past a falling sphere in a fluid (e.g., a droplet of fog falling through the air): streamlines, drag force F_d and force by gravity F_g.

At terminal (or settling) velocity, the excess force F_g due to the difference between the weight and buoyancy of the sphere (both caused by gravity) is given by:

$$F_g = \left(\rho_p - \rho_f\right) g \frac{4}{3} \pi R^3,$$

with ρ_p and ρ_f the mass densities of the sphere and fluid, respectively, and g the gravitational acceleration. Requiring the force balance $F_d = F_g$ and solving for the velocity V gives the terminal velocity V_s. Note that since buoyant force increases as R^3 and Stokes' drag increases as R, the terminal velocity increases as R^2 and thus varies greatly with particle size as shown below. If the particle is falling in the viscous fluid under its own weight, then a terminal velocity, or settling velocity, is reached when this frictional force combined with the buoyant force exactly balances the gravitational force. This velocity V (m/s) is given by:

$$V = \frac{2}{9} \frac{\left(\rho_p - \rho_f\right)}{\mu} g R^2$$

(vertically downwards if $\rho_p > \rho_f$, upwards if $\rho_p < \rho_f$), where:

- g is the gravitational acceleration (m/s²)

- ρ_p is the mass density of the particles (kg/m³)

- ρ_f is the mass density of the fluid (kg/m³)

- μ is the dynamic viscosity (kg/m*s).

Steady Stokes Flow

In Stokes flow, at very low Reynolds number, the convective acceleration terms in the Navier–Stokes equations are neglected. Then the flow equations become, for an incompressible steady flow:

$$\nabla p = \mu \nabla^2 \mathbf{u} = -\mu \nabla \times \omega, \nabla \cdot \mathbf{u} = 0,$$

where:

- p is the fluid pressure (in Pa),

- u is the flow velocity (in m/s), and

- ω is the vorticity (in s⁻¹), defined as $\grave{u} = \nabla \times \mathbf{u}$.

By using some vector calculus identities, these equations can be shown to result in Laplace's equations for the pressure and each of the components of the vorticity vector:

$$\nabla^2 \omega = 0 \quad \text{and} \quad \nabla^2 p = 0.$$

Additional forces like those by gravity and buoyancy have not been taken into account, but can easily be added since the above equations are linear, so linear superposition of solutions and associated forces can be applied.

Flow Around a Sphere

For the case of a sphere in a uniform far field flow, it is advantageous to use a cylindrical coordinate system (r , φ , z). The z–axis is through the centre of the sphere and aligned with the mean flow direction, while r is the radius as measured perpendicular to the z–axis. The origin is at the sphere centre. Because the flow is axisymmetric around the z–axis, it is independent of the azimuth φ.

In this cylindrical coordinate system, the incompressible flow can be described with a Stokes stream function ψ, depending on r and z:

$$u_z = \frac{1}{r}\frac{\partial \psi}{\partial r}, \qquad u_r = -\frac{1}{r}\frac{\partial \psi}{\partial z},$$

with u_r and u_z the flow velocity components in the r and z direction, respectively. The azimuthal velocity component in the φ–direction is equal to zero, in this axisymmetric case. The volume flux, through a tube bounded by a surface of some constant value ψ, is equal to $2\pi\,\psi$ and is constant.

For this case of an axisymmetric flow, the only non-zero component of the vorticity vector ω is the azimuthal φ–component ω_φ

$$\omega_\varphi = \frac{\partial u_r}{\partial z} - \frac{\partial u_z}{\partial r} = -\frac{\partial}{\partial r}\left(\frac{1}{r}\frac{\partial \psi}{\partial r}\right) - \frac{1}{r}\frac{\partial^2 \psi}{\partial z^2}.$$

The Laplace operator, applied to the vorticity ω_φ, becomes in this cylindrical coordinate system with axisymmetry:

$$\nabla^2 \omega_\varphi = \frac{1}{r}\frac{\partial}{\partial r}\left(r\frac{\partial \omega_\varphi}{\partial r}\right) + \frac{\partial^2 \omega_\varphi}{\partial z^2} - \frac{\omega_\varphi}{r^2} = 0.$$

From the previous two equations, and with the appropriate boundary conditions, for a far-field uniform-flow velocity u in the z–direction and a sphere of radius R, the solution is found to be

$$\psi = -\frac{1}{2}ur^2\left[1 - \frac{3}{2}\frac{R}{\sqrt{r^2+z^2}} + \frac{1}{2}\left(\frac{R}{\sqrt{r^2+z^2}}\right)^3\right]$$

The viscous force per unit area σ, exerted by the flow on the surface on the sphere, is in the z–direction everywhere. More strikingly, it has also the same value everywhere on the sphere:

$$\sigma = \frac{3\mu u}{2R}e_z$$

with e_z the unit vector in the z–direction. For other shapes than spherical, σ is not constant along the body surface. Integration of the viscous force per unit area σ over the sphere surface gives the frictional force F_d according to Stokes' law.

Faxén's Law

In fluid dynamics, Faxén's laws relate a sphere's velocity $\mathbf{U}$ and angular velocity $\dot{\mathbf{U}}$ to the forces, torque, stresslet and flow it experiences under low Reynolds number (creeping flow) conditions.

First Law

Faxen's first law was introduced in 1922 by Swedish physicist Hilding Faxén, who at the time was active at Uppsala University, and is given by

$$\mathbf{F} = 6\pi\mu a\left[\left(1+\frac{a^2}{6}\nabla^2\right)\mathbf{u}'-(\mathbf{U}-\mathbf{u}^\infty)\right],$$

where

- $\mathbf{F}$ is the force exerted by the fluid on the sphere

- μ is the Newtonian viscosity of the solvent in which the sphere is placed

- a is the sphere's radius

- $\mathbf{U}$ is the (translational) velocity of the sphere

- $\mathbf{u}'$ is the disturbance velocity caused by the other spheres in suspension (not by the background impressed flow), evaluated at the sphere centre

- $\mathbf{u}^\infty$ is the background impressed flow, evaluated at the sphere centre (set to zero in some references).

It can also be written in the form

$$\mathbf{U}-\mathbf{u}^\infty = \mathbf{u}'+b_0\mathbf{F}+\frac{a^2}{6}\nabla^2\mathbf{u}',$$

where $b_0 = -\dfrac{1}{6\pi\mu a}$ is the mobility.

In the case that the pressure gradient is small compared with the length scale of the sphere's diameter, and when there is no external force, the last two terms of this form may be neglected. In this case the external fluid flow simply advects the sphere.

Second Law

Faxen's second law is given by

$$T = 8\pi\mu a^3\left[\frac{1}{2}(\nabla\times\mathbf{u}')-(\Omega-\Omega^\infty)\right],$$

where

- **T** is the torque exerted by the fluid on the sphere

- **U̇** is the angular velocity of the sphere

- **U̇**$^\infty$ is the angular velocity of the background flow, evaluated at the sphere centre (set to zero in some references).

'Third Law'

Batchelor and Green derived an equation for the stresslet, given by

$$S = \frac{10}{3}\pi\mu a^3\left[-2E^\infty + \left(1+\frac{1}{10}a^2\nabla^2\right)\left(\nabla \mathbf{u}' + (\nabla \mathbf{u}')^\mathrm{T}\right)\right],$$

where

- **S** is the stresslet (symmetric part of the first moment of force) exerted by the fluid on the sphere,

- $\nabla \mathbf{u}'$ is the velocity gradient tensor; $^\mathrm{T}$ represents transpose; and so $\frac{1}{2}\left[\nabla \mathbf{u}' + (\nabla \mathbf{u}')^\mathrm{T}\right]$ is the rate of strain, or deformation, tensor.

- $E^\infty = \frac{1}{2}\left[\nabla \mathbf{u}^\infty + (\nabla \mathbf{u}^\infty)^\mathrm{T}\right]$ is the rate of strain of the background flow, evaluated at the sphere centre (set to zero in some references).

Note there is no rate of strain on the sphere (no **E**) since the spheres are assumed to be rigid.

Faxén's law is a correction to Stokes' law for the friction on spherical objects in a viscous fluid, valid where the object moves close to a wall of the container.

Hagen–Poiseuille Equation

In nonideal fluid dynamics, the Hagen–Poiseuille equation, also known as the Hagen–Poiseuille law, Poiseuille law or Poiseuille equation, is a physical law that gives the pressure drop in an incompressible and Newtonian fluid in laminar flow flowing through a long cylindrical pipe of constant cross section. It can be successfully applied to air flow in lung alveoli, for the flow through a drinking straw or through a hypodermic needle. It was experimentally derived independently by Jean Léonard Marie Poiseuille in 1838 and Gotthilf Heinrich Ludwig Hagen, and published by Poiseuille in 1840 and 1846.

The assumptions of the equation are that the fluid is incompressible and Newtonian; the flow is laminar through a pipe of constant circular cross-section that is substantially longer than its diameter; and there is no acceleration of fluid in the pipe. For velocities and pipe diameters above a threshold, actual fluid flow is not laminar but turbulent, leading to larger pressure drops than calculated by the Hagen–Poiseuille equation.

Equation
Standard Fluid-dynamics Notation

In standard fluid-dynamics notation:

$$\Delta P = \frac{8\mu LQ}{\pi r^4},$$

or

$$\Delta P = \frac{128\mu LQ}{\pi d^4},$$

or

$$\Delta P = \frac{32\mu Lv}{d^2},$$

where:

ΔP is the pressure reduction,

L is the length of pipe,

μ is the dynamic viscosity,

Q is the volumetric flow rate,

r is the pipe radius,

d is the pipe diameter,

π is the mathematical constant π,

v is the velocity.

Physics Notation

$$\Phi = \frac{dV}{dt} = v\pi R^2 = \frac{\pi R^4}{8\eta}\left(\frac{-\Delta P}{\Delta x}\right) = \frac{\pi R^4}{8\eta}\frac{|\Delta P|}{L},$$

where in compatible units (e.g., SI):

Φ is the volumetric flow rate (denoted as Q above),

$V(t)$ is the volume of the liquid transferred as a function of time, t,

v is mean fluid velocity along the length of the tube,

x is distance in direction of flow,

R is the internal radius of the tube,

ΔP is the pressure difference between the two ends,

η is the dynamic fluid viscosity,

L is the length of the tube.

The equation does not hold close to the pipe entrance.

The equation fails in the limit of low viscosity, wide and/or short pipe. Low viscosity or a wide pipe may result in turbulent flow, making it necessary to use more complex models, such as Darcy–Weisbach equation. If the pipe is too short, the Hagen–Poiseuille equation may result in unphysically high flow rates; the flow is bounded by Bernoulli's principle, under less restrictive conditions, by

$$\Phi_{max} = \pi R^2 \sqrt{2\Delta P / \rho}.$$

Relation to Darcy–Weisbach

Normally, Hagen-Poiseuille flow implies not just the relation for the pressure drop, above, but also the full solution for the laminar flow profile, which is parabolic. However, the result for the pressure drop can be extended to turbulent flow by inferring an effective turbulent viscosity in the case of turbulent flow, even though the flow profile in turbulent flow is strictly speaking not actually parabolic. In both cases, laminar or turbulent, the pressure drop is related to the stress at the wall, which determines the so-called friction factor. The wall stress can be determined phenomenologically by the Darcy–Weisbach equation in the field of hydraulics, given a relationship for the friction factor in terms of the Reynolds number. In the case of laminar flow:

$$\Lambda = \frac{64}{Re}, \quad Re = \frac{2\rho vr}{\eta},$$

where Re is the Reynolds number, ρ is the fluid density, v is the mean flow velocity, which is half the maximal flow velocity in the case of laminar flow. It proves more useful to define the Reynolds number in terms of the mean flow velocity because this quantity remains well defined even in the case of turbulent flow, whereas the maximal flow velocity may not be, or in any case, it may be difficult to infer. In this form the law approximates the *Darcy friction factor*, the *energy (head) loss factor, friction loss factor* or *Darcy (friction) factor* Λ in the laminar flow at very low velocities in cylindrical tube. The theoretical derivation of a slightly different form of the law was made independently by Wiedman in 1856 and Neumann and E. Hagenbach in 1858 (1859, 1860). Hagenbach was the first who called this law the Poiseuille's law.

The law is also very important in hemorheology and hemodynamics, both fields of physiology.

The Poiseuilles' law was later in 1891 extended to turbulent flow by L. R. Wilberforce, based on Hagenbach's work.

Derivation

The Hagen–Poiseuille equation can be derived from the Navier–Stokes equations. Although more lengthy than directly using the Navier–Stokes equations, an alternative method of deriving the Hagen–Poiseuille equation is as follows.

Liquid Flow Through a Pipe

Assume the liquid exhibits laminar flow. Laminar flow in a round pipe prescribes that there are a bunch of circular layers (lamina) of liquid, each having a velocity determined only by their radial distance from the center of the tube. Also assume the center is moving fastest while the liquid touching the walls of the tube is stationary (due to the no-slip condition).

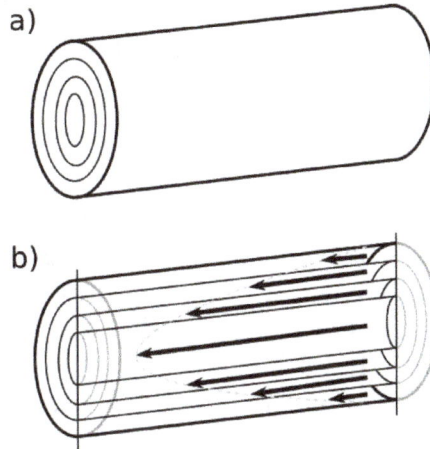

a)

b)

a) A tube showing the imaginary lamina. b) A cross section of the tube shows the lamina moving at different speeds. Those closest to the edge of the tube are moving slowly while those near the center are moving quickly.

To figure out the motion of the liquid, all forces acting on each lamina must be known:

1. The pressure force pushing the liquid through the tube is the change in pressure multiplied by the area: $F = -A\Delta P$. This force is in the direction of the motion of the liquid. The negative sign comes from the conventional way we define $\Delta P = P_{end} - P_{top} < 0..$

2. Viscosity effects will pull from the faster lamina immediately closer to the center of the tube.

3. Viscosity effects will drag from the slower lamina immediately closer to the walls of the tube.

Viscosity

Two fluids moving past each other in the x direction. The liquid on top is moving faster and will be pulled in the negative direction by the bottom liquid while the bottom liquid will be pulled in the positive direction by the top liquid.

When two layers of liquid in contact with each other move at different speeds, there will be a shear force between them. This force is proportional to the area of contact A, the velocity gradient in the direction of flow $\Delta v_x / \Delta y$, and a proportionality constant η (viscosity) and is given by

$$F_{viscosity, top} = -\eta A \frac{\Delta v_x}{\Delta y}.$$

The negative sign is in there because we are concerned with the faster moving liquid (top in figure), which is being slowed by the slower liquid (bottom in figure). By Newton's third law of motion, the force on the slower liquid is equal and opposite (no negative sign) to the force on the faster liquid. This equation assumes that the area of contact is so large that we can ignore any effects from the edges and that the fluids behave as Newtonian fluids.

Faster Lamina

Assume that we are figuring out the force on the lamina with radius r. From the equation above, we need to know the area of contact and the velocity gradient. Think of the lamina as a ring of radius r, thickness dr, and length Δx. The area of contact between the lamina and the faster one is simply the area of the inside of the cylinder: $A = 2\pi r \Delta x$. We don't know the exact form for the velocity of the liquid within the tube yet, but we do know (from our assumption above) that it is dependent on the radius. Therefore, the velocity gradient is the change of the velocity with respect to the change in the radius at the intersection of these two laminae. That intersection is at a radius of r. So, considering that this force will be positive with respect to the movement of the liquid (but the derivative of the velocity is negative), the final form of the equation becomes

$$F_{\text{viscosity, fast}} = -2\pi r \eta \Delta x \frac{dv}{dr}\Big|_r$$

where the vertical bar and subscript r following the derivative indicates that it should be taken at a radius of r.

Slower Lamina

Next let's find the force of drag from the slower lamina. We need to calculate the same values that we did for the force from the faster lamina. In this case, the area of contact is at $r+dr$ instead of r. Also, we need to remember that this force opposes the direction of movement of the liquid and will therefore be negative (and that the derivative of the velocity is negative).

$$F_{\text{viscosity, slow}} = 2\pi(r+dr)\eta \Delta x \frac{dv}{dr}\Big|_{r+dr}$$

Putting it all Together

To find the solution for the flow of a laminar layer through a tube, we need to make one last assumption. There is no acceleration of liquid in the pipe, and by Newton's first law, there is no net force. If there is no net force then we can add all of the forces together to get zero

$$0 = F_{\text{pressure}} + F_{\text{viscosity, fast}} + F_{\text{viscosity, slow}}$$

or

$$0 = \Delta P 2\pi r\, dr - 2\pi r \eta \Delta x \frac{dv}{dr}\Big|_r + 2\pi(r+dr)\eta \Delta x \frac{dv}{dr}\Big|_{r+dr}$$

First, to get everything happening at the same point, use the first two terms of a Taylor series expansion of the velocity gradient:

$$\frac{dv}{dr}\bigg|_{r+dr} = \frac{dv}{dr}\bigg|_{r} + \frac{d^2v}{dr^2}\bigg|_{r} dr.$$

The expression is valid for all laminae. Grouping like terms and dropping the vertical bar since all derivatives are assumed to be at radius r,

$$0 = \Delta P 2\pi r dr + 2\pi\eta\, dr\, \Delta x \frac{dv}{dr} + 2\pi r\eta\, dr\, \Delta x \frac{d^2v}{dr^2} + 2\pi\eta (dr)^2 \Delta x \frac{d^2v}{dr^2}$$

Finally, put this expression in the form of a differential equation, dropping the term quadratic in dr.

$$-\frac{1}{\eta}\frac{\Delta P}{\Delta x} = \frac{d^2v}{dr^2} + \frac{1}{r}\frac{dv}{dr}$$

It can be seen that both sides of the equations are negative: there is a drop of pressure along the tube (left side) and both first and second derivatives of the velocity are negative (velocity has a maximum value at the center of the tube, where $r = 0$). Using the product rule, the equation may be re-arranged to:

$$-\frac{1}{\eta}\frac{\Delta P}{\Delta x} = \frac{1}{r}\frac{d}{dr}\left(r\frac{dv}{dr}\right).$$

The right-hand side is the radial term of the Laplace operator ∇^2, so this differential equation is a special case of the Poisson equation. It is subject to the following boundary conditions:

$v(r) = 0$ at $r = R$ -- "no-slip" boundary condition at the wall

$\frac{dv}{dr} = 0$ at $r = 0$ -- axial symmetry.

Axial symmetry means that the velocity v(r) is maximum at the center of the tube, therefore the first derivative $\frac{dv}{dr}$ is zero at r = 0.

The differential equation can be integrated to:

$$v(r) = -\frac{1}{4\eta}r^2\frac{\Delta P}{\Delta x} + A\ln(r) + B.$$

To find A and B, we use the boundary conditions.

First, the symmetry boundary condition indicates:

$$\frac{dv}{dr} = -\frac{1}{2\eta}r\frac{\Delta P}{\Delta x} + \frac{A}{r} = 0 \text{ at r = 0.}$$

A solution possible only if A = 0. Next the no-slip boundary condition is applied to the remaining equation:

$$v(R) = -\frac{1}{4\eta}R^2\frac{\Delta P}{\Delta x} + B = 0$$

so therefore

$$B = \frac{1}{4\eta}R^2\frac{\Delta P}{\Delta x}.$$

Now we have a formula for the velocity of liquid moving through the tube as a function of the distance from the center of the tube

$$v = \frac{1}{4\eta}\frac{\Delta P}{\Delta x}(R^2 - r^2)$$

or, at the center of the tube where the liquid is moving fastest ($r = 0$) with R being the radius of the tube,

$$v_{max} = \frac{1}{4\eta}\frac{\Delta P}{\Delta x}R^2.$$

Poiseuille's Law

To get the total volume that flows through the tube, we need to add up the contributions from each lamina. To calculate the flow through each lamina, we multiply the velocity (from above) and the area of the lamina.

$$\Phi(r)dr = \frac{1}{4\eta}\frac{|\Delta P|}{\Delta x}(R^2 - r^2)2\pi r\,dr = \frac{\pi}{2\eta}\frac{|\Delta P|}{\Delta x}(rR^2 - r^3)dr$$

Finally, we integrate over all lamina via the radius variable r.

$$\Phi = \frac{\pi}{2\eta}\frac{|\Delta P|}{\Delta x}\int_0^R (rR^2 - r^3)dr = \frac{|\Delta P|\pi R^4}{8\eta\Delta x}$$

Poiseuille's Equation for Compressible Fluids

For a compressible fluid in a tube the volumetric flow rate and the linear velocity is not constant along the tube. The flow is usually expressed at outlet pressure. As fluid is compressed or expands, work is done and the fluid is heated or cooled. This means that the flow rate depends on the heat transfer to and from the fluid. For an ideal gas in the isothermal case, where the temperature of the fluid is permitted to equilibrate with its surroundings, and when the pressure difference between ends of the pipe is small, the volumetric flow rate at the pipe outlet is given by

$$\Phi = \frac{dV}{dt} = v\pi R^2 = \frac{\pi R^4\left(P_i - P_o\right)}{8\eta L}\times\frac{P_i + P_o}{2P_o} = \frac{\pi R^4}{16\eta L}\left(\frac{P_i^2 - P_o^2}{P_o}\right)$$

where:

P_i inlet pressure

P_o outlet pressure

L is the length of tube

η is the viscosity

R is the radius

V is the volume of the fluid at outlet pressure

v is the velocity of the fluid at outlet pressure

This is usually a good approximation when the flow velocity is less than mach 0.3

This equation can be seen as Poiseuille's law with an extra correction factor $\dfrac{P_i + P_o}{2} \times \dfrac{1}{P_o}$ expressing the average pressure relative to the outlet pressure.

Electrical Circuits Analogy

Electricity was originally understood to be a kind of fluid. This hydraulic analogy is still conceptually useful for understanding circuits. This analogy is also used to study the frequency response of fluid-mechanical networks using circuit tools, in which case the fluid network is termed a hydraulic circuit.

Poiseuille's law corresponds to Ohm's law for electrical circuits, $V = IR$.

Since the net force acting on the fluid is equal to

$$\Delta F = S\Delta P,$$

where $S = \pi r^2$, i.e. $\Delta F = \pi r^2 \Delta P$, then from Poiseuille's law

$$\Delta P = \frac{8\mu L Q}{\pi r^4}$$

it follows that

$$\Delta F = \frac{8\mu L Q}{r^2}.$$

For electrical circuits, let n be the concentration of free charged particles, (in m⁻³) and let q^* be the charge of each particle (in coulombs). (For electrons, $q^* = e = 1,6.10^{-19} C$.)

Then nQ is the number of particles in the volume Q, and nQq^* is their total charge. This is the charge that flows through the cross section per unit time, i.e. the current I. Therefore, $I = nQq^*$.

Consequently, $Q = \dfrac{I}{nq^*}$, and

$$\Delta F = \frac{8\mu L I}{nr^2 q^*}$$

But $\Delta F = Eq,$, where q is the total charge in the volume of the tube. The volume of the tube is equal to $\pi r^2 L$, so the number of charged particles in this volume is equal to $n\pi r^2 L$, and their total charge is

$$q = n\pi r^2 L q^*.$$

Now,

$$E = \frac{\Delta F}{q} = \frac{8\mu I}{n^2 \pi r^4 (q^*)^2}.$$

Since the voltage $V = EL,$, we get

$$V = \frac{8\mu LI}{n^2 \pi r^4 (q^*)^2}$$

This is exactly Ohm's law, where the resistance $R = \dfrac{V}{I}$ is described by the formula

$$R = \frac{8\mu L}{n^2 \pi r^4 (q^*)^2}.$$

It follows that the resistance R is proportional to the length R of the resistor, which is true. However, it also follows that the resistance R is inversely proportional to the fourth power of the radius r, i.e. the resistance R is inversely proportional to the second power of the cross section area $S = \pi r^2$ of the resistor, which is wrong (according to the electrical analogy)!

The correct relation is

$$R = \frac{\rho L}{S},$$

where ρ is the specific resistance; i.e. the resistance R is inversely proportional to the first power of the cross section area S of the resistor.

The reason why Poiseuille's law leads to a wrong formula for the resistance R is the difference between the fluid flow and the electric current. Electron gas is inviscid, so its velocity does not depend on the distance to the walls of the conductor. The resistance is due to the interaction between the flowing electrons and the atoms of the conductor. Therefore, Poiseuille's law and the hydraulic analogy are useful only within certain limits when applied to electricity.

Both Ohm's law and Poiseuille's law illustrate transport phenomena.

Medical Applications – Intravenous Access and Fluid Delivery

The Hagen Poiseuille's equation is useful in determining the flow rate of IV fluids that may be achieved using various sizes of peripheral and central cannulas. The equation states that flow rate is proportional to the radius to the fourth power, meaning that a small increase in the internal diameter of the cannula yields a significant increase in flow rate of IV fluids. The radius of IV cannulas is typically measured in "gauge", which is inversely proportional to the radius. Peripheral IV cannulas are typically available (from large to small) 14g, 16g, 18g, 20g, 22g, 24g. As an example, the flow of a 14g cannula is typically twice that of a 16g, and ten times that of a 20g. It also states that flow is inversely proportional to length, meaning that longer lines have lower flow rates. This is important to remember as in an emergency, many clinicians would intuitively favour a long central venous catheter over a short peripherally sited cannula. While of less clinical importance, the change in pressure can be used to speed up flow rate by pressurizing the bag of fluid, squeezing the bag, or hanging the bag higher from the level of the cannula. It is also useful to understand that viscous fluids will flow slower (e.g. blood transfusion).

References

- Morin, David (2008). Introduction to Classical Mechanics. New York: Cambridge University Press. ISBN 9780521876223.

- Barut, Asim O. (1980) [1964]. Introduction to Classical Mechanics. New York: Dover Publications. ISBN 9780486640389.

- Einstein, Albert (2004) [1920]. Relativity. Translated by Robert W. Lawson. New York: Barnes & Noble. ISBN 9780760759219.

- Goldstein, Herbert; Poole, Charles P., Jr.; Safko, John L. (2002), Classical Mechanics (3rd ed.), San Francisco, CA: Addison Wesley, pp. 347–349, ISBN 0-201-65702-3

- Chanson, Hubert (2004). Hydraulics of Open Channel Flow: An Introduction. Butterworth-Heinemann. ISBN 0-7506-5978-5.

- Oertel, Herbert; Prandtl, Ludwig; Böhle, M.; Mayes, Katherine (2004). Prandtl's Essentials of Fluid Mechanics. Springer. pp. 70–71. ISBN 0-387-40437-6.

- Feynman, R.P.; Leighton, R.B.; Sands, M. (1963). The Feynman Lectures on Physics. ISBN 0-201-02116-1., Vol. 2, §40–3, pp. 40–6 – 40–9.

- Tipler, Paul (1991). Physics for Scientists and Engineers: Mechanics (3rd extended ed.). W. H. Freeman. ISBN 0-87901-432-6.

- Chanson, H. (2009). Applied Hydrodynamics: An Introduction to Ideal and Real Fluid Flows. CRC Press, Taylor & Francis Group, Leiden, The Netherlands, 478 pages. ISBN 978-0-415-49271-3.

- Daryn Lehoux (2012). What Did the Romans Know? An Inquiry into Science and Worldmaking. University of Chicago Press. (ISBN 9780226471143)

- Toro, E.F. (1999). "Chapter 2. Notions on Hyperbolic PDEs". Riemann Solvers and Numerical Methods for Fluid Dynamics. Springer-Verlag. ISBN 3-540-65966-8.

- Lamb, H. (1994). Hydrodynamics (6th ed.). Cambridge University Press. ISBN 978-0-521-45868-9. Originally published in 1879, the 6th extended edition appeared first in 1932.

Motion: A Comprehensive Study

The movement of any object from one point to another is because of motion. Velocity of an object is the rate of this change from one point to another. Momentum, frame of reference and Newton's laws of motion have also been explained in the following section. Physics is best understood in confluence with the major topics listed in the following chapter.

Motion (Physics)

In physics, motion is a change in position of an object with respect to time. Motion is typically described in terms of displacement, distance, velocity, acceleration, time and speed. Motion of a body is observed by attaching a frame of reference to an observer and measuring the change in position of the body relative to that frame.

If the position of a body is not changing with respect to a given frame of reference, the body is said to be *at rest*, *motionless*, *immobile*, *stationary*, or to have constant (time-invariant) position. An object's motion cannot change unless it is acted upon by a force, as described. Momentum is a quantity which is used for measuring motion of an object. An object's momentum is directly related to the object's mass and velocity, and the total momentum of all objects in an isolated system (one not affected by external forces) does not change with time, as described by the law of conservation of momentum.

As there is no absolute frame of reference, *absolute motion* cannot be determined. Thus, everything in the universe can be considered to be moving.

Motion involves a change in position, such as in this perspective of rapidly leaving Yongsan Station.

More generally, motion is a concept that applies to objects, bodies, and matter particles, to radiation, radiation fields and radiation particles, and to space, its curvature and space-time. One can also speak of motion of shapes and boundaries. So, the term motion in general signifies a continuous change in the configuration of a physical system. For example, one can talk about motion of a wave or about motion of a quantum particle, where the configuration consists of probabilities of occupying specific positions.

Laws of Motion

In physics, motion is described through two sets of apparently contradictory laws of mechanics. Motions of all large scale and familiar objects in the universe (such as projectiles, planets, cells, and humans) are described by classical mechanics. Whereas the motion of very small atomic and sub-atomic objects is described by quantum mechanics.

Classical Mechanics

Classical mechanics is used for describing the motion of macroscopic objects, from projectiles to parts of machinery, as well as astronomical objects, such as spacecraft, planets, stars, and galaxies. It produces very accurate results within these domains, and is one of the oldest and largest in science, engineering, and technology.

Classical mechanics is fundamentally based on Newton's laws of motion. These laws describe the relationship between the forces acting on a body and the motion of that body. They were first compiled by Sir Isaac Newton in his work *Philosophiæ Naturalis Principia Mathematica*, first published on July 5, 1687. His three laws are:

1. A body either is at rest or moves with constant velocity, until and unless an outer force is applied to it.

2. An object will travel in one direction only until an outer force changes its direction.

3. Whenever one body exerts a force F onto a second body,(in some cases, which is standing still) the second body exerts the force −F on the first body. F and −F are equal in magnitude and opposite in sense. So, the body which exerts F will go backwards.

Newton's three laws of motion, along with his Newton's law of motion, which were the first to accurately provide a mathematical model for understanding orbiting bodies in outer space. This explanation unified the motion of celestial bodies and motion of objects on earth.

Classical mechanics was later further enhanced by Albert Einstein's special relativity and general relativity. Motion of objects with a high velocity, approaching the speed of light; general relativity is employed to handle gravitational motion at a deeper level.

Quantum Mechanics

Quantum mechanics is a set of principles describing physical reality at the atomic level of matter (molecules and atoms) and the subatomic particles (electrons, protons, and even smaller particles). These descriptions include the simultaneous wave-like and particle-like behavior of both matter and radiation energy, this is described in the wave–particle duality.

In classical mechanics, accurate measurements and predictions of the state of objects can be calculated, such as location and velocity. In the quantum mechanics, due to the Heisenberg uncertainty principle), the complete state of a subatomic particle, such as its location and velocity, cannot be simultaneously determined.

In addition to describing the motion of atomic level phenomena, quantum mechanics is useful in understanding some large scale phenomenon such as superfluidity, superconductivity, and biological systems, including the function of smell receptors and the structures of proteins.

List of "Imperceptible" Human Motions

Humans, like all known things in the universe, are in constant motion, however, aside from obvious movements of the various external body parts and locomotion, humans are in motion in a variety of ways which are more difficult to perceive. Many of these "imperceptible motions" are only perceivable with the help of special tools and careful observation. The larger scales of "imperceptible motions" are difficult for humans to perceive for two reasons: 1) Newton's laws of motion (particularly Inertia) which prevent humans from feeling motions of a mass to which they are connected, and 2) the lack of an obvious frame of reference which would allow individuals to easily see that they are moving. The smaller scales of these motions are too small for humans to sense.

Universe

- Spacetime (the fabric of the universe) is actually expanding. Essentially, everything in the universe is stretching like a rubber band. This motion is the most obscure as it is not physical motion as such, but rather a change in the very nature of the universe. The primary source of verification of this expansion was provided by Edwin Hubble who demonstrated that all galaxies and distant astronomical objects were moving away from us (*"Hubble's law"*) as predicted by a universal expansion.

Galaxy

- The Milky Way Galaxy, is moving through space. Many astronomers believe the Milky Way is moving at approximately 600 km/s relative to the observed locations of other nearby galaxies. Another reference frame is provided by the Cosmic microwave background. This frame of reference indicates that The Milky Way is moving at around 582 km/s.

Sun and Solar System

- The Milky Way is rotating around its dense galactic center, thus the sun is moving in a circle within the galaxy's gravity. Away from the central bulge or outer rim, the typical stellar velocity is between 210 and 240 km/s. All planets and their moons move with the sun. Thus the solar system is moving.

Earth

- The Earth is rotating or spinning around its axis, this is evidenced by day and night, at the equator the earth has an eastward velocity of 0.4651 km/s (1040 mi/h).

- The Earth is orbiting around the Sun in an orbital revolution. A complete orbit around the sun takes one year or about 365 days; it averages a speed of about 30 km/s (67,000 mi/h).

Continents

- The Theory of Plate tectonics tells us that the continents are drifting on convection currents within the mantle causing them to move across the surface of the planet at the slow speed of approximately 1 inch (2.54 cm) per year. However, the velocities of plates range widely. The fastest-moving plates are the oceanic plates, with the Cocos Plate advancing at a rate of 75 mm/yr (3.0 in/yr) and the Pacific Plate moving 52–69 mm/yr (2.1–2.7 in/yr). At the other extreme, the slowest-moving plate is the Eurasian Plate, progressing at a typical rate of about 21 mm/yr (0.8 in/yr).

Internal Body

- The human heart is constantly contracting to move blood throughout the body. Through larger veins and arteries in the body blood has been found to travel at approximately 0.33 m/s. Though considerable variation exists, and peak flows in the venae cavae have been found between 0.1 m/s and 0.45 m/s.

- The smooth muscles of hollow internal organs are moving. The most familiar would be peristalsis which is where digested food is forced throughout the digestive tract. Though different foods travel through the body at rates, an average speed through the human small intestine is 2.16 m/h (0.036 m/s).

- Typically some sound is audible at any given moment, when the vibration of these sound waves reaches the ear drum it moves in response and allows the sense of hearing.

- The human lymphatic system is constantly moving excess fluids, lipids, and immune system related products around the body. The lymph fluid has been found to move through a lymph capillary of the skin at approximately 0.0000097 m/s.

Cells

The cells of the human body have many structures which move throughout them.

- Cytoplasmic streaming is a way which cells move molecular substances throughout the cytoplasm.

- Various motor proteins work as molecular motors within a cell and move along the surface of various cellular substrates such as microtubuless. Motor proteins are typically powered by the hydrolysis of adenosine triphosphate (ATP), and convert chemical energy into mechanical work. Vesicles propelled by motor proteins have been found to have a velocity of approximately 0.00000152 m/s.

Particles

- According to the laws of thermodynamics all particles of matter are in constant random motion as long as the temperature is above absolute zero. Thus the molecules and atoms which make up the human body are vibrating, colliding, and moving. This motion can be detected as temperature; higher temperatures, which represent greater kinetic energy in the particles, feel warm to humans whom sense the thermal energy transferring from the object being touched to their nerves. Similarly, when lower temperature objects are touched, the senses perceive the transfer of heat away from the body as feeling cold.

Subatomic Particles

- Within each atom, electrons exist in an area around the nucleus. This area is called the electron cloud. According to Bohr's model of the atom, electrons have a high velocity, and the larger the nucleus they are orbiting the faster they would need to move. If electrons 'move' about the electron cloud in strict paths the same way planets orbit the sun, then electrons would be required to do so at speeds which far exceed the speed of light. However, there is no reason that one must confine one's self to this strict conceptualization, that electrons move in paths the same way macroscopic objects do. Rather one can conceptualize electrons to be 'particles' that capriciously exist within the bounds of the electron cloud.

- Inside the atomic nucleus the protons and neutrons are also probably moving around due the electrical repulsion of the protons and the presence of angular momentum of both particles.

Light

Light propagates at 299,792,458 m/s, often approximated as 300,000 kilometres per second or 186,000 miles per second. The speed of light (or c) is also the speed of all massless particles and associated fields in a vacuum, and it is the upper limit on the speed at which energy, matter, and information can travel. The speed of light is the limit speed for physical systems.

In addition, the speed of light is an invariant quantity: it has the same value, irrespective of the position or speed of the observer. This property makes the speed of light c the natural measurement unit for speed.

Velocity

The velocity of an object is the rate of change of its position with respect to a frame of reference, and is a function of time. Velocity is equivalent to a specification of its speed and direction of motion (e.g. 60 km/h to the north). Velocity is an important concept in kinematics, the branch of classical mechanics that describes the motion of bodies.

Velocity is a physical vector quantity; both magnitude and direction are needed to define it. The scalar absolute value (magnitude) of velocity is called "speed", being a coherent derived unit whose quantity is measured in the SI (metric) system as metres per second (m/s) or as the SI base unit of (m·s⁻¹). For example, "5 metres per second" is a scalar, whereas "5 metres per second east" is a vector.

If there is a change in speed, direction or both, then the object has a changing velocity and is said to be undergoing an *acceleration*.

Constant Velocity vs Acceleration

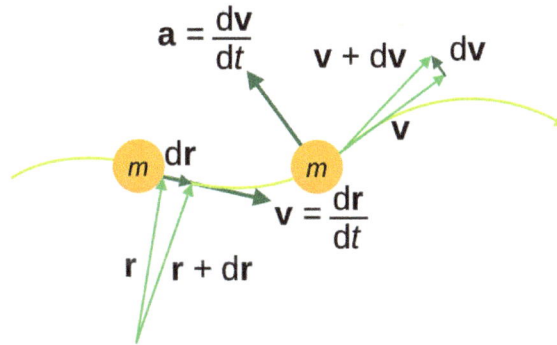

Kinematic quantities of a classical particle: mass m, position $\mathbf{r}$, velocity $\mathbf{v}$, acceleration $\mathbf{a}$.

To have a constant velocity, an object must have a constant speed in a constant direction. Constant direction constrains the object to motion in a straight path thus, a constant velocity means motion in a straight line at a constant speed.

For example, a car moving at a constant 20 kilometres per hour in a circular path has a constant speed, but does not have a constant velocity because its direction changes. Hence, the car is considered to be undergoing an acceleration.

Distinction Between Speed and Velocity

Speed describes only how fast an object is moving, whereas velocity gives both how fast and in what direction the object is moving. If a car is said to travel at 60 km/h, its speed has been specified. However, if the car is said to move at 60 km/h to the north, its velocity has now been specified.

The big difference can be noticed when we consider movement around a circle. When something moves in a circle and returns to its starting point, its average velocity is zero but its average speed is found by dividing the circumference of the circle by the time taken to move around the circle. This is because the average velocity is calculated by only considering the displacement between the starting and the end points while the average speed considers only the total distance traveled.

Equation of Motion

Average Velocity

Velocity is defined as the rate of change of position with respect to time, which may also be referred to as the *instantaneous velocity* to emphasize the distinction from the average velocity. In some applications the "average velocity" of an object might be needed, that is to say, the constant velocity that would provide the same resultant displacement as a variable velocity in the same time interval, v(t), over some time period Δt. Average velocity can be calculated as:

$$\overline{\mathbf{v}} = \frac{\Delta \mathbf{x}}{\Delta t}.$$

The average velocity is always less than or equal to the average speed of an object. This can be seen by realizing that while distance is always strictly increasing, displacement can increase or decrease in magnitude as well as change direction.

In terms of a displacement-time (x vs. t) graph, the instantaneous velocity (or, simply, velocity) can be thought of as the slope of the tangent line to the curve at any point, and the average velocity as the slope of the secant line between two points with t coordinates equal to the boundaries of the time period for the average velocity.

The average velocity is the same as the velocity averaged over time – that is to say, its time-weighted average, which may be calculated as the time integral of the velocity:

$$\overline{\mathbf{v}} = \frac{1}{t_1 - t_0} \int_{t_0}^{t_1} \mathbf{v}(t)\, dt,$$

where we may identify

$$\Delta \mathbf{x} = \int_{t_0}^{t_1} \mathbf{v}(t)\, dt$$

and

$$\Delta t = t_1 - t_0.$$

Instantaneous Velocity

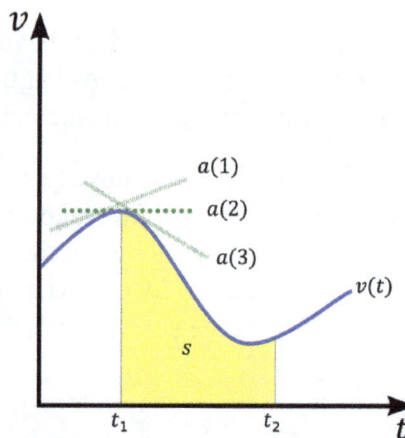

Example of a velocity vs. time graph, and the relationship between velocity v on the y-axis, acceler-ation a (the three green tangent lines represent the values for acceleration at different points along the curve) and displacement s (the yellow area under the curve.)

If we consider v as velocity and x as the displacement (change in position) vector, then we can express the (instantaneous) velocity of a particle or object, at any particular time t, as the derivative of the position with respect to time:

$$\mathbf{v} = \lim_{\Delta t \to 0} \frac{\Delta \mathbf{x}}{\Delta t} = \frac{d\mathbf{x}}{dt}.$$

From this derivative equation, in the one-dimensional case it can be seen that the area under a velocity vs. time (v vs. t graph) is the displacement, x. In calculus terms, the integral of the velocity function $v(t)$ is the displacement function $x(t)$. In the figure, this corresponds to the yellow area under the curve labeled s (s being an alternative notation for displacement).

$$\mathbf{x} = \int \mathbf{v}\, dt.$$

Since the derivative of the position with respect to time gives the change in position (in metres) divided by the change in time (in seconds), velocity is measured in metres per second (m/s). Although the concept of an instantaneous velocity might at first seem counter-intuitive, it may be thought of as the velocity that the object would continue to travel at if it stopped accelerating at that moment.

Relationship to Acceleration

Although velocity is defined as the rate of change of position, it is often common to start with an expression for an object's acceleration. As seen by the three green tangent lines in the figure, an object's instantaneous acceleration at a point in time is the slope of the line tangent to the curve of a (v vs. t graph at that point. In other words, acceleration is defined as the derivative of velocity with respect to time:

$$\mathbf{a} = \frac{d\mathbf{v}}{dt}.$$

From there, we can obtain an expression for velocity as the area under an acceleration vs. time (a vs. t) graph. As above, this is done using the concept of the integral:

$$\mathbf{v} = \int \mathbf{a}\, dt.$$

Constant Acceleration

In the special case of constant acceleration, velocity can be studied using the suvat equations. By considering a as being equal to some arbitrary constant vector, it is trivial to show that

$$\mathbf{v} = \mathbf{u} + \mathbf{a}t$$

with v as the velocity at time t and u as the velocity at time $t = 0$. By combining this equation with the suvat equation $x = ut + at^2/2$, it is possible to relate the displacement and the average velocity by

$$\mathbf{x} = \frac{(\mathbf{u} + \mathbf{v})}{2}t = \overline{\mathbf{v}}t.$$

It is also possible to derive an expression for the velocity independent of time, known as the Torricelli equation, as follows:

$$v^2 = \mathbf{v} \cdot \mathbf{v} = (\mathbf{u} + \mathbf{a}t) \cdot (\mathbf{u} + \mathbf{a}t) = u^2 + 2t(\mathbf{a} \cdot \mathbf{u}) + a^2 t^2$$

$$(2\mathbf{a}) \cdot \mathbf{x} = (2\mathbf{a}) \cdot (\mathbf{u}t + \frac{1}{2}\mathbf{a}t^2) = 2t(\mathbf{a} \cdot \mathbf{u}) + a^2 t^2 = v^2 - u^2$$

$$\therefore v^2 = u^2 + 2(\mathbf{a} \cdot \mathbf{x})$$

where $v = |v|$ etc.

The above equations are valid for both Newtonian mechanics and special relativity. Where Newtonian mechanics and special relativity differ is in how different observers would describe the same situation. In particular, in Newtonian mechanics, all observers agree on the value of t and the transformation rules for position create a situation in which all non-accelerating observers would describe the acceleration of an object with the same values. Neither is true for special relativity. In other words, only relative velocity can be calculated.

Quantities that are Dependent on Velocity

The kinetic energy of a moving object is dependent on its velocity and is given by the equation

$$E_k = \frac{1}{2}mv^2$$

ignoring special relativity, where E_k is the kinetic energy and m is the mass. Kinetic energy is a scalar quantity as it depends on the square of the velocity, however a related quantity, momentum, is a vector and defined by

$$\mathbf{p} = m\mathbf{v}$$

In special relativity, the dimensionless Lorentz Factor appears frequently, and is given by

$$\gamma = \frac{1}{\sqrt{1 - \frac{v^2}{c^2}}}$$

where γ is the Lorentz factor and c is the speed of light.

Escape velocity is the minimum speed a ballistic object needs to escape from a massive body such as Earth. It represents the kinetic energy that, when added to the object's gravitational potential energy, (which is always negative) is equal to zero. The general formula for the escape velocity of an object at a distance r from the center of a planet with mass M is

$$v_e = \sqrt{\frac{2GM}{r}} = \sqrt{2gr},$$

where G is the Gravitational constant and g is the Gravitational acceleration. The escape velocity from Earth's surface is about 11 200 m/s, and is irrespective of the direction of the object.

Relative Velocity

Relative velocity is a measurement of velocity between two objects as determined in a single coordinate system. Relative velocity is fundamental in both classical and modern physics, since many systems in physics deal with the relative motion of two or more particles. In Newtonian mechanics, the relative velocity is independent of the chosen inertial reference frame. This is not the case anymore with special relativity in which velocities depend on the choice of reference frame.

If an object A is moving with velocity vector v and an object B with velocity vector w, then the velocity of object A *relative to* object B is defined as the difference of the two velocity vectors:

$$\mathbf{v}_{A \text{ relative to } B} = \mathbf{v} - \mathbf{w}$$

Similarly the relative velocity of object B moving with velocity w, relative to object A moving with velocity v is:

$$\mathbf{v}_{B \text{ relative to } A} = \mathbf{w} - \mathbf{v}$$

Usually the inertial frame is chosen in which the latter of the two mentioned objects is in rest.

Scalar Velocities

In the one-dimensional case, the velocities are scalars and the equation is either:

$v_{rel} = v - (-w),$, if the two objects are moving in opposite directions, or:

$v_{rel} = v - (+w),$, if the two objects are moving in the same direction.

Polar Coordinates

In polar coordinates, a two-dimensional velocity is described by a radial velocity, defined as the component of velocity away from or toward the origin (also known as *velocity made good*), and an angular velocity, which is the rate of rotation about the origin (with positive quantities representing counter-clockwise rotation and negative quantities representing clockwise rotation, in a right-handed coordinate system).

The radial and angular velocities can be derived from the Cartesian velocity and displacement vectors by decomposing the velocity vector into radial and transverse components. The transverse velocity is the component of velocity along a circle centered at the origin.

$$\mathbf{v} = \mathbf{v}_T + \mathbf{v}_R$$

where

$\mathbf{v}_T$ is the transverse velocity

$\mathbf{v}_R$ is the radial velocity.

The *magnitude of the radial velocity* is the dot product of the velocity vector and the unit vector in the direction of the displacement.

$$v_R = \frac{\mathbf{v} \cdot \mathbf{r}}{|\mathbf{r}|}$$

where

$\mathbf{r}$ is displacement.

The *magnitude of the transverse velocity* is that of the cross product of the unit vector in the direction of the displacement and the velocity vector. It is also the product of the angular speed ω and the magnitude of the displacement.

$$v_T = \frac{|\mathbf{r} \times \mathbf{v}|}{|\mathbf{r}|} = \omega|\mathbf{r}|$$

such that

$$\omega = \frac{|\boldsymbol{r} \times \boldsymbol{v}|}{|\boldsymbol{r}|^2}.$$

Angular momentum in scalar form is the mass times the distance to the origin times the transverse velocity, or equivalently, the mass times the distance squared times the angular speed. The sign convention for angular momentum is the same as that for angular velocity.

$$L = mrv_T = mr^2\omega$$

where

m is mass

$r = \|\mathbf{r}\|$.

The expression mr^2 is known as moment of inertia. If forces are in the radial direction only with an inverse square dependence, as in the case of a gravitational orbit, angular momentum is constant, and transverse speed is inversely proportional to the distance, angular speed is inversely proportional to the distance squared, and the rate at which area is swept out is constant. These relations are known as Kepler's laws of planetary motion.

Momentum

In classical mechanics, linear momentum, translational momentum, or simply momentum (pl. momenta; SI unit kg · m/s) is the product of the mass and velocity of an object, quantified in kilogram-meters per second. It is dimensionally equivalent to impulse, the product of force and time, quantified in newton-seconds. Newton's second law of motion states that the change in linear momentum of a body is equal to the net impulse acting on it. For example, a heavy truck moving rapidly has a large momentum, and it takes a large or prolonged force to get the truck up to this speed, and would take a similarly large or prolonged force to bring it to a stop. If the truck were lighter, or moving more slowly, then it would have less momentum and therefore require less impulse to start or stop.

Like velocity, linear momentum is a vector quantity, possessing a direction as well as a magnitude:

$$\mathbf{p} = m\mathbf{v},$$

where p is the three-dimensional vector stating the object's momentum in the three directions of three-dimensional space, v is the three-dimensional velocity vector giving the object's rate of movement in each direction, and m is the object's mass.

Linear momentum is also a *conserved* quantity, meaning that if a closed system is not affected by external forces, its total linear momentum cannot change.

In classical mechanics, conservation of linear momentum is implied by Newton's laws. It also holds in special relativity (with a modified formula) and, with appropriate definitions, a (generalized) linear momentum conservation law holds in electrodynamics, quantum mechanics, quantum field theory, and general relativity. It is ultimately an expression of one of the fundamental symmetries of space and time, that of translational symmetry.

Linear momentum depends on frame of reference. Observers in different frames would find different values of linear momentum of a system. But each would observe that the value of linear momentum does not change with time, provided the system is isolated.

Newtonian Mechanics

Momentum has a direction as well as magnitude. Quantities that have both a magnitude and a direction are known as vector quantities. Because momentum has a direction, it can be used to predict the resulting direction of objects after they collide, as well as their speeds. Below, the basic properties of momentum are described in one dimension. The vector equations are almost identical to the scalar equations.

Single Particle

The momentum of a particle is traditionally represented by the letter p. It is the product of two quantities, the mass (represented by the letter m) and velocity (v):

$$p = mv.$$

The units of momentum are the product of the units of mass and velocity. In SI units, if the mass is in kilograms and the velocity in meters per second then the momentum is in kilogram meters/second (kg m/s). In cgs units, if the mass is in grams and the velocity in centimeters per second, then the momentum is in gram centimeters/second (g cm/s).

Being a vector, momentum has magnitude and direction. For example, a 1 kg model airplane, traveling due north at 1 m/s in straight and level flight, has a momentum of 1 kg m/s due north measured from the ground.

Many Particles

The momentum of a system of particles is the sum of their momenta. If two particles have masses m_1 and m_2, and velocities v_1 and v_2, the total momentum is

$$p = p_1 + p_2$$
$$= m_1 v_1 + m_2 v_2.$$

The momenta of more than two particles can be added more generally with the following:

$$p = \sum_i m_i v_i$$

A system of particles has a center of mass, a point determined by the weighted sum of their positions:

$$r_{cm} = \frac{m_1 r_1 + m_2 r_2 + \cdots}{m_1 + m_2 + \cdots} = \frac{\sum_i m_i r_i}{\sum_i m_i}.$$

If all the particles are moving, the center of mass will generally be moving as well (unless the system is in pure rotation around it). If the center of mass is moving at velocity v_{cm}, the momentum is:

$$p = m v_{cm}.$$

This is known as Euler's first law.

Relation to Force

If a force F is applied to a particle for a time interval Δt, the momentum of the particle changes by an amount

$$\Delta p = F \Delta t.$$

In differential form, this is Newton's second law; the rate of change of the momentum of a particle is proportional to the force F acting on it,

$$F = \frac{dp}{dt}.$$

If the force depends on time, the change in momentum (or impulse J) between times t_1 and t_2 is

$$\Delta p = J = \int_{t_1}^{t_2} F(t) dt.$$

Impulse is measured in the derived units of the newton second (1 N s = 1 kg m/s) or dyne second (1 dyne s = 1 g m/s)

Under the assumption of constant mass m, it is equivalent to write

$$F = m \frac{dv}{dt} = ma,$$

so the force is equal to mass times acceleration.

Example: A model airplane of 1 kg accelerates from rest to a velocity of 6 m/s due north in 2 s. The net force required to produce this acceleration is 3 newtons due north. The change in momentum is 6 kg m/s. The rate of change of momentum is 3 (kg m/s)/s = 3 N.

Conservation

A Newton's cradle demonstrates conservation of momentum.

In a closed system (one that does not exchange any matter with its surroundings and is not acted on by external forces) the total momentum is constant. This fact, known as the *law of conservation of momentum*, is implied by Newton's laws of motion. Suppose, for example, that two particles interact. Because of the third law, the forces between them are equal and opposite. If the particles are numbered 1 and 2, the second law states that $F_1 = dp_1/dt$ and $F_2 = dp_2/dt$. Therefore,

$$\frac{dp_1}{dt} = -\frac{dp_2}{dt},$$

with the negative sign indicating that the forces oppose. Equivalently,

$$\frac{d}{dt}(p_1 + p_2) = 0.$$

If the velocities of the particles are u_1 and u_2 before the interaction, and afterwards they are v_1 and v_2, then

$$m_1 u_1 + m_2 u_2 = m_1 v_1 + m_2 v_2$$

This law holds no matter how complicated the force is between particles. Similarly, if there are several particles, the momentum exchanged between each pair of particles adds up to zero, so the total change in momentum is zero. This conservation law applies to all interactions, including collisions and separations caused by explosive forces. It can also be generalized to situations where Newton's laws do not hold, for example in the theory of relativity and in electrodynamics.

Dependence on Reference Frame

Momentum is a measurable quantity, and the measurement depends on the motion of the observer. For example: if an apple is sitting in a glass elevator that is descending, an outside observer, looking into the elevator, sees the apple moving, so, to that observer, the apple has a non-zero momentum. To someone inside the elevator, the apple does not move, so, it has zero momentum. The two observers each have a frame of reference, in which, they observe motions, and, if the elevator is descending steadily, they will see behavior that is consistent with those same physical laws.

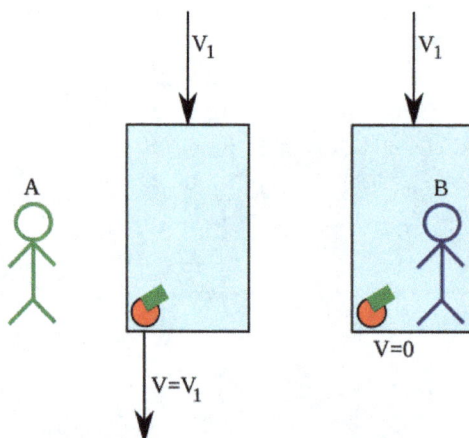

Newton's apple in Einstein's elevator. In person A's frame of reference, the apple has non-zero velocity and momentum. In the elevator's and person B's frames of reference, it has zero velocity and momentum.

Suppose a particle has position x in a stationary frame of reference. From the point of view of another frame of reference, moving at a uniform speed u, the position (represented by a primed coordinate) changes with time as

$$x' = x - ut.$$

This is called a Galilean transformation. If the particle is moving at speed $dx/dt = v$ in the first frame of reference, in the second, it is moving at speed

$$v' = \frac{dx'}{dt} = v - u.$$

Since u does not change, the accelerations are the same:

$$a' = \frac{dv'}{dt} = a.$$

Thus, momentum is conserved in both reference frames. Moreover, as long as the force has the same form, in both frames, Newton's second law is unchanged. Forces such as Newtonian gravity, which depend only on the scalar distance between objects, satisfy this criterion. This independence of reference frame is called Newtonian relativity or Galilean invariance.

A change of reference frame, can, often, simplify calculations of motion. For example, in a collision of two particles, a reference frame can be chosen, where, one particle begins at rest. Another, commonly used reference frame, is the center of mass frame - one that is moving with the center of mass. In this frame, the total momentum is zero.

Application to Collisions

By itself, the law of conservation of momentum is not enough to determine the motion of particles after a collision. Another property of the motion, kinetic energy, must be known. This is not necessarily conserved. If it is conserved, the collision is called an *elastic collision*; if not, it is an *inelastic collision*.

Elastic Collisions

Elastic collision of equal masses

Elastic collision of unequal masses

An elastic collision is one in which no kinetic energy is lost. Perfectly elastic "collisions" can occur when the objects do not touch each other, as for example in atomic or nuclear scattering where electric repulsion keeps them apart. A slingshot maneuver of a satellite around a planet can also be viewed as a perfectly elastic collision from a distance. A collision between two pool balls is a good example of an *almost* totally elastic collision, due to their high rigidity; but when bodies come in contact there is always some dissipation.

A head-on elastic collision between two bodies can be represented by velocities in one dimension, along a line passing through the bodies. If the velocities are u_1 and u_2 before the collision and v_1 and v_2 after, the equations expressing conservation of momentum and kinetic energy are:

$$m_1 u_1 + m_2 u_2 = m_1 v_1 + m_2 v_2$$

$$\frac{1}{2} m_1 u_1^2 + \frac{1}{2} m_2 u_2^2 = \frac{1}{2} m_1 v_1^2 + \frac{1}{2} m_2 v_2^2.$$

A change of reference frame can often simplify the analysis of a collision. For example, suppose there are two bodies of equal mass m, one stationary and one approaching the other at a speed v (as in the figure). The center of mass is moving at speed $v/2$ and both bodies are moving towards it at speed $v/2$. Because of the symmetry, after the collision both must be moving away from the center of mass at the same speed. Adding the speed of the center of mass to both, we find that the body that was moving is now stopped and the other is moving away at speed v. The bodies have exchanged their velocities. Regardless of the velocities of the bodies, a switch to the center of mass frame leads us to the same conclusion. Therefore, the final velocities are given by

$$v_1 = u_2$$

$$v_2 = u_1.$$

In general, when the initial velocities are known, the final velocities are given by

$$v_1 = \left(\frac{m_1 - m_2}{m_1 + m_2}\right)u_1 + \left(\frac{2m_2}{m_1 + m_2}\right)u_2$$

$$v_2 = \left(\frac{m_2 - m_1}{m_1 + m_2}\right)u_2 + \left(\frac{2m_1}{m_1 + m_2}\right)u_1.$$

If one body has much greater mass than the other, its velocity will be little affected by a collision while the other body will experience a large change.

Inelastic Collisions

a perfectly inelastic collision between equal masses

In an inelastic collision, some of the kinetic energy of the colliding bodies is converted into other forms of energy such as heat or sound. Examples include traffic collisions, in which the effect of lost kinetic energy can be seen in the damage to the vehicles; electrons losing some of their energy to atoms (as in the Franck–Hertz experiment); and particle accelerators in which the kinetic energy is converted into mass in the form of new particles.

In a perfectly inelastic collision (such as a bug hitting a windshield), both bodies have the same motion afterwards. If one body is motionless to begin with, the equation for conservation of momentum is

$$m_1 u_1 = (m_1 + m_2)v,$$

so

$$v = \frac{m_1}{m_1 + m_2}u_1.$$

In a frame of reference moving at the speed v), the objects are brought to rest by the collision and 100% of the kinetic energy is converted.

One measure of the inelasticity of the collision is the coefficient of restitution C_R, defined as the ratio of relative velocity of separation to relative velocity of approach. In applying this measure to ball sports, this can be easily measured using the following formula:

$$C_R = \sqrt{\frac{\text{bounce height}}{\text{drop height}}}.$$

The momentum and energy equations also apply to the motions of objects that begin together and then move apart. For example, an explosion is the result of a chain reaction that transforms potential energy stored in chemical, mechanical, or nuclear form into kinetic energy, acoustic energy, and electromagnetic radiation. Rockets also make use of conservation of momentum: propellant is thrust outward, gaining momentum, and an equal and opposite momentum is imparted to the rocket.

Multiple Dimensions

Two-dimensional elastic collision. There is no motion perpendicular to the image,
so only two components are needed to represent the velocities and momenta.
The two blue vectors represent velocities after the collision and add vectorially to get the initial (red) velocity.

Real motion has both direction and velocity and must be represented by a vector. In a coordinate system with x, y, z axes, velocity has components v_x in the x direction, v_y in the y direction, v_z in the z direction. The vector is represented by a boldface symbol:

$$\mathbf{v} = \left(v_x, v_y, v_z \right).$$

Similarly, the momentum is a vector quantity and is represented by a boldface symbol:

$$\mathbf{p} = \left(p_x, p_y, p_z \right).$$

The equations in the previous sections, work in vector form if the scalars p and v are replaced by vectors p and v. Each vector equation represents three scalar equations. For example,

$$\mathbf{p} = m\mathbf{v}$$

represents three equations:

$$p_x = mv_x$$

$$p_y = mv_y$$

$$p_z = mv_z.$$

The kinetic energy equations are exceptions to the above replacement rule. The equations are still one-dimensional, but each scalar represents the magnitude of the vector, for example,

$$v^2 = v_x^2 + v_y^2 + v_z^2.$$

Each vector equation represents three scalar equations. often coordinates can be chosen so that only two components are needed, as in the figure. Each component can be obtained separately and the results combined to produce a vector result.

A simple construction involving the center of mass frame can be used to show that if a stationary elastic sphere is struck by a moving sphere, the two will head off at right angles after the collision (as in the figure).

Objects of Variable Mass

The concept of momentum plays a fundamental role in explaining the behavior of variable-mass objects such as a rocket ejecting fuel or a star accreting gas. In analyzing such an object, one treats the object's mass as a function that varies with time: $m(t)$. The momentum of the object at time t is therefore $p(t) = m(t)v(t)$. One might then try to invoke Newton's second law of motion by saying that the external force F on the object is related to its momentum $p(t)$ by $F = dp/dt$, but this is incorrect, as is the related expression found by applying the product rule to $d(mv)/dt$:

$$F = m(t)\frac{dv}{dt} + v(t)\frac{dm}{dt}. \qquad \text{(incorrect)}$$

This equation does not correctly describe the motion of variable-mass objects. The correct equation is

$$F = m(t)\frac{dv}{dt} - u\frac{dm}{dt},$$

where u is the velocity of the ejected/accreted mass *as seen in the object's rest frame*. This is distinct from v, which is the velocity of the object itself as seen in an inertial frame.

This equation is derived by keeping track of both the momentum of the object as well as the momentum of the ejected/accreted mass (dm). When considered together, the object and the mass (dm) constitute a closed system in which total momentum is conserved.

$$P(t+dt) = (m-dm)(v+dv) + dm(v-u) = mv + mdv - udm = P(t) + mdv - udm$$

Relativistic Mechanics

Lorentz Invariance

Newtonian physics assumes that absolute time and space exist outside of any observer; this gives rise to the Galilean invariance described earlier. It also results in a prediction that the speed of light can vary from one reference frame to another. This is contrary to observation. In the special theory of relativity, Einstein keeps the postulate that the equations of motion do not depend on the reference frame, but assumes that the speed of light c is invariant. As a result, position and time in two reference frames are related by the Lorentz transformation instead of the Galilean transformation.

Consider, for example, a reference frame moving relative to another at velocity v in the x direction. The Galilean transformation gives the coordinates of the moving frame as

$$t' = t$$
$$x' = x - vt$$

while the Lorentz transformation gives

$$t' = \gamma\left(t - \frac{vx}{c^2}\right)$$
$$x' = \gamma\left(x - vt\right)$$

where γ is the Lorentz factor:

$$\gamma = \frac{1}{\sqrt{1 - v^2/c^2}}.$$

Newton's second law, with mass fixed, is not invariant under a Lorentz transformation. However, it can be made invariant by making the *inertial mass* m of an object a function of velocity:

$$m = \gamma m_0;$$

m_0 is the object's invariant mass.

The modified momentum,

$$\mathbf{p} = \gamma m_0 \mathbf{v},$$

obeys Newton's second law:

$$\mathbf{F} = \frac{d\mathbf{p}}{dt}.$$

Within the domain of classical mechanics, relativistic momentum closely approximates Newtonian momentum: at low velocity, $\gamma m_0 \mathbf{v}$ is approximately equal to $m_0 \mathbf{v}$, the Newtonian expression for momentum.

Four-vector Formulation

In the theory of special relativity, physical quantities are expressed in terms of four-vectors that include time as a fourth coordinate along with the three space coordinates. These vectors are generally represented by capital letters, for example R for position. The expression for the *four-momentum* depends on how the coordinates are expressed. Time may be given in its normal units or multiplied by the speed of light so that all the components of the four-vector have dimensions of length. If the latter scaling is used, an interval of proper time, τ, defined by

$$c^2 d\tau^2 = c^2 dt^2 - dx^2 - dy^2 - dz^2,$$

is invariant under Lorentz transformations (in this expression and in what follows the (+ − − −) metric signature has been used, different authors use different conventions). Mathematically this invariance can be ensured in one of two ways: by treating the four-vectors as Euclidean vectors and multiplying time by $\sqrt{-1}$; or by keeping time a real quantity and embedding the vectors in a Minkowski space. In a Minkowski space, the scalar product of two four-vectors $U = (U_0, U_1, U_2, U_3)$ and $V = (V_0, V_1, V_2, V_3)$ is defined as

$$\mathbf{U} \cdot \mathbf{V} = U_0 V_0 - U_1 V_1 - U_2 V_2 - U_3 V_3.$$

In all the coordinate systems, the (contravariant) relativistic four-velocity is defined by

$$\mathbf{U} \equiv \frac{d\mathbf{R}}{d\tau} = \gamma \frac{d\mathbf{R}}{dt},$$

and the (contravariant) four-momentum is

$$\mathbf{P} = m_0 \mathbf{U},$$

where m_0 is the invariant mass. If R = (ct, x, y, z) (in Minkowski space), then

$$\mathbf{P} = \gamma m_0 (c, \mathbf{v}) = (mc, \mathbf{p}).$$

Using Einstein's mass-energy equivalence, $E = mc^2$, this can be rewritten as

$$\mathbf{P} = \left(\frac{E}{c}, \mathbf{P} \right).$$

Thus, conservation of four-momentum is Lorentz-invariant and implies conservation of both mass and energy.

The 4-Momentum is related to the 4-WaveVector in Special Relativity

$$\mathbf{P} = \left(\frac{E}{c}, \vec{\mathbf{p}}\right) = \hbar\mathbf{K} = \hbar\left(\frac{\omega}{c}, \vec{\mathbf{k}}\right)$$ which is the full 4-vector version of:

The (temporal component) Planck–Einstein relation $= \hbar\omega$

The (spatial components) de Broglie matter wave relation $\vec{\mathbf{p}} = \hbar\vec{\mathbf{k}}$

The magnitude of the momentum four-vector is equal to $m_0 c$:

$$\| \mathbf{P} \|^2 = \mathbf{P} \cdot \mathbf{P} = \gamma^2 m_0^2 (c^2 - v^2) = (m_0 c)^2,$$

and is invariant across all reference frames.

The relativistic energy–momentum relationship holds even for massless particles such as photons; by setting $m_0 = 0$ it follows that

$$E = pc.$$

In a game of relativistic "billiards", if a stationary particle is hit by a moving particle in an elastic collision, the paths formed by the two afterwards will form an acute angle. This is unlike the non-relativistic case where they travel at right angles.

Generalized Coordinates

Newton's laws can be difficult to apply to many kinds of motion because the motion is limited by *constraints*. For example, a bead on an abacus is constrained to move along its wire and a pendulum bob is constrained to swing at a fixed distance from the pivot. Many such constraints can be incorporated by changing the normal Cartesian coordinates to a set of *generalized coordinates* that may be fewer in number. Refined mathematical methods have been developed for solving mechanics problems in generalized coordinates. They introduce a *generalized momentum*, also known as the *canonical* or *conjugate momentum*, that extends the concepts of both linear momentum and angular momentum. To distinguish it from generalized momentum, the product of mass and velocity is also referred to as *mechanical, kinetic* or *kinematic momentum*. The two main methods are described below.

Lagrangian Mechanics

In Lagrangian mechanics, a Lagrangian is defined as the difference between the kinetic energy T and the potential energy V:

$$\mathcal{L} = T - V.$$

If the generalized coordinates are represented as a vector $q = (q_1, q_2, \dots, q_N)$ and time differentiation is represented by a dot over the variable, then the equations of motion (known as the Lagrange or Euler–Lagrange equations) are a set of N equations:

$$\frac{d}{dt}\left(\frac{\partial \mathcal{L}}{\partial \dot{q}_j}\right) - \frac{\partial \mathcal{L}}{\partial q_j} = 0.$$

If a coordinate q_i is not a Cartesian coordinate, the associated generalized momentum component p_i does not necessarily have the dimensions of linear momentum. Even if q_i is a Cartesian coordinate, p_i will not be the same as the mechanical momentum if the potential depends on velocity. Some sources represent the kinematic momentum by the symbol Π.

In this mathematical framework, a generalized momentum is associated with the generalized coordinates. Its components are defined as

$$p_j = \frac{\partial \mathcal{L}}{\partial \dot{q}_j}.$$

Each component p_j is said to be the *conjugate momentum* for the coordinate q_j.

Now if a given coordinate q_i does not appear in the Lagrangian (although its time derivative might appear), then

$$p_j = \text{constant}.$$

This is the generalization of the conservation of momentum.

Even if the generalized coordinates are just the ordinary spatial coordinates, the conjugate momenta are not necessarily the ordinary momentum coordinates. An example is found in the section on electromagnetism.

Hamiltonian Mechanics

In Hamiltonian mechanics, the Lagrangian (a function of generalized coordinates and their derivatives) is replaced by a Hamiltonian that is a function of generalized coordinates and momentum. The Hamiltonian is defined as

$$\mathcal{H}(\mathbf{q}, \mathbf{p}, t) = \mathbf{p} \cdot \dot{\mathbf{q}} - \mathcal{L}(\mathbf{q}, \dot{\mathbf{q}}, t),$$

where the momentum is obtained by differentiating the Lagrangian as above. The Hamiltonian equations of motion are

$$\dot{q}_i = \frac{\partial \mathcal{H}}{\partial p_i}$$

$$-\dot{p}_i = \frac{\partial \mathcal{H}}{\partial q_i}$$

$$-\frac{\partial \mathcal{L}}{\partial t} = \frac{d\mathcal{H}}{dt}.$$

As in Lagrangian mechanics, if a generalized coordinate does not appear in the Hamiltonian, its conjugate momentum component is conserved.

Symmetry and Conservation

Conservation of momentum is a mathematical consequence of the homogeneity (shift symmetry) of space (position in space is the canonical conjugate quantity to momentum). That is, conservation of momentum is a consequence of the fact that the laws of physics do not depend on position; this is a special case of Noether's theorem.

Electromagnetism

In Newtonian mechanics, the law of conservation of momentum can be derived from the law of action and reaction, which states that every force has a reciprocating equal and opposite force. Under some circumstances, moving charged particles can exert forces on each other in non-opposite directions. Moreover, Maxwell's equations, the foundation of classical electrodynamics, are Lorentz-invariant. Nevertheless, the combined momentum of the particles and the electromagnetic field is conserved.

Vacuum

In Maxwell's equations, the forces between particles are mediated by electric and magnetic fields. The electromagnetic force (*Lorentz force*) on a particle with charge q due to a combination of electric field E and magnetic field (as given by the "B-field" B) is

$$\mathbf{F} = q(\mathbf{E} + \mathbf{v} \times \mathbf{B}).$$

This force imparts a momentum to the particle, so by Newton's second law the particle must impart a momentum to the electromagnetic fields.

In a vacuum, the momentum per unit volume is

$$\mathbf{g} = \frac{1}{\mu_0 c^2} \mathbf{E} \times \mathbf{B},$$

where μ_0 is the vacuum permeability and c is the speed of light. The momentum density is proportional to the Poynting vector S which gives the directional rate of energy transfer per unit area:

$$\mathbf{g} = \frac{\mathbf{S}}{c^2}.$$

If momentum is to be conserved over the volume V over a region Q, changes in the momentum of matter through the Lorentz force must be balanced by changes in the momentum of the electromagnetic field and outflow of momentum. If P_{mech} is the momentum of all the particles in Q, and the particles are treated as a continuum, then Newton's second law gives

$$\frac{d\mathbf{P}_{mech}}{dt} = \iiint_Q (\rho \mathbf{E} + \mathbf{J} \times \mathbf{B}) dV$$

The electromagnetic momentum is

$$\mathbf{P}_{field} = \frac{1}{\mu_0 c^2} \iiint_Q \mathbf{E} \times \mathbf{B} \, dv$$

and the equation for conservation of each component i of the momentum is

$$\frac{d}{dt} (\mathbf{P}_{mech} + \mathbf{P}_{field})_i = \iint_\sigma \left(\sum_j T_{ij} n_j \right) d\Sigma.$$

The term on the right is an integral over the surface area Σ of the surface σ representing momentum flow into and out of the volume, and n_j is a component of the surface normal of S. The quantity T_{ij} is called the Maxwell stress tensor, defined as

$$T_{ij} \equiv \epsilon_0 \left(E_i E_j - \frac{1}{2} \delta_{ij} E^2 \right) + \frac{1}{\mu_0} \left(B_i B_j - \frac{1}{2} \delta_{ij} B^2 \right).$$

Media

The above results are for the *microscopic* Maxwell equations, applicable to electromagnetic forces in a vacuum (or on a very small scale in media). It is more difficult to define momentum density in media because the division into electromagnetic and mechanical is arbitrary. The definition of electromagnetic momentum density is modified to

$$\mathbf{g} = \frac{1}{c^2} \mathbf{E} \times \mathbf{H} = \frac{\mathbf{S}}{c^2},$$

where the H-field H is related to the B-field and the magnetization M by

$$\mathbf{B} = \mu_0 \left(\mathbf{H} + \mathbf{M} \right).$$

The electromagnetic stress tensor depends on the properties of the media.

Particle in Field

If a charged particle q moves in an electromagnetic field, neither its kinetic momentum $m\mathbf{v}$ nor its canonical momentum is conserved.

Canonical Commutation Relations

The kinetic momentum (p above) satisfies the commutation relation:

$$\left[p_j, p_k \right] = \frac{i \hbar e}{c} \varepsilon_{jkl} B_l$$

where: j, k, l are indices labelling vector components, B_l is a component of the magnetic field, and ε_{kjl} is the Levi-Civita symbol, here in 3 dimensions.

Quantum Mechanics

In quantum mechanics, momentum is defined as a self-adjoint operator on the wave function. The Heisenberg uncertainty principle defines limits on how accurately the momentum and position of a single observable system can be known at once. In quantum mechanics, position and momentum are conjugate variables.

For a single particle described in the position basis the momentum operator can be written as

$$\mathbf{p} = \frac{\hbar}{i}\nabla = -i\hbar\nabla,$$

where ∇ is the gradient operator, $\hbar$ is the reduced Planck constant, and i is the imaginary unit. This is a commonly encountered form of the momentum operator, though the momentum operator in other bases can take other forms. For example, in momentum space the momentum operator is represented as

$$\mathbf{p}\psi(p) = p\psi(p),$$

where the operator p acting on a wave function $\psi(p)$ yields that wave function multiplied by the value p, in an analogous fashion to the way that the position operator acting on a wave function $\psi(x)$ yields that wave function multiplied by the value x.

For both massive and massless objects, relativistic momentum is related to the phase constant β by

$$p = \hbar\beta$$

Electromagnetic radiation (including visible light, ultraviolet light, and radio waves) is carried by photons.Even though photons (the particle aspect of light) have no mass, they still carry momentum. This leads to applications such as the solar sail. The calculation of the momentum of light within dielectric media is somewhat controversial.

Deformable Bodies and Fluids

Conservation in a Continuum

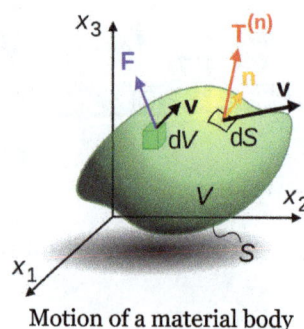

Motion of a material body

In fields such as fluid dynamics and solid mechanics, it is not feasible to follow the motion of individual atoms or molecules. Instead, the materials must be approximated by a continuum in which there is a particle or fluid parcel at each point that is assigned the average of the properties of atoms in a small region nearby. In particular, it has a density ρ and velocity v that depend on time t and position r. The momentum per unit volume is ρv.

Consider a column of water in hydrostatic equilibrium. All the forces on the water are in balance and the water is motionless. On any given drop of water, two forces are balanced. The first is gravity, which acts directly on each atom and molecule inside. The gravitational force per unit volume is ρg, where g is the gravitational acceleration. The second force is the sum of all the forces exerted on its surface by the surrounding water. The force from below is greater than the force from above by just the amount needed to balance gravity. The normal force per unit area is the pressure p. The average force per unit volume inside the droplet is the gradient of the pressure, so the force balance equation is

$$-\nabla p + \rho \mathbf{g} = 0.$$

If the forces are not balanced, the droplet accelerates. This acceleration is not simply the partial derivative $\partial v/\partial t$ because the fluid in a given volume changes with time. Instead, the material derivative is needed:

$$\frac{D}{Dt} \equiv \frac{\partial}{\partial t} + \mathbf{v} \cdot \nabla.$$

Applied to any physical quantity, the material derivative includes the rate of change at a point and the changes due to advection as fluid is carried past the point. Per unit volume, the rate of change in momentum is equal to $\rho D v/Dt$. This is equal to the net force on the droplet.

Forces that can change the momentum of a droplet include the gradient of the pressure and gravity, as above. In addition, surface forces can deform the droplet. In the simplest case, a shear stress τ, exerted by a force parallel to the surface of the droplet, is proportional to the rate of deformation or strain rate. Such a shear stress occurs if the fluid has a velocity gradient because the fluid is moving faster on one side than another. If the speed in the x direction varies with z, the tangential force in direction x per unit area normal to the z direction is

$$\sigma_{zx} = -\mu \frac{\partial v_x}{\partial z},$$

where μ is the viscosity. This is also a flux, or flow per unit area, of x-momentum through the surface.

Including the effect of viscosity, the momentum balance equations for the incompressible flow of a Newtonian fluid are

$$\rho \frac{D\mathbf{v}}{Dt} = -\nabla p + \mu \nabla^2 \mathbf{v} + \rho \mathbf{g}.$$

These are known as the Navier–Stokes equations.

The momentum balance equations can be extended to more general materials, including solids. For each surface with normal in direction i and force in direction j, there is a stress component σ_{ij}. The nine components make up the Cauchy stress tensor σ, which includes both pressure and shear. The local conservation of momentum is expressed by the Cauchy momentum equation:

$$\rho \frac{\mathbf{Dv}}{\mathbf{Dt}} = \nabla \cdot \sigma + f,$$

where f is the body force.

The Cauchy momentum equation is broadly applicable to deformations of solids and liquids. The relationship between the stresses and the strain rate depends on the properties of the material.

Acoustic Waves

A disturbance in a medium gives rise to oscillations, or waves, that propagate away from their source. In a fluid, small changes in pressure p can often be described by the acoustic wave equation:

$$\frac{\partial^2 p}{\partial t^2} = c^2 \nabla^2 p,$$

where c is the speed of sound. In a solid, similar equations can be obtained for propagation of pressure (P-waves) and shear (S-waves).

The flux, or transport per unit area, of a momentum component ρv_j by a velocity v_i is equal to $\rho\, v_j v_j$. In the linear approximation that leads to the above acoustic equation, the time average of this flux is zero. However, nonlinear effects can give rise to a nonzero average. It is possible for momentum flux to occur even though the wave itself does not have a mean momentum.

History of the Concept

In about 530 A.D., working in Alexandria, Byzantine philosopher John Philoponus developed a concept of momentum in his commentary to Aristotle's *Physics*. Aristotle claimed that everything that is moving must be kept moving by something. For example, a thrown ball must be kept moving by motions of the air. Most writers continued to accept Aristotle's theory until the time of Galileo, but a few were skeptical. Philoponus pointed out the absurdity in Aristotle's claim that motion of an object is promoted by the same air that is resisting its passage. He proposed instead that an impetus was imparted to the object in the act of throwing it. Ibn Sīnā (also known by his Latinized name Avicenna) read Philoponus and published his own theory of motion in *The Book of Healing* in 1020. He agreed that an impetus is imparted to a projectile by the thrower; but unlike Philoponus, who believed that it was a temporary virtue that would decline even in a vacuum, he viewed it as a persistent, requiring external forces such as air resistance to dissipate it. The work of Philoponus, and possibly that of Ibn Sīnā, was read and refined by the European philosophers Peter Olivi and Jean Buridan. Buridan, who in about 1350 was made rector of the University of Paris, referred to impetus being proportional to the weight times the speed. Moreover, Buridan's theory was different from his predecessor's in that he did not consider impetus to be self-dissipating, asserting that a body would be arrested by the forces of air resistance and gravity which might be opposing its impetus.

René Descartes believed that the total "quantity of motion" in the universe is conserved, where the quantity of motion is understood as the product of size and speed. This should not be read as a statement of the modern law of momentum, since he had no concept of mass as distinct from weight and size, and more importantly he believed that it is speed rather than velocity that is conserved. So for Descartes if a moving object were to bounce off a surface, changing its direction but not its speed, there would be no change in its quantity of motion. Galileo, later, in his *Two New Sciences*, used the Italian word *impeto*.

Leibniz, in his "Discourse on Metaphysics", gave an argument against Descartes' construction of the conservation of the "quantity of motion" using an example of dropping blocks of different sizes different distances. He points out that force is conserved but quantity of motion, construed as the product of size and speed of an object, is not conserved.

The first correct statement of the law of conservation of momentum was by English mathematician John Wallis in his 1670 work, *Mechanica sive De Motu, Tractatus Geometricus*: "the initial state of the body, either of rest or of motion, will persist" and "If the force is greater than the resistance, motion will result". Wallis uses *momentum* and *vis* for force. Newton's *Philosophiæ Naturalis Principia Mathematica*, when it was first published in 1687, showed a similar casting around for words to use for the mathematical momentum. His Definition II defines *quantitas motus*, "quantity of motion", as "arising from the velocity and quantity of matter conjointly", which identifies it as momentum. Thus when in Law II he refers to *mutatio motus*, "change of motion", being proportional to the force impressed, he is generally taken to mean momentum and not motion. It remained only to assign a standard term to the quantity of motion. The first use of "momentum" in its proper mathematical sense is not clear but by the time of Jenning's *Miscellanea* in 1721, five years before the final edition of Newton's *Principia Mathematica*, momentum M or "quantity of motion" was being defined for students as "a rectangle", the product of Q and V, where Q is "quantity of material" and V is "velocity", s/t.

Frame of Reference

In physics, a frame of reference (or reference frame) consists of an abstract coordinate system and the set of physical reference points that uniquely fix (locate and orient) the coordinate system and standardize measurements.

In n dimensions, n+1 reference points are sufficient to fully define a reference frame. Using rectangular (Cartesian) coordinates, a reference frame may be defined with a reference point at the origin and a reference point at one unit distance along each of the n coordinate axes.

In Einsteinian relativity, reference frames are used to specify the relationship between a moving observer and the phenomenon or phenomena under observation. In this context, the phrase often becomes "observational frame of reference" (or "observational reference frame"), which implies that the observer is at rest in the frame, although not necessarily located at its origin. A relativistic reference frame includes (or implies) the coordinate time, which does not correspond across different frames moving relatively to each other. The situation thus differs from Galilean relativity, where all possible coordinate times are essentially equivalent.

Different Aspects of "Frame of Reference"

The need to distinguish between the various meanings of "frame of reference" has led to a variety of terms. For example, sometimes the type of coordinate system is attached as a modifier, as in *Cartesian frame of reference*. Sometimes the state of motion is emphasized, as in *rotating frame of reference.* Sometimes the way it transforms to frames considered as related is emphasized as in *Galilean frame of reference.* Sometimes frames are distinguished by the scale of their observations, as in *macroscopic* and *microscopic frames of reference.*

In this article, the term *observational frame of reference* is used when emphasis is upon the *state of motion* rather than upon the coordinate choice or the character of the observations or observational apparatus. In this sense, an observational frame of reference allows study of the effect of motion upon an entire family of coordinate systems that could be attached to this frame. On the other hand, a *coordinate system* may be employed for many purposes where the state of motion is not the primary concern. For example, a coordinate system may be adopted to take advantage of the symmetry of a system. In a still broader perspective, the formulation of many problems in physics employs *generalized coordinates, normal modes* or *eigenvectors*, which are only indirectly related to space and time. It seems useful to divorce the various aspects of a reference frame for the discussion below. We therefore take observational frames of reference, coordinate systems, and observational equipment as independent concepts, separated as below:

- An observational frame (such as an inertial frame or non-inertial frame of reference) is a physical concept related to state of motion.

- A coordinate system is a mathematical concept, amounting to a choice of language used to describe observations. Consequently, an observer in an observational frame of reference can choose to employ any coordinate system (Cartesian, polar, curvilinear, generalized, ...) to describe observations made from that frame of reference. A change in the choice of this coordinate system does not change an observer's state of motion, and so does not entail a change in the observer's *observational* frame of reference. This viewpoint can be found elsewhere as well. Which is not to dispute that some coordinate systems may be a better choice for some observations than are others.

- Choice of what to measure and with what observational apparatus is a matter separate from the observer's state of motion and choice of coordinate system.

Here is a quotation applicable to moving observational frames $\mathfrak{R}$ and various associated Euclidean three-space coordinate systems [R, R', etc.]:

> " We first introduce the notion of *reference frame,* itself related to the idea of *observer*: the reference frame is, in some sense, the "Euclidean space carried by the observer". Let us give a more mathematical definition:... the reference frame is... the set of all points in the Euclidean space with the rigid body motion of the observer. The frame, denoted $\mathfrak{R}$, is said to move with the observer.... The spatial positions of particles are labelled relative to a frame $\mathfrak{R}$ by establishing a *coordinate system* R with origin O. The corresponding set of axes, sharing the rigid body motion of the frame $\mathfrak{R}$, can be considered to give a physical realization of $\mathfrak{R}$. In a frame $\mathfrak{R}$, coordinates are changed from R to R' by carrying out, at each instant of time, the same coordinate transformation on the components of *intrinsic* objects (vectors and tensors) introduced to represent physical quantities *in this frame.* "

and this on the utility of separating the notions of $\mathfrak{R}$ and [R, R', etc.]:

" As noted by Brillouin, a distinction between mathematical sets of coordinates and physical frames of reference must be made. The ignorance of such distinction is the source of much confusion... the dependent functions such as velocity for example, are measured with respect to a physical reference frame, but one is free to choose any mathematical coordinate system in which the equations are specified. "

and this, also on the distinction between $\mathfrak{R}$ and $[R, R', etc.]$:

" The idea of a reference frame is really quite different from that of a coordinate system. Frames differ just when they define different *spaces* (sets of *rest* points) or times (sets of simultaneous events). So the ideas of a space, a time, of rest and simultaneity, go inextricably together with that of frame. However, a mere shift of origin, or a purely spatial rotation of space coordinates results in a new coordinate system. So frames correspond at best to *classes* of coordinate systems. "

and from J. D. Norton:

" In traditional developments of special and general relativity it has been customary not to distinguish between two quite distinct ideas. The first is the notion of a coordinate system, understood simply as the smooth, invertible assignment of four numbers to events in spacetime neighborhoods. The second, the frame of reference, refers to an idealized system used to assign such numbers ... To avoid unnecessary restrictions, we can divorce this arrangement from metrical notions. ... Of special importance for our purposes is that each frame of reference has a definite state of motion at each event of spacetime....Within the context of special relativity and as long as we restrict ourselves to frames of reference in inertial motion, then little of importance depends on the difference between an inertial frame of reference and the inertial coordinate system it induces. This comfortable circumstance ceases immediately once we begin to consider frames of reference in nonuniform motion even within special relativity....More recently, to negotiate the obvious ambiguities of Einstein's treatment, the notion of frame of reference has reappeared as a structure distinct from a coordinate system. "

The discussion is taken beyond simple space-time coordinate systems by Brading and Castellani. Extension to coordinate systems using generalized coordinates underlies the Hamiltonian and Lagrangian formulations of quantum field theory, classical relativistic mechanics, and quantum gravity.

Coordinate Systems

Although the term "coordinate system" is often used (particularly by physicists) in a nontechnical sense, the term "coordinate system" does have a precise meaning in mathematics, and sometimes that is what the physicist means as well.

A coordinate system in mathematics is a facet of geometry or of algebra, in particular, a property of manifolds (for example, in physics, configuration spaces or phase spaces). The coordinates of a point r in an n-dimensional space are simply an ordered set of n numbers:

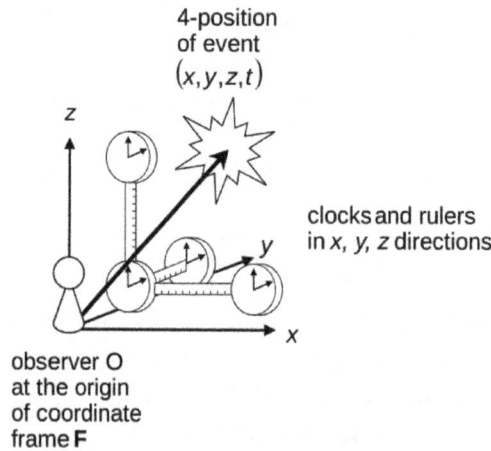

An observer O, situated at the origin of a local set of coordinates – a frame of reference **F**. The observer in this frame uses the coordinates (x, y, z, t) to describe a spacetime event, shown as a star.

$$\mathbf{r} = [x^1, x^2, \ldots, x^n].$$

In a general Banach space, these numbers could be (for example) coefficients in a functional expansion like a Fourier series. In a physical problem, they could be spacetime coordinates or normal mode amplitudes. In a robot design, they could be angles of relative rotations, linear displacements, or deformations of joints. Here we will suppose these coordinates can be related to a Cartesian coordinate system by a set of functions:

$$x^j = x^j(x, y, z, \ldots), \quad j = 1, \ldots, n$$

where x, y, z, *etc.* are the n Cartesian coordinates of the point. Given these functions, coordinate surfaces are defined by the relations:

$$x^j(x, y, z, \ldots) = \text{constant} \quad j = 1, \ldots, n$$

The intersection of these surfaces define coordinate lines. At any selected point, tangents to the intersecting coordinate lines at that point define a set of basis vectors $\{e_1, e_2, \ldots, e_n\}$ at that point. That is:

$$\mathbf{e}_i(\mathbf{r}) = \lim_{\epsilon \to 0} \frac{\mathbf{r}\left(x^1, \ldots, x^i + \epsilon, \ldots, x^n\right) - \mathbf{r}\left(x^1, \ldots, x^i, \ldots, x^n\right)}{\epsilon}, \quad i = 1, \ldots, n$$

which can be normalized to be of unit length.

Coordinate surfaces, coordinate lines, and basis vectors are components of a coordinate system. If the basis vectors are orthogonal at every point, the coordinate system is an orthogonal coordinate system.

An important aspect of a coordinate system is its metric tensor g_{ik}, which determines the arc length ds in the coordinate system in terms of its coordinates:

$$(ds)^2 = g_{ik} \, dx^i \, dx^k,$$

where repeated indices are summed over.

As is apparent from these remarks, a coordinate system is a mathematical construct, part of an axiomatic system. There is no necessary connection between coordinate systems and physical motion (or any other aspect of reality). However, coordinate systems can include time as a coordinate, and can be used to describe motion. Thus, Lorentz transformations and Galilean transformations may be viewed as coordinate transformations.

Observational Frames of Reference

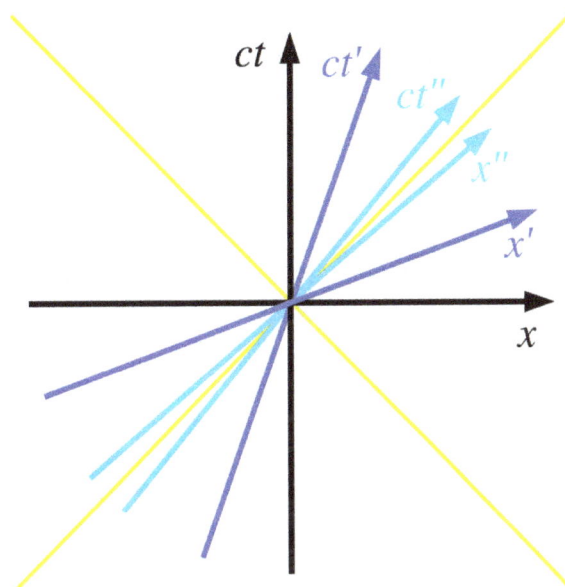

Three frames of reference in special relativity. The black frame is at rest. The primed frame moves at 40% of light speed, and the double primed frame at 80%. Note the scissors-like change as speed increases.

An observational frame of reference, often referred to as a *physical frame of reference*, a *frame of reference*, or simply a *frame*, is a physical concept related to an observer and the observer's state of motion. Here we adopt the view expressed by Kumar and Barve: an observational frame of reference is characterized *only by its state of motion*. However, there is lack of unanimity on this point. In special relativity, the distinction is sometimes made between an *observer* and a *frame*. According to this view, a *frame* is an *observer* plus a coordinate lattice constructed to be an orthonormal right-handed set of spacelike vectors perpendicular to a timelike vector. This restricted view is not used here, and is not universally adopted even in discussions of relativity. In general relativity the use of general coordinate systems is common

There are two types of observational reference frame: inertial and non-inertial. An inertial frame of reference is defined as one in which all laws of physics take on their simplest form. In special relativity these frames are related by Lorentz transformations, which are parametrized by rapidity. In Newtonian mechanics, a more restricted definition requires only that Newton's first law holds true; that is, a Newtonian inertial frame is one in which a free particle travels in a straight line at constant speed, or is at rest. These frames are related by Galilean transformations. These relativistic and Newtonian transformations are expressed in spaces of general dimension in terms of representations of the Poincaré group and of the Galilean group.

In contrast to the inertial frame, a non-inertial frame of reference is one in which fictitious forces must be invoked to explain observations. An example is an observational frame of reference centered at a point on the Earth's surface. This frame of reference orbits around the center of the Earth, which introduces the fictitious forces known as the Coriolis force, centrifugal force, and gravitational force. (All of these forces including gravity disappear in a truly inertial reference frame, which is one of free-fall.)

Measurement Apparatus

A further aspect of a frame of reference is the role of the measurement apparatus (for example, clocks and rods) attached to the frame. This question is not addressed in this article, and is of particular interest in quantum mechanics, where the relation between observer and measurement is still under discussion.

In physics experiments, the frame of reference in which the laboratory measurement devices are at rest is usually referred to as the laboratory frame or simply "lab frame." An example would be the frame in which the detectors for a particle accelerator are at rest. The lab frame in some experiments is an inertial frame, but it is not required to be (for example the laboratory on the surface of the Earth in many physics experiments is not inertial). In particle physics experiments, it is often useful to transform energies and momenta of particles from the lab frame where they are measured, to the center of momentum frame "COM frame" in which calculations are sometimes simplified, since potentially all kinetic energy still present in the COM frame may be used for making new particles.

In this connection it may be noted that the clocks and rods often used to describe observers' measurement equipment in thought, in practice are replaced by a much more complicated and indirect metrology that is connected to the nature of the vacuum, and uses atomic clocks that operate according to the standard model and that must be corrected for gravitational time dilation.

In fact, Einstein felt that clocks and rods were merely expedient measuring devices and they should be replaced by more fundamental entities based upon, for example, atoms and molecules.

Types

- Body-fixed frames of reference
- Space-fixed frames of reference
- Inertial frames of reference
- Non-Inertial frames of reference

Examples of Inertial Frames of Reference

Simple Example

Consider a situation common in everyday life. Two cars travel along a road, both moving at constant velocities. See Figure. At some particular moment, they are separated by 200 metres. The car in front is travelling at 22 metres per second and the car behind is travelling at 30 metres per second. If we want to find out how long it will take the second car to catch up with the first, there are three obvious "frames of reference" that we could choose.

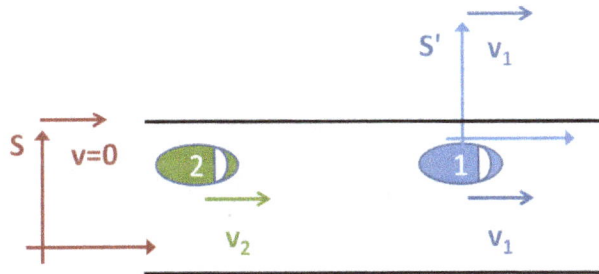

Figure: Two cars moving at different but constant velocities observed from stationary inertial frame S attached to the road and moving inertial frame S' attached to the first car.

First, we could observe the two cars from the side of the road. We define our "frame of reference" S as follows. We stand on the side of the road and start a stop-clock at the exact moment that the second car passes us, which happens to be when they are a distance d = 200 m apart. Since neither of the cars is accelerating, we can determine their positions by the following formulas, where $x_1(t)$ is the position in meters of car one after time t in seconds and $x_2(t)$ is the position of car two after time t.

$$x_1(t) = d + v_1 t = 200 + 22t; \quad x_2(t) = v_2 t = 30t$$

Notice that these formulas predict at t = 0 s the first car is 200 m down the road and the second car is right beside us, as expected. We want to find the time at which $x_1 = x_2$. Therefore, we set $x_1 = x_2$ and solve for t, that is:

$$200 + 22t = 30t$$

$$8t = 200$$

$$t = 25 \quad \text{seconds}$$

Alternatively, we could choose a frame of reference S' situated in the first car. In this case, the first car is stationary and the second car is approaching from behind at a speed of $v_2 - v_1$ = 8 m / s. In order to catch up to the first car, it will take a time of $d/v_2 - v_1$ = 200/8 s, that is, 25 seconds, as before. Note how much easier the problem becomes by choosing a suitable frame of reference. The third possible frame of reference would be attached to the second car. That example resembles the case just discussed, except the second car is stationary and the first car moves backward towards it at 8 m / s.

It would have been possible to choose a rotating, accelerating frame of reference, moving in a complicated manner, but this would have served to complicate the problem unnecessarily. It is also necessary to note that one is able to convert measurements made in one coordinate system to another. For example, suppose that your watch is running five minutes fast compared to the local standard time. If you know that this is the case, when somebody asks you what time it is, you are able to deduct five minutes from the time displayed on your watch in order to obtain the correct time. The measurements that an observer makes about a system depend therefore on the observer's frame of reference (you might say that the bus arrived at 5 past three, when in fact it arrived at three).

Additional Example

Figure : Simple-minded frame-of-reference example

For a simple example involving only the orientation of two observers, consider two people standing, facing each other on either side of a north-south street. See Figure. A car drives past them heading south. For the person facing east, the car was moving toward the right. However, for the person facing west, the car was moving toward the left. This discrepancy is because the two people used two different frames of reference from which to investigate this system.

For a more complex example involving observers in relative motion, consider Alfred, who is standing on the side of a road watching a car drive past him from left to right. In his frame of reference, Alfred defines the spot where he is standing as the origin, the road as the x-axis and the direction in front of him as the positive y-axis. To him, the car moves along the x axis with some velocity v in the positive x-direction. Alfred's frame of reference is considered an inertial frame of reference because he is not accelerating (ignoring effects such as Earth's rotation and gravity).

Now consider Betsy, the person driving the car. Betsy, in choosing her frame of reference, defines her location as the origin, the direction to her right as the positive x-axis, and the direction in front of her as the positive y-axis. In this frame of reference, it is Betsy who is stationary and the world around her that is moving – for instance, as she drives past Alfred, she observes him moving with velocity v in the negative y-direction. If she is driving north, then north is the positive y-direction; if she turns east, east becomes the positive y-direction.

Finally, as an example of non-inertial observers, assume Candace is accelerating her car. As she passes by him, Alfred measures her acceleration and finds it to be a in the negative x-direction. Assuming Candace's acceleration is constant, what acceleration does Betsy measure? If Betsy's velocity v is constant, she is in an inertial frame of reference, and she will find the acceleration to be the same as Alfred in her frame of reference, a in the negative y-direction. However, if she is accelerating at rate A in the negative y-direction (in other words, slowing down), she will find Candace's acceleration to be $a' = a - A$ in the negative y-direction - a smaller value than Alfred has measured. Similarly, if she is accelerating at rate A in the positive y-direction (speeding up), she will observe Candace's acceleration as $a' = a + A$ in the negative y-direction – a larger value than Alfred's measurement.

Frames of reference are especially important in special relativity, because when a frame of reference is moving at some significant fraction of the speed of light, then the flow of time in that frame does not necessarily apply in another frame. The speed of light is considered to be the only true constant between moving frames of reference.

Remarks

It is important to note some assumptions made above about the various inertial frames of reference. Newton, for instance, employed universal time, as explained by the following example. Suppose that you own two clocks, which both tick at exactly the same rate. You synchronize them so that they both display exactly the same time. The two clocks are now separated and one clock is on a fast moving train, traveling at constant velocity towards the other. According to Newton, these two clocks will still tick at the same rate and will both show the same time. Newton says that the rate of time as measured in one frame of reference should be the same as the rate of time in another. That is, there exists a "universal" time and all other times in all other frames of reference will run at the same rate as this universal time irrespective of their position and velocity. This concept of time and simultaneity was later generalized by Einstein in his special theory of relativity (1905) where he developed transformations between inertial frames of reference based upon the universal nature of physical laws and their economy of expression (Lorentz transformations).

It is also important to note that the definition of inertial reference frame can be extended beyond three-dimensional Euclidean space. Newton's assumed a Euclidean space, but general relativity uses a more general geometry. As an example of why this is important, let us consider the geometry of an ellipsoid. In this geometry, a "free" particle is defined as one at rest or traveling at constant speed on a geodesic path. Two free particles may begin at the same point on the surface, traveling with the same constant speed in different directions. After a length of time, the two particles collide at the opposite side of the ellipsoid. Both "free" particles traveled with a constant speed, satisfying the definition that no forces were acting. No acceleration occurred and so Newton's first law held true. This means that the particles were in inertial frames of reference. Since no forces were acting, it was the geometry of the situation which caused the two particles to meet each other again. In a similar way, it is now common to describe that we exist in a four-dimensional geometry known as spacetime. In this picture, the curvature of this 4D space is responsible for the way in which two bodies with mass are drawn together even if no forces are acting. This curvature of spacetime replaces the force known as gravity in Newtonian mechanics and special relativity.

Non-inertial Frames

Here the relation between inertial and non-inertial observational frames of reference is considered. The basic difference between these frames is the need in non-inertial frames for fictitious forces, as described below.

An accelerated frame of reference is often delineated as being the "primed" frame, and all variables that are dependent on that frame are notated with primes, e.g. x', y', a'.

The vector from the origin of an inertial reference frame to the origin of an accelerated reference frame is commonly notated as R. Given a point of interest that exists in both frames, the vector from the inertial origin to the point is called r, and the vector from the accelerated origin to the point is called r'. From the geometry of the situation, we get

$$\mathbf{r} = \mathbf{R} + \mathbf{r}'$$

Taking the first and second derivatives of this with respect to time, we obtain

$$\mathbf{v} = \mathbf{V} + \mathbf{v}'$$

$$\mathbf{a} = \mathbf{A} + \mathbf{a}'$$

where V and A are the velocity and acceleration of the accelerated system with respect to the inertial system and v and a are the velocity and acceleration of the point of interest with respect to the inertial frame.

These equations allow transformations between the two coordinate systems; for example, we can now write Newton's second law as

$$\mathbf{F} = m\mathbf{a} = m\mathbf{A} + m\mathbf{a}'$$

When there is accelerated motion due to a force being exerted there is manifestation of inertia. If an electric car designed to recharge its battery system when decelerating is switched to braking, the batteries are recharged, illustrating the physical strength of manifestation of inertia. However, the manifestation of inertia does not prevent acceleration (or deceleration), for manifestation of inertia occurs in response to change in velocity due to a force. Seen from the perspective of a rotating frame of reference the manifestation of inertia appears to exert a force (either in centrifugal direction, or in a direction orthogonal to an object's motion, the Coriolis effect).

A common sort of accelerated reference frame is a frame that is both rotating and translating (an example is a frame of reference attached to a CD which is playing while the player is carried). This arrangement leads to the equation:

$$a = a' + \dot{\omega} \times r' + 2\omega \times v' + \omega \times (\omega \times r') + A_0$$

or, to solve for the acceleration in the accelerated frame,

$$a = a' - \dot{\omega} \times r' - 2\omega \times v' - \omega \times (\omega \times r') - A_0$$

Multiplying through by the mass m gives

$$\mathbf{F}' = \mathbf{F}_{physical} + \mathbf{F}'_{Euler} + \mathbf{F}'_{Coriolis} + \mathbf{F}'_{centripetal} - m\mathbf{A}_0$$

where

$$F'_{Euler} = -m\,\dot{\omega} \times r' \quad \text{(Euler force)}$$

$$F'_{Coriolis} = -2m\omega \times v' \quad \text{(Coriolis force)}$$

$$F'_{centrifugal} = -m\omega \times (\omega \times r') = m(\omega^2 r' - (\omega \cdot r')\omega) \quad \text{(centrifugal force)}$$

Particular Frames of Reference in Common Use

- International Terrestrial Reference Frame

- International Celestial Reference Frame

- In fluid mechanics, Lagrangian and Eulerian specification of the flow field

Other Frames

- Frame fields in general relativity

- Linguistic frame of reference

- Moving frame in Mathematics

Newton's Laws of Motion

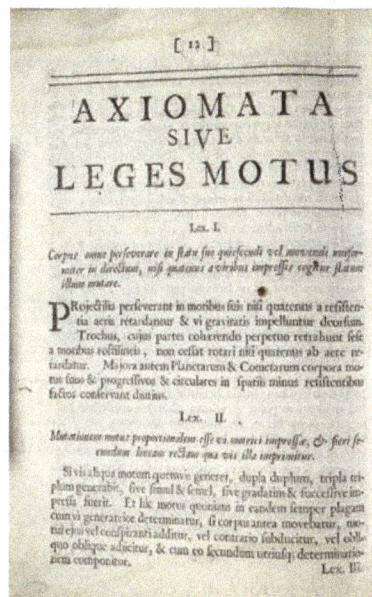

Newton's First and Second laws, in Latin, from the original 1687 *Principia Mathematica*.

Newton's laws of motion are three physical laws that, together, laid the foundation for classical mechanics. They describe the relationship between a body and the forces acting upon it, and its motion in response to those forces. They have been expressed in several different ways, over nearly three centuries, and can be summarised as follows.

First law:	In an inertial reference frame, an object either remains at rest or continues to move at a constant velocity, unless acted upon by a net force.
Second law:	In an inertial reference frame, the vector sum of the forces F on an object is equal to the mass m of that object multiplied by the acceleration a of the object: F = ma.

Third law:	When one body exerts a force on a second body, the second body simultaneously exerts a force equal in magnitude and opposite in direction on the first body.

The three laws of motion were first compiled by Isaac Newton in his *Philosophiæ Naturalis Principia Mathematica* (*Mathematical Principles of Natural Philosophy*), first published in 1687. Newton used them to explain and investigate the motion of many physical objects and systems. For example, in the third volume of the text, Newton showed that these laws of motion, combined with his law of universal gravitation, explained Kepler's laws of planetary motion.

Overview

Isaac Newton (1643–1727), the physicist who formulated the laws

Newton's laws are applied to objects which are idealised as single point masses, in the sense that the size and shape of the object's body are neglected to focus on its motion more easily. This can be done when the object is small compared to the distances involved in its analysis, or the deformation and rotation of the body are of no importance. In this way, even a planet can be idealised as a particle for analysis of its orbital motion around a star.

In their original form, Newton's laws of motion are not adequate to characterise the motion of rigid bodies and deformable bodies. Leonhard Euler in 1750 introduced a generalisation of Newton's laws of motion for rigid bodies called Euler's laws of motion, later applied as well for deformable bodies assumed as a continuum. If a body is represented as an assemblage of discrete particles, each governed by Newton's laws of motion, then Euler's laws can be derived from Newton's laws. Euler's laws can, however, be taken as axioms describing the laws of motion for extended bodies, independently of any particle structure.

Newton's laws hold only with respect to a certain set of frames of reference called Newtonian or inertial reference frames. Some authors interpret the first law as defining what an inertial reference frame is; from this point of view, the second law only holds when the observation is made from an inertial reference frame, and therefore the first law cannot be proved as a special case of the second. Other authors do treat the first law as a corollary of the second. The explicit concept of an inertial frame of reference was not developed until long after Newton's death.

In the given interpretation mass, acceleration, momentum, and (most importantly) force are assumed to be externally defined quantities. This is the most common, but not the only interpretation of the way one can consider the laws to be a definition of these quantities.

Newtonian mechanics has been superseded by special relativity, but it is still useful as an approximation when the speeds involved are much slower than the speed of light.

Newton's First Law

Explanation of Newton's first law and reference frames. (MIT Course 8.01)

The first law states that if the net force (the vector sum of all forces acting on an object) is zero, then the velocity of the object is constant. Velocity is a vector quantity which expresses both the object's speed and the direction of its motion; therefore, the statement that the object's velocity is constant is a statement that both its speed and the direction of its motion are constant.

The first law can be stated mathematically when the mass is a non-zero constant, as,

$$\sum \mathbf{F} = 0 \Leftrightarrow \frac{d\mathbf{v}}{dt} = 0.$$

Consequently,

- An object that is at rest will stay at rest unless a force acts upon it.

- An object that is in motion will not change its velocity unless a force acts upon it.

This is known as *uniform motion*. An object *continues* to do whatever it happens to be doing unless a force is exerted upon it. If it is at rest, it continues in a state of rest (demonstrated when a tablecloth is skilfully whipped from under dishes on a tabletop and the dishes remain in their initial state of rest). If an object is moving, it continues to move without turning or changing its speed. This is evident in space probes that continually move in outer space. Changes in motion must be imposed against the tendency of an object to retain its state of motion. In the absence of net forces, a moving object tends to move along a straight line path indefinitely.

Newton placed the first law of motion to establish frames of reference for which the other laws are applicable. The first law of motion postulates the existence of at least one frame of reference called a Newtonian or inertial reference frame, relative to which the motion of a particle not subject to forces is a straight line at a constant speed. Newton's first law is often referred to as the *law of inertia*. Thus, a condition necessary for the uniform motion of a particle relative to an inertial reference frame is that the total net force acting on it is zero. In this sense, the first law can be restated as:

In every material universe, the motion of a particle in a preferential reference frame Φ is determined by the action of forces whose total vanished for all times when and only when the velocity of the particle is constant in Φ. That is, a particle initially at rest or in uniform motion in the preferential frame Φ continues in that state unless compelled by forces to change it.

Newton's laws are valid only in an inertial reference frame. Any reference frame that is in uniform motion with respect to an inertial frame is also an inertial frame, i.e. Galilean invariance or the principle of Newtonian relativity.

Newton's Second Law

Explanation of Newton's second law, using gravity as an example. (MIT OCW)

The second law states that the rate of change of momentum of a body, is directly proportional to the force applied and this change in momentum takes place in the direction of the applied force.

$$\mathbf{F} = \frac{d\mathbf{p}}{dt} = \frac{d(m\mathbf{v})}{dt}.$$

The second law can also be stated in terms of an object's acceleration. Since Newton's second law is only valid for constant-mass systems, it can be taken outside the differentiation operator by the constant factor rule in differentiation. Thus,

$$\mathbf{F} = m\frac{d\mathbf{v}}{dt} = m\mathbf{a},$$

where F is the net force applied, m is the mass of the body, and a is the body's acceleration. Thus, the net force applied to a body produces a proportional acceleration. In other words, if a body is accelerating, then there is a force on it.

Consistent with the first law, the time derivative of the momentum is non-zero when the momentum changes direction, even if there is no change in its magnitude; such is the case with uniform circular motion. The relationship also implies the conservation of momentum: when the net force on the body is zero, the momentum of the body is constant. Any net force is equal to the rate of change of the momentum.

Any mass that is gained or lost by the system will cause a change in momentum that is not the result of an external force. A different equation is necessary for variable-mass systems.

Newton's second law requires modification if the effects of special relativity are to be taken into account, because at high speeds the approximation that momentum is the product of rest mass and velocity is not accurate.

Impulse

An impulse J occurs when a force F acts over an interval of time Δt, and it is given by

$$\mathbf{J} = \int_{\Delta t} \mathbf{F} \, dt.$$

Since force is the time derivative of momentum, it follows that

$$\mathbf{J} = \Delta \mathbf{p} = m\Delta \mathbf{v}.$$

This relation between impulse and momentum is closer to Newton's wording of the second law.

Impulse is a concept frequently used in the analysis of collisions and impacts.

Variable-mass Systems

Variable-mass systems, like a rocket burning fuel and ejecting spent gases, are not closed and cannot be directly treated by making mass a function of time in the second law; that is, the following formula is wrong:

$$\mathbf{F}_{net} = \frac{d}{dt}\left[m(t)\mathbf{v}(t)\right] = m(t)\frac{d\mathbf{v}}{dt} + \mathbf{v}(t)\frac{dm}{dt}. \qquad \text{(wrong)}$$

The falsehood of this formula can be seen by noting that it does not respect Galilean invariance: a variable-mass object with F = 0 in one frame will be seen to have F ≠ 0 in another frame. The correct equation of motion for a body whose mass m varies with time by either ejecting or accreting mass is obtained by applying the second law to the entire, constant-mass system consisting of the body and its ejected/accreted mass; the result is

$$\mathbf{F} + \mathbf{u}\frac{dm}{dt} = m\frac{d\mathbf{v}}{dt}$$

where u is the velocity of the escaping or incoming mass relative to the body. From this equation one can derive the equation of motion for a varying mass system, for example, the Tsiolkovsky rocket equation. Under some conventions, the quantity $u\,dm/dt$ on the left-hand side, which represents the advection of momentum, is defined as a force (the force exerted on the body by the changing mass, such as rocket exhaust) and is included in the quantity F. Then, by substituting the definition of acceleration, the equation becomes F = ma.

Newton's Third Law

An illustration of Newton's third law in which two skaters push against each other. The first skater on the left exerts a normal force N_{12} on the second skater directed towards the right, and the second skater exerts a normal force N_{21} on the first skater directed towards the left. The magnitudes of both forces are equal, but they have opposite directions, as dictated by Newton's third law.

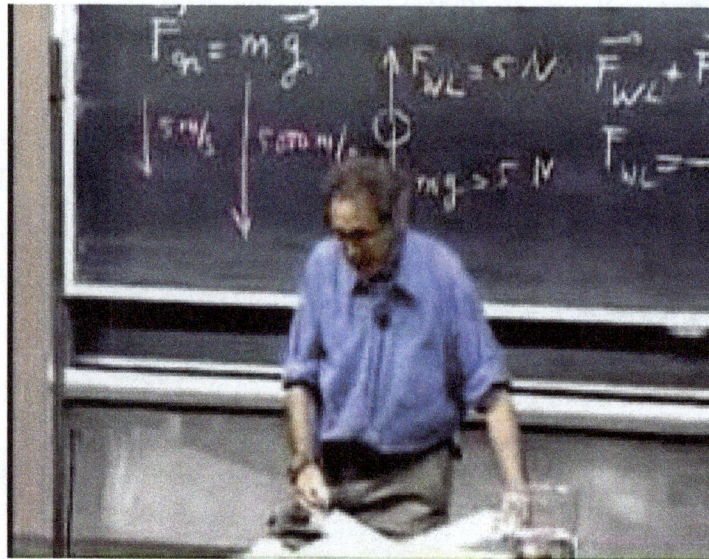

A description of Newton's third law and contact forces

The third law states that all forces between two objects exist in equal magnitude and opposite direction: if one object A exerts a force F_A on a second object B, then B simultaneously exerts a force F_B on A, and the two forces are equal in magnitude and opposite in direction: $F_A = -F_B$. The third law means that all forces are *interactions* between different bodies, or different regions within one body, and thus that there is no such thing as a force that is not accompanied by an equal and opposite force. In some situations, the magnitude and direction of the forces are determined entirely by one of the two bodies, say Body A; the force exerted by Body A on Body B is called the "action", and the force exerted by Body B on Body A is called the "reaction". This law is sometimes referred to as the *action-reaction law*, with F_A called the "action" and F_B the "reaction". In other situations the magnitude and directions of the forces are determined jointly by both bodies and it isn't necessary to identify one force as the "action" and the other as the "reaction". The action and the reaction are simultaneous, and it does not matter which is called the *action* and which is called *reaction*; both forces are part of a single interaction, and neither force exists without the other.

The two forces in Newton's third law are of the same type (e.g., if the road exerts a forward frictional force on an accelerating car's tires, then it is also a frictional force that Newton's third law predicts for the tires pushing backward on the road).

From a conceptual standpoint, Newton's third law is seen when a person walks: they push against the floor, and the floor pushes against the person. Similarly, the tires of a car push against the road while the road pushes back on the tires—the tires and road simultaneously push against each other. In swimming, a person interacts with the water, pushing the water backward, while the water simultaneously pushes the person forward—both the person and the water push against each other. The reaction forces account for the motion in these examples. These forces depend on friction; a person or car on ice, for example, may be unable to exert the action force to produce the needed reaction force.

History

Newton's 1st Law

From the original Latin of Newton's *Principia*:

> "   *Lex I: Corpus omne perseverare in statu suo quiescendi vel movendi uniformiter in directum, nisi quatenus a viribus impressis cogitur statum illum mutare.*   "

Translated to English, this reads:

> "   Law I: Every body persists in its state of being at rest or of moving uniformly straight forward, except insofar as it is compelled to change its state by force impressed.   "

The ancient Greek philosopher Aristotle had the view that all objects have a natural place in the universe: that heavy objects (such as rocks) wanted to be at rest on the Earth and that light objects like smoke wanted to be at rest in the sky and the stars wanted to remain in the heavens. He thought that a body was in its natural state when it was at rest, and for the body to move in a straight line at a constant speed an external agent was needed continually to propel it, otherwise it would stop moving. Galileo Galilei, however, realised that a force is necessary to change the velocity of a body, i.e., acceleration, but no force is needed to maintain its velocity. In other words, Galileo stated that, in the *absence* of a force, a moving object will continue moving. (The tendency of objects to resist changes in motion was what Johannes Kepler had called *inertia*.) This insight was refined by Newton, who made it into his first law, also known as the "law of inertia"—no force means no acceleration, and hence the body will maintain its velocity. As Newton's first law is a restatement of the law of inertia which Galileo had already described, Newton appropriately gave credit to Galileo.

The law of inertia apparently occurred to several different natural philosophers and scientists independently, including Thomas Hobbes in his *Leviathan*. The 17th century philosopher and mathematician René Descartes also formulated the law, although he did not perform any experiments to confirm it.

Newton's 2nd Law

> " *Lex II: Mutationem motus proportionalem esse vi motrici impressae, et fieri secundum lineam rectam qua vis illa imprimitur.* "

This was translated quite closely in Motte's 1729 translation as:

> " Law II: The alteration of motion is ever proportional to the motive force impress'd; and is made in the direction of the right line in which that force is impress'd. "

According to modern ideas of how Newton was using his terminology, this is understood, in modern terms, as an equivalent of:

The change of momentum of a body is proportional to the impulse impressed on the body, and happens along the straight line on which that impulse is impressed.

This may be expressed by the formula F = p', where p' is the time derivative of the momentum p. This equation can be seen clearly in the Wren Library of Trinity College, Cambridge, in a glass case in which Newton's manuscript is open to the relevant page.

Motte's 1729 translation of Newton's Latin continued with Newton's commentary on the second law of motion, reading:

If a force generates a motion, a double force will generate double the motion, a triple force triple the motion, whether that force be impressed altogether and at once, or gradually and successively. And this motion (being always directed the same way with the generating force), if the body moved before, is added to or subtracted from the former motion, according as they directly conspire with or are directly contrary to each other; or obliquely joined, when they are oblique, so as to produce a new motion compounded from the determination of both.

The sense or senses in which Newton used his terminology, and how he understood the second law and intended it to be understood, have been extensively discussed by historians of science, along with the relations between Newton's formulation and modern formulations.

Newton's 3rd Law

> " *Lex III: Actioni contrariam semper et æqualem esse reactionem: sive corporum duorum actiones in se mutuo semper esse æquales et in partes contrarias dirigi.* "

Translated to English, this reads:

> " Law III: To every action there is always opposed an equal reaction: or the mutual actions of two bodies upon each other are always equal, and directed to contrary parts. "

Newton's Scholium (explanatory comment) to this law:

Whatever draws or presses another is as much drawn or pressed by that other. If you press a stone with your finger, the finger is also pressed by the stone. If a horse draws a stone tied to a rope, the horse (if I may so say) will be equally drawn back towards the stone: for the distended rope, by the same endeavour to relax or unbend itself, will draw the horse as much towards the stone, as it does the stone towards the horse, and will obstruct the progress of the one as much as it advances that of the other. If a body impinges upon another, and by its force changes the motion of the other, that body also (because of the equality of the mutual pressure) will undergo an equal change, in its own motion, toward the contrary part. The changes made by these actions are equal, not in the velocities but in the motions of the bodies; that is to say, if the bodies are not hindered by any other impediments. For, as the motions are equally changed, the changes of the velocities made toward contrary parts are reciprocally proportional to the bodies. This law takes place also in attractions, as will be proved in the next scholium.

In the above, as usual, *motion* is Newton's name for momentum, hence his careful distinction between motion and velocity.

Newton used the third law to derive the law of conservation of momentum; from a deeper perspective, however, conservation of momentum is the more fundamental idea (derived via Noether's theorem from Galilean invariance), and holds in cases where Newton's third law appears to fail, for instance when force fields as well as particles carry momentum, and in quantum mechanics.

Importance and Range of Validity

Newton's laws were verified by experiment and observation for over 200 years, and they are excellent approximations at the scales and speeds of everyday life. Newton's laws of motion, together with his law of universal gravitation and the mathematical techniques of calculus, provided for the first time a unified quantitative explanation for a wide range of physical phenomena.

These three laws hold to a good approximation for macroscopic objects under everyday conditions. However, Newton's laws (combined with universal gravitation and classical electrodynamics) are inappropriate for use in certain circumstances, most notably at very small scales, very high speeds (in special relativity, the Lorentz factor must be included in the expression for momentum along with the rest mass and velocity) or very strong gravitational fields. Therefore, the laws cannot be used to explain phenomena such as conduction of electricity in a semiconductor, optical properties of substances, errors in non-relativistically corrected GPS systems and superconductivity. Explanation of these phenomena requires more sophisticated physical theories, including general relativity and quantum field theory.

In quantum mechanics, concepts such as force, momentum, and position are defined by linear operators that operate on the quantum state; at speeds that are much lower than the speed of light, Newton's laws are just as exact for these operators as they are for classical objects. At speeds comparable to the speed of light, the second law holds in the original form $F = dp/dt$, where F and p are four-vectors.

Relationship to the Conservation Laws

In modern physics, the laws of conservation of momentum, energy, and angular momentum are of more general validity than Newton's laws, since they apply to both light and matter, and to both classical and non-classical physics.

This can be stated simply, "Momentum, energy and angular momentum cannot be created or destroyed."

Because force is the time derivative of momentum, the concept of force is redundant and subordinate to the conservation of momentum, and is not used in fundamental theories (e.g., quantum mechanics, quantum electrodynamics, general relativity, etc.). The standard model explains in detail how the three fundamental forces known as gauge forces originate out of exchange by virtual particles. Other forces, such as gravity and fermionic degeneracy pressure, also arise from the momentum conservation. Indeed, the conservation of 4-momentum in inertial motion via curved space-time results in what we call gravitational force in general relativity theory. The application of the space derivative (which is a momentum operator in quantum mechanics) to the overlapping wave functions of a pair of fermions (particles with half-integer spin) results in shifts of maxima of compound wavefunction away from each other, which is observable as the "repulsion" of the fermions.

Newton stated the third law within a world-view that assumed instantaneous action at a distance between material particles. However, he was prepared for philosophical criticism of this action at a distance, and it was in this context that he stated the famous phrase "I feign no hypotheses". In modern physics, action at a distance has been completely eliminated, except for subtle effects involving quantum entanglement. However, in modern engineering, in all practical applications involving the motion of vehicles and satellites, the concept of action at a distance is used extensively.

The discovery of the second law of thermodynamics by Carnot in the 19th century showed that every physical quantity is not conserved over time, thus disproving the validity of inducing the opposite metaphysical view from Newton's laws. Hence, a "steady-state" worldview based solely on Newton's laws and the conservation laws does not take entropy into account.

References

- Scott, J.F. (1981). The Mathematical Work of John Wallis, D.D., F.R.S. Chelsea Publishing Company. p. 111. ISBN 0-8284-0314-7.

- Wahlin, Lars (1997). "9.1 Relative and absolute motion". The Deadbeat Universe (PDF). Boulder, CO: Coultron Research. pp. 121–129. ISBN 0-933407-03-3. Retrieved 25 January 2013.

- Tyson, Neil de Grasse; Charles Tsun-Chu Liu; Robert Irion (2000). The universe : at home in the cosmos. Washington, DC: National Academy Press. ISBN 0-309-06488-0.

- Serway, Raymond A.; John W. Jewett, Jr (2012). Principles of physics : a calculus-based text (5th ed.). Boston, MA: Brooks/Cole, Cengage Learning. p. 245. ISBN 9781133104261.

- McGinnis, Peter M. (2005). Biomechanics of sport and exercise (2nd ed.). Champaign, IL [u.a.]: Human Kinetics. p. 85. ISBN 9780736051019.

- Misner, Charles W.; Kip S. Thorne; John Archibald Wheeler (1973). Gravitation. 24th printing. New York: W. H. Freeman. p. 51. ISBN 9780716703440.

- Rindler, Wolfgang (1991). Introduction to Special Relativity (2nd ed.). Oxford Science Publications. pp. 82–84.

ISBN 0-19-853952-5.

- Lerner, Rita G.; Trigg, George L., eds. (2005). Encyclopedia of physics (3rd ed.). Weinheim: Wiley-VCH-Verl. ISBN 978-3527405541.

- Hand, Louis N.; Finch, Janet D. (1998). Analytical mechanics (7th print ed.). Cambridge, England: Cambridge University Press. Chapter 4. ISBN 9780521575720.

- Bird, R. Byron; Warren Stewart; Edwin N. Lightfoot (2007). Transport phenomena (2nd ed.). New York: Wiley. p. 13. ISBN 9780470115398.

- Gubbins, David (1992). Seismology and plate tectonics (Repr. (with corr.) ed.). Cambridge [England]: Cambridge University Press. p. 59. ISBN 0521379954.

- LeBlond, Paul H.; Mysak, Lawrence A. (1980). Waves in the ocean (2. impr. ed.). Amsterdam [u.a.]: Elsevier. p. 258. ISBN 9780444419262.

- Seyyed Hossein Nasr & Mehdi Amin Razavi (1996). The Islamic intellectual tradition in Persia. Routledge. p. 72. ISBN 0-7007-0314-4.

- Park, David (1990). The how and the why : an essay on the origins and development of physical theory. With drawings by Robin Brickman (3rd print ed.). Princeton, N.J.: Princeton University Press. pp. 139–141. ISBN 9780691025087.

- Daniel Garber (1992). "Descartes' Physics". In John Cottingham. The Cambridge Companion to Descartes. Cambridge: Cambridge University Press. pp. 310–319. ISBN 0-521-36696-8.

- Rothman, Milton A. (1989). Discovering the natural laws : the experimental basis of physics (2nd ed.). New York: Dover Publications. pp. 83–88. ISBN 9780486261782.

- G. W. Leibniz (1989). "Discourse on Metaphysics". In Roger Ariew; Daniel Garber. Philosophical Essays. Indianapolis, IN: Hackett Publishing Company, Inc. pp. 49–51. ISBN 0-87220-062-0.

- Rescigno, Aldo (2003). Foundation of Pharmacokinetics. New York: Kluwer Academic/Plenum Publishers. p. 19. ISBN 0306477041.

- J X Zheng-Johansson; Per-Ivar Johansson (2006). Unification of Classical, Quantum and Relativistic Mechanics and of the Four Forces. Nova Publishers. p. 13. ISBN 1-59454-260-0.

- Jean Salençon; Stephen Lyle (2001). Handbook of Continuum Mechanics: General Concepts, Thermoelasticity. Springer. p. 9. ISBN 3-540-41443-6.

- Patrick Cornille (Akhlesh Lakhtakia, editor) (1993). Essays on the Formal Aspects of Electromagnetic Theory. World Scientific. p. 149. ISBN 981-02-0854-5.

- Graham Nerlich (1994). What Spacetime Explains: Metaphysical essays on space and time. Cambridge University Press. p. 64. ISBN 0-521-45261-9.

- Katherine Brading; Elena Castellani (2003). Symmetries in Physics: Philosophical Reflections. Cambridge University Press. p. 417. ISBN 0-521-82137-1.

- Oliver Davis Johns (2005). Analytical Mechanics for Relativity and Quantum Mechanics. Oxford University Press. Chapter 16. ISBN 0-19-856726-X.

- Donald T Greenwood (1997). Classical dynamics (Reprint of 1977 edition by Prentice-Hall ed.). Courier Dover Publications. p. 313. ISBN 0-486-69690-1.

Work and Energy: An Integrated Study

In physics, force can change the velocity of any object. It has both magnitude and direction. The section on work and energy also covers themes such as friction, power, energy, forms of energy, conservation of energy etc. The major concepts of work and energy are explained in this chapter.

Force

In physics, a force is any interaction that, when unopposed, will change the motion of an object. In other words, a force can cause an object with mass to change its velocity (which includes to begin moving from a state of rest), i.e., to accelerate. Force can also be described by intuitive concepts such as a push or a pull. A force has both magnitude and direction, making it a vector quantity. It is measured in the SI unit of newtons and represented by the symbol F.

The original form of Newton's second law states that the net force acting upon an object is equal to the rate at which its momentum changes with time. If the mass of the object is constant, this law implies that the acceleration of an object is directly proportional to the net force acting on the object, is in the direction of the net force, and is inversely proportional to the mass of the object

Related concepts to force include: thrust, which increases the velocity of an object; drag, which decreases the velocity of an object; and torque, which produces changes in rotational speed of an object. In an extended body, each part usually applies forces on the adjacent parts; the distribution of such forces through the body is the so-called mechanical stress. Pressure is a simple type of stress. Stress usually causes deformation of solid materials, or flow in fluids.

Development of the Concept

Philosophers in antiquity used the concept of force in the study of stationary and moving objects and simple machines, but thinkers such as Aristotle and Archimedes retained fundamental errors in understanding force. In part this was due to an incomplete understanding of the sometimes non-obvious force of friction, and a consequently inadequate view of the nature of natural motion. A fundamental error was the belief that a force is required to maintain motion, even at a constant velocity. Most of the previous misunderstandings about motion and force were eventually corrected by Galileo Galilei and Sir Isaac Newton. With his mathematical insight, Sir Isaac Newton formulated laws of motion that were not improved-on for nearly three hundred years. By the early 20th century, Einstein developed a theory of relativity that correctly predicted the action of forces on objects with increasing momenta near the speed of light, and also provided insight into the forces produced by gravitation and inertia.

With modern insights into quantum mechanics and technology that can accelerate particles close to the speed of light, particle physics has devised a Standard Model to describe forces between particles smaller than atoms. The Standard Model predicts that exchanged particles called gauge bosons are the fundamental means by which forces are emitted and absorbed. Only four main interactions are known: in order of decreasing strength, they are: strong, electromagnetic, weak, and gravitational. High-energy particle physics observations made during the 1970s and 1980s confirmed that the weak and electromagnetic forces are expressions of a more fundamental electroweak interaction.

Pre-newtonian Concepts

Aristotle famously described a force as anything that causes an object to undergo "unnatural motion"

Since antiquity the concept of force has been recognized as integral to the functioning of each of the simple machines. The mechanical advantage given by a simple machine allowed for less force to be used in exchange for that force acting over a greater distance for the same amount of work. Analysis of the characteristics of forces ultimately culminated in the work of Archimedes who was especially famous for formulating a treatment of buoyant forces inherent in fluids.

Aristotle provided a philosophical discussion of the concept of a force as an integral part of Aristotelian cosmology. In Aristotle's view, the terrestrial sphere contained four elements that come to rest at different "natural places" therein. Aristotle believed that motionless objects on Earth, those composed mostly of the elements earth and water, to be in their natural place on the ground and that they will stay that way if left alone. He distinguished between the innate tendency of objects to find their "natural place" (e.g., for heavy bodies to fall), which led to "natural motion", and unnatural or forced motion, which required continued application of a force. This theory, based on the everyday experience of how objects move, such as the constant application of a force needed to keep a cart moving, had conceptual trouble accounting for the behavior of projectiles, such as the flight of arrows. The place where the archer moves the projectile was at the start of the flight, and while the projectile sailed through the air, no discernible efficient cause acts on it. Aristotle was aware of this problem and proposed that the air displaced through the projectile's path carries the projectile to its target. This explanation demands a continuum like air for change of place in general.

Aristotelian physics began facing criticism in Medieval science, first by John Philoponus in the 6th century.

The shortcomings of Aristotelian physics would not be fully corrected until the 17th century work of Galileo Galilei, who was influenced by the late Medieval idea that objects in forced motion carried an innate force of impetus. Galileo constructed an experiment in which stones and cannonballs were both rolled down an incline to disprove the Aristotelian theory of motion early in the 17th century. He showed that the bodies were accelerated by gravity to an extent that was independent of their mass and argued that objects retain their velocity unless acted on by a force, for example friction.

Newtonian Mechanics

Sir Isaac Newton sought to describe the motion of all objects using the concepts of inertia and force, and in doing so he found that they obey certain conservation laws. In 1687, Newton went on to publish his thesis *Philosophiæ Naturalis Principia Mathematica*. In this work Newton set out three laws of motion that to this day are the way forces are described in physics.

First Law

Newton's First Law of Motion states that objects continue to move in a state of constant velocity unless acted upon by an external net force or *resultant force*. This law is an extension of Galileo's insight that constant velocity was associated with a lack of net force. Newton proposed that every object with mass has an innate inertia that functions as the fundamental equilibrium "natural state" in place of the Aristotelian idea of the "natural state of rest". That is, the first law contradicts the intuitive Aristotelian belief that a net force is required to keep an object moving with constant velocity. By making *rest* physically indistinguishable from *non-zero constant velocity*, Newton's First Law directly connects inertia with the concept of relative velocities. Specifically, in systems where objects are moving with different velocities, it is impossible to determine which object is "in motion" and which object is "at rest". In other words, to phrase matters more technically, the laws of physics are the same in every inertial frame of reference, that is, in all frames related by a Galilean transformation.

For instance, while traveling in a moving vehicle at a constant velocity, the laws of physics do not change from being at rest. A person can throw a ball straight up in the air and catch it as it falls down without worrying about applying a force in the direction the vehicle is moving. This is true even though another person who is observing the moving vehicle pass by also observes the ball follow a curving parabolic path in the same direction as the motion of the vehicle. It is the inertia of the ball associated with its constant velocity in the direction of the vehicle's motion that ensures the ball continues to move forward even as it is thrown up and falls back down. From the perspective of the person in the car, the vehicle and everything inside of it is at rest: It is the outside world that is moving with a constant speed in the opposite direction. Since there is no experiment that can distinguish whether it is the vehicle that is at rest or the outside world that is at rest, the two situations are considered to be physically indistinguishable. Inertia therefore applies equally well to constant velocity motion as it does to rest.

The concept of inertia can be further generalized to explain the tendency of objects to continue in many different forms of constant motion, even those that are not strictly constant velocity. The rotational inertia of planet Earth is what fixes the constancy of the length of a day and the length of a year. Albert Einstein extended the principle of inertia further when he explained that reference frames subject to constant acceleration, such as those free-falling toward a gravitating object, were physically equivalent to inertial reference frames. This is why, for example, astronauts experience weightlessness when in free-fall orbit around the Earth, and why Newton's Laws of Motion are more easily discernible in such environments. If an astronaut places an object with mass in mid-air next to himself, it will remain stationary with respect to the astronaut due to its inertia. This is the same thing that would occur if the astronaut and the object were in intergalactic space with no net force of gravity acting on their shared reference frame. This principle of equivalence was one of the foundational underpinnings for the development of the general theory of relativity.

Second Law

A modern statement of Newton's Second Law is a vector equation:

$$\vec{F} = \frac{d\vec{p}}{dt},$$

where $\vec{p}$ is the momentum of the system, and $\vec{F}$ is the net (vector sum) force. In equilibrium, there is zero *net* force by definition, but (balanced) forces may be present nevertheless. In contrast, the second law states an *unbalanced* force acting on an object will result in the object's momentum changing over time.

By the definition of momentum,

$$\vec{F} = \frac{d\vec{p}}{dt} = \frac{d(m\vec{v})}{dt},$$

where m is the mass and $\vec{v}$ is the velocity.

Newton's second law applies only to a system of constant mass, and hence m may be moved outside the derivative operator. The equation then becomes

$$\vec{F} = m\frac{d\vec{v}}{dt}.$$

By substituting the definition of acceleration, the algebraic version of Newton's Second Law is derived:

$$\vec{F} = m\vec{a}.$$

Newton never explicitly stated the formula in the reduced form above.

Newton's Second Law asserts the direct proportionality of acceleration to force and the inverse proportionality of acceleration to mass. Accelerations can be defined through kinematic measurements. However, while kinematics are well-described through reference frame analysis in advanced physics, there are still deep questions that remain as to what is the proper definition of mass. General relativity offers an equivalence between space-time and mass, but lacking a coherent theory of quantum gravity, it is unclear as to how or whether this connection is relevant on microscales. With some justification, Newton's second law can be taken as a quantitative definition of *mass* by writing the law as an equality; the relative units of force and mass then are fixed.

The use of Newton's Second Law as a *definition* of force has been disparaged in some of the more rigorous textbooks, because it is essentially a mathematical truism. Notable physicists, philosophers and mathematicians who have sought a more explicit definition of the concept of force include Ernst Mach, Clifford Truesdell and Walter Noll.

Newton's Second Law can be used to measure the strength of forces. For instance, knowledge of the masses of planets along with the accelerations of their orbits allows scientists to calculate the gravitational forces on planets.

Third Law

Newton's Third Law is a result of applying symmetry to situations where forces can be attributed to the presence of different objects. The third law means that all forces are *interactions* between different bodies, and thus that there is no such thing as a unidirectional force or a force that acts on only one body. Whenever a first body exerts a force F on a second body, the second body exerts a force $-F$ on the first body. F and $-F$ are equal in magnitude and opposite in direction. This law is sometimes referred to as the *action-reaction law*, with F called the "action" and $-F$ the "reaction". The action and the reaction are simultaneous:

$$\vec{F}_{1,2} = -\vec{F}_{2,1}.$$

If object 1 and object 2 are considered to be in the same system, then the net force on the system due to the interactions between objects 1 and 2 is zero since

$$\vec{F}_{1,2} + \vec{F}_{2,1} = 0$$
$$\sum \vec{F} = 0.$$

This means that in a closed system of particles, there are no internal forces that are unbalanced. That is, the action-reaction force shared between any two objects in a closed system will not cause the center of mass of the system to accelerate. The constituent objects only accelerate with respect to each other, the system itself remains unaccelerated. Alternatively, if an external force acts on the system, then the center of mass will experience an acceleration proportional to the magnitude of the external force divided by the mass of the system.

Combining Newton's Second and Third Laws, it is possible to show that the linear momentum of a system is conserved. Using

$$\vec{F}_{1,2} = \frac{d\vec{p}_{1,2}}{dt} = -\vec{F}_{2,1} = -\frac{d\vec{p}_{2,1}}{dt}$$

and integrating with respect to time, the equation:

$$\Delta \vec{p}_{1,2} = -\Delta \vec{p}_{2,1}$$

is obtained. For a system that includes objects 1 and 2,

$$\sum \Delta \vec{p} = \Delta \vec{p}_{1,2} + \Delta \vec{p}_{2,1} = 0,$$

which is the conservation of linear momentum. Using the similar arguments, it is possible to generalize this to a system of an arbitrary number of particles. This shows that exchanging momentum between constituent objects will not affect the net momentum of a system. In general, as long as all forces are due to the interaction of objects with mass, it is possible to define a system such that net momentum is never lost nor gained.

Special Theory of Relativity

In the special theory of relativity, mass and energy are equivalent (as can be seen by calculating the work required to accelerate an object). When an object's velocity increases, so does its energy and hence its mass equivalent (inertia). It thus requires more force to accelerate it the same amount than it did at a lower velocity. Newton's Second Law

$$\vec{F} = d\vec{p}/dt$$

remains valid because it is a mathematical definition. But in order to be conserved, relativistic momentum must be redefined as:

$$\vec{p} = \frac{m_0 \vec{v}}{\sqrt{1-v^2/c^2}}$$

where

v is the velocity and

c is the speed of light

m_0 is the rest mass.

The relativistic expression relating force and acceleration for a particle with constant non-zero rest mass m moving in the x direction is:

$$F_x = \gamma^3 m a_x$$
$$F_y = \gamma m a_y$$
$$F_z = \gamma m a_z$$

where the Lorentz factor

$$\gamma = \frac{1}{\sqrt{1-v^2/c^2}}.$$

In the early history of relativity, the expressions $\gamma^3 m$ and γm were called longitudinal and transverse mass. Relativistic force does not produce a constant acceleration, but an ever decreasing acceleration as the object approaches the speed of light. Note that γ is undefined for an object with a non-zero rest mass at the speed of light, and the theory yields no prediction at that speed.

If v is very small compared to c, then γ is very close to 1 and

$$F = ma$$

is a close approximation. Even for use in relativity, however, one can restore the form of

$$F^\mu = mA^\mu$$

through the use of four-vectors. This relation is correct in relativity when F^μ is the four-force, m is the invariant mass, and A^μ is the four-acceleration.

Descriptions

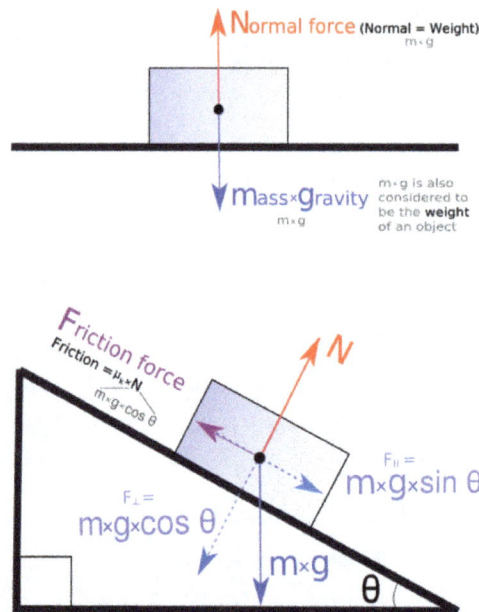

Free body diagrams of a block on a flat surface and an inclined plane.
Forces are resolved and added together to determine their magnitudes and the net force.

Since forces are perceived as pushes or pulls, this can provide an intuitive understanding for describing forces. As with other physical concepts (e.g. temperature), the intuitive understanding of forces is quantified using precise operational definitions that are consistent with direct observations and compared to a standard measurement scale. Through experimentation, it is determined that laboratory measurements of forces are fully consistent with the conceptual definition of force offered by Newtonian mechanics.

Forces act in a particular direction and have sizes dependent upon how strong the push or pull is. Because of these characteristics, forces are classified as "vector quantities". This means that forces follow a different set of mathematical rules than physical quantities that do not have direction (denoted scalar quantities). For example, when determining what happens when two forces act on the same object, it is necessary to know both the magnitude and the direction of both forces to calculate the result. If both of these pieces of information are not known for each force, the situation is ambiguous. For example, if you know that two people are pulling on the same rope with known magnitudes of force but you do not know which direction either person is pulling, it is impossible to determine what the acceleration of the rope will be. The two people could be pulling against each other as in tug of war or the two people could be pulling in the same direction. In this simple one-dimensional example, without knowing the direction of the forces it is impossible to decide whether the net force is the result of adding the two force magnitudes or subtracting one from the other. Associating forces with vectors avoids such problems.

Historically, forces were first quantitatively investigated in conditions of static equilibrium where several forces canceled each other out. Such experiments demonstrate the crucial properties that forces are additive vector quantities: they have magnitude and direction. When two forces act on a point particle, the resulting force, the *resultant* (also called the *net force*), can be determined by following the parallelogram rule of vector addition: the addition of two vectors represented by sides of a parallelogram, gives an equivalent resultant vector that is equal in magnitude and direction to the transversal of the parallelogram. The magnitude of the resultant varies from the difference of the magnitudes of the two forces to their sum, depending on the angle between their lines of action. However, if the forces are acting on an extended body, their respective lines of application must also be specified in order to account for their effects on the motion of the body.

Free-body diagrams can be used as a convenient way to keep track of forces acting on a system. Ideally, these diagrams are drawn with the angles and relative magnitudes of the force vectors preserved so that graphical vector addition can be done to determine the net force.

As well as being added, forces can also be resolved into independent components at right angles to each other. A horizontal force pointing northeast can therefore be split into two forces, one pointing north, and one pointing east. Summing these component forces using vector addition yields the original force. Resolving force vectors into components of a set of basis vectors is often a more mathematically clean way to describe forces than using magnitudes and directions. This is because, for orthogonal components, the components of the vector sum are uniquely determined by the scalar addition of the components of the individual vectors. Orthogonal components are independent of each other because forces acting at ninety degrees to each other have no effect on the magnitude or direction of the other. Choosing a set of orthogonal basis vectors is often done by considering what set of basis vectors will make the mathematics most convenient. Choosing a basis vector that is in the same direction as one of the forces is desirable, since that force would then have only one non-zero component. Orthogonal force vectors can be three-dimensional with the third component being at right-angles to the other two.

Equilibrium

Equilibrium occurs when the resultant force acting on a point particle is zero (that is, the vector sum of all forces is zero). When dealing with an extended body, it is also necessary that the net torque in it is 0.

There are two kinds of equilibrium: static equilibrium and dynamic equilibrium.

Static

Static equilibrium was understood well before the invention of classical mechanics. Objects that are at rest have zero net force acting on them.

The simplest case of static equilibrium occurs when two forces are equal in magnitude but opposite in direction. For example, an object on a level surface is pulled (attracted) downward toward the center of the Earth by the force of gravity. At the same time, surface forces resist the downward force with equal upward force (called the normal force). The situation is one of zero net force and no acceleration.

Pushing against an object on a frictional surface can result in a situation where the object does not move because the applied force is opposed by static friction, generated between the object and the table surface. For a situation with no movement, the static friction force *exactly* balances the applied force resulting in no acceleration. The static friction increases or decreases in response to the applied force up to an upper limit determined by the characteristics of the contact between the surface and the object.

A static equilibrium between two forces is the most usual way of measuring forces, using simple devices such as weighing scales and spring balances. For example, an object suspended on a vertical spring scale experiences the force of gravity acting on the object balanced by a force applied by the "spring reaction force", which equals the object's weight. Using such tools, some quantitative force laws were discovered: that the force of gravity is proportional to volume for objects of constant density (widely exploited for millennia to define standard weights); Archimedes' principle for buoyancy; Archimedes' analysis of the lever; Boyle's law for gas pressure; and Hooke's law for springs. These were all formulated and experimentally verified before Isaac Newton expounded his Three Laws of Motion.

Dynamic

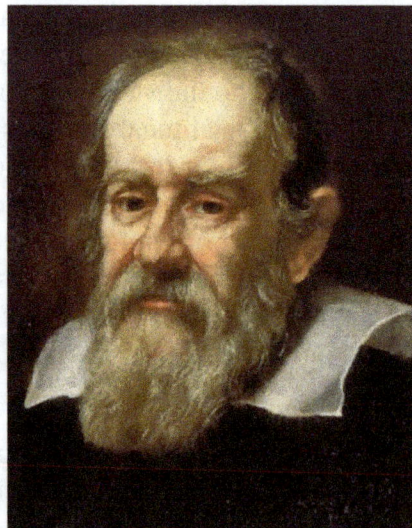

Galileo Galilei was the first to point out the inherent contradictions contained in Aristotle's description of forces.

Dynamic equilibrium was first described by Galileo who noticed that certain assumptions of Aristotelian physics were contradicted by observations and logic. Galileo realized that simple velocity addition demands that the concept of an "absolute rest frame" did not exist. Galileo concluded that motion in a constant velocity was completely equivalent to rest. This was contrary to Aristotle's notion of a "natural state" of rest that objects with mass naturally approached. Simple experiments showed that Galileo's understanding of the equivalence of constant velocity and rest were correct. For example, if a mariner dropped a cannonball from the crow's nest of a ship moving at a constant velocity, Aristotelian physics would have the cannonball fall straight down while the ship moved beneath it. Thus, in an Aristotelian universe, the falling cannonball would land behind the foot of the mast of a moving ship. However, when this experiment is actually conducted, the cannonball always falls at the foot of the mast, as if the cannonball knows to travel with the ship despite being separated from it. Since there is no forward horizontal force being applied on the cannonball as it falls, the only conclusion left is that the cannonball continues to move with the same velocity as the boat as it falls. Thus, no force is required to keep the cannonball moving at the constant forward velocity.

Moreover, any object traveling at a constant velocity must be subject to zero net force (resultant force). This is the definition of dynamic equilibrium: when all the forces on an object balance but it still moves at a constant velocity.

A simple case of dynamic equilibrium occurs in constant velocity motion across a surface with kinetic friction. In such a situation, a force is applied in the direction of motion while the kinetic friction force exactly opposes the applied force. This results in zero net force, but since the object started with a non-zero velocity, it continues to move with a non-zero velocity. Aristotle misinterpreted this motion as being caused by the applied force. However, when kinetic friction is taken into consideration it is clear that there is no net force causing constant velocity motion.

Forces in Quantum Mechanics

The notion "force" keeps its meaning in quantum mechanics, though one is now dealing with operators instead of classical variables and though the physics is now described by the Schrödinger equation instead of Newtonian equations. This has the consequence that the results of a measurement are now sometimes "quantized", i.e. they appear in discrete portions. This is, of course, difficult to imagine in the context of "forces". However, the potentials V(x,y,z) or fields, from which the forces generally can be derived, are treated similar to classical position variables, i.e., $V(x, y, z) \rightarrow \hat{V}(\hat{x}, \hat{y}, \hat{z})$..

This becomes different only in the framework of quantum field theory, where these fields are also quantized.

However, already in quantum mechanics there is one "caveat", namely the particles acting onto each other do not only possess the spatial variable, but also a discrete intrinsic angular momentum-like variable called the "spin", and there is the Pauli principle relating the space and the spin variables. Depending on the value of the spin, identical particles split into two different classes, fermions and bosons. If two identical fermions (e.g. electrons) have a *symmetric* spin function (e.g. parallel spins) the spatial variables must be *antisymmetric* (i.e. they exclude each other from their places much as if there was a repulsive force), and vice versa, i.e. for antiparallel *spins* the *position variables* must be symmetric (i.e. the apparent force must be attractive). Thus in the case of two fermions there is a strictly negative correlation between spatial and spin variables, whereas for two bosons (e.g. quanta of electromagnetic waves, photons) the correlation is strictly positive.

Thus the notion "force" loses already part of its meaning.

Feynman Diagrams

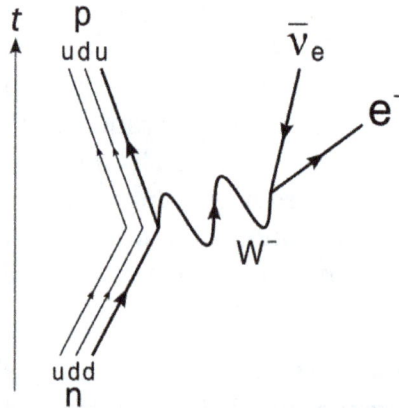

Feynman diagram for the decay of a neutron into a proton.
The W boson is between two vertices indicating a repulsion.

In modern particle physics, forces and the acceleration of particles are explained as a mathematical by-product of exchange of momentum-carrying gauge bosons. With the development of quantum field theory and general relativity, it was realized that force is a redundant concept arising from conservation of momentum (4-momentum in relativity and momentum of virtual particles in quantum electrodynamics). The conservation of momentum can be directly derived from the homogeneity or symmetry of space and so is usually considered more fundamental than the concept of a force. Thus the currently known fundamental forces are considered more accurately to be "fundamental interactions". When particle A emits (creates) or absorbs (annihilates) virtual particle B, a momentum conservation results in recoil of particle A making impression of repulsion or attraction between particles A A' exchanging by B. This description applies to all forces arising from fundamental interactions. While sophisticated mathematical descriptions are needed to predict, in full detail, the accurate result of such interactions, there is a conceptually simple way to describe such interactions through the use of Feynman diagrams. In a Feynman diagram, each matter particle is represented as a straight line traveling through time, which normally increases up or to the right in the diagram. Matter and anti-matter particles are identical except for their direction of propagation through the Feynman diagram. World lines of particles intersect at interaction vertices, and the Feynman diagram represents any force arising from an interaction as occurring at the vertex with an associated instantaneous change in the direction of the particle world lines. Gauge bosons are emitted away from the vertex as wavy lines and, in the case of virtual particle exchange, are absorbed at an adjacent vertex.

The utility of Feynman diagrams is that other types of physical phenomena that are part of the general picture of fundamental interactions but are conceptually separate from forces can also be described using the same rules. For example, a Feynman diagram can describe in succinct detail how a neutron decays into an electron, proton, and neutrino, an interaction mediated by the same gauge boson that is responsible for the weak nuclear force.

Fundamental Forces

All of the forces in the universe are based on four fundamental interactions. The strong and weak forces are nuclear forces that act only at very short distances, and are responsible for the interactions between subatomic particles, including nucleons and compound nuclei. The electromagnetic force acts between electric charges, and the gravitational force acts between masses. All other forces in nature derive from these four fundamental interactions. For example, friction is a manifestation of the electromagnetic force acting between the atoms of two surfaces, and the Pauli exclusion principle, which does not permit atoms to pass through each other. Similarly, the forces in springs, modeled by Hooke's law, are the result of electromagnetic forces and the Exclusion Principle acting together to return an object to its equilibrium position. Centrifugal forces are acceleration forces that arise simply from the acceleration of rotating frames of reference.

The development of fundamental theories for forces proceeded along the lines of unification of disparate ideas. For example, Isaac Newton unified the force responsible for objects falling at the surface of the Earth with the force responsible for the orbits of celestial mechanics in his universal theory of gravitation. Michael Faraday and James Clerk Maxwell demonstrated that electric and magnetic forces were unified through one consistent theory of electromagnetism. In the 20th century, the development of quantum mechanics led to a modern understanding that the first three fundamental forces (all except gravity) are manifestations of matter (fermions) interacting by exchanging virtual particles called gauge bosons. This standard model of particle physics posits a similarity between the forces and led scientists to predict the unification of the weak and electromagnetic forces in electroweak theory subsequently confirmed by observation. The complete formulation of the standard model predicts an as yet unobserved Higgs mechanism, but observations such as neutrino oscillations indicate that the standard model is incomplete. A Grand Unified Theory allowing for the combination of the electroweak interaction with the strong force is held out as a possibility with candidate theories such as supersymmetry proposed to accommodate some of the outstanding unsolved problems in physics. Physicists are still attempting to develop self-consistent unification models that would combine all four fundamental interactions into a theory of everything. Einstein tried and failed at this endeavor, but currently the most popular approach to answering this question is string theory.

The four fundamental forces of nature					
Property/Interaction	**Gravitation**	**Weak**	**Electromagnetic**	**Strong**	
		(Electroweak)		**Fundamental**	**Residual**
Acts on:	Mass - Energy	Flavor	Electric charge	Color charge	Atomic nuclei
Particles experiencing:	All	Quarks, leptons	Electrically charged	Quarks, Gluons	Hadrons
Particles mediating:	Graviton (not yet observed)	$W^+ W^- Z^0$	γ	Gluons	Mesons
Strength in the scale of quarks:	10^{-41}	10^{-4}	1	60	Not applicable to quarks
Strength in the scale of protons/neutrons:	10^{-36}	10^{-7}	1	Not applicable to hadrons	20

Gravitational

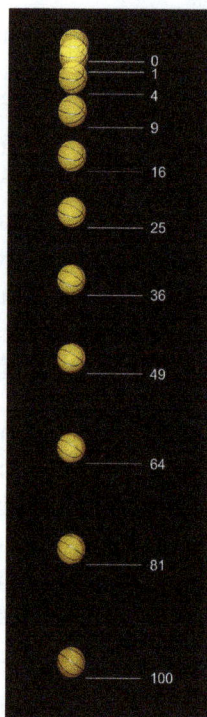

Images of a freely falling basketball taken with a stroboscope at 20 flashes per second. The distance units on the right are multiples of about 12 millimetres. The basketball starts at rest. At the time of the first flash (distance zero) it is released, after which the number of units fallen is equal to the square of the number of flashes.

What we now call gravity was not identified as a universal force until the work of Isaac Newton. Before Newton, the tendency for objects to fall towards the Earth was not understood to be related to the motions of celestial objects. Galileo was instrumental in describing the characteristics of falling objects by determining that the acceleration of every object in free-fall was constant and independent of the mass of the object. Today, this acceleration due to gravity towards the surface of the Earth is usually designated as $\vec{g}$ and has a magnitude of about 9.81 meters per second squared (this measurement is taken from sea level and may vary depending on location), and points toward the center of the Earth. This observation means that the force of gravity on an object at the Earth's surface is directly proportional to the object's mass. Thus an object that has a mass of m will experience a force:

$$\vec{F} = m\vec{g}$$

In free-fall, this force is unopposed and therefore the net force on the object is its weight. For objects not in free-fall, the force of gravity is opposed by the reactions of their supports. For example, a person standing on the ground experiences zero net force, since his weight is balanced by a normal force exerted by the ground.

Newton's contribution to gravitational theory was to unify the motions of heavenly bodies, which Aristotle had assumed were in a natural state of constant motion, with falling motion observed on the Earth. He proposed a law of gravity that could account for the celestial motions that had been described earlier using Kepler's laws of planetary motion.

Newton came to realize that the effects of gravity might be observed in different ways at larger distances. In particular, Newton determined that the acceleration of the Moon around the Earth could be ascribed to the same force of gravity if the acceleration due to gravity decreased as an inverse square law. Newton realized that the acceleration due to gravity is proportional to the mass of the attracting body. Combining these ideas gives a formula that relates the mass ($m_\oplus$) and the radius ($R_\oplus$) of the Earth to the gravitational acceleration:

$$\vec{g} = -\frac{Gm_\oplus}{R_\oplus^2}\hat{r}$$

where the vector direction is given by , the unit vector directed outward from the center of the Earth.

In this equation, a dimensional constant G is used to describe the relative strength of gravity. This constant has come to be known as Newton's Universal Gravitation Constant, though its value was unknown in Newton's lifetime. Not until 1798 was Henry Cavendish able to make the first measurement of G using a torsion balance; this was widely reported in the press as a measurement of the mass of the Earth since knowing G could allow one to solve for the Earth's mass given the above equation. Newton, however, realized that since all celestial bodies followed the same laws of motion, his law of gravity had to be universal. Succinctly stated, Newton's Law of Gravitation states that the force on a spherical object of mass m_1 due to the gravitational pull of mass m_2 is

$$\vec{F} = -\frac{Gm_1m_2}{r^2}\hat{r}$$

where r is the distance between the two objects' centers of mass and $\hat{r}$ is the unit vector pointed in the direction away from the center of the first object toward the center of the second object.

This formula was powerful enough to stand as the basis for all subsequent descriptions of motion within the solar system until the 20th century. During that time, sophisticated methods of perturbation analysis were invented to calculate the deviations of orbits due to the influence of multiple bodies on a planet, moon, comet, or asteroid. The formalism was exact enough to allow mathematicians to predict the existence of the planet Neptune before it was observed.

Instruments like GRAVITY provide a powerful probe for gravity force detection.

It was only the orbit of the planet Mercury that Newton's Law of Gravitation seemed not to fully explain. Some astrophysicists predicted the existence of another planet (Vulcan) that would explain the discrepancies; however, despite some early indications, no such planet could be found. When Albert Einstein formulated his theory of general relativity (GR) he turned his attention to the problem of Mercury's orbit and found that his theory added a correction, which could account for the discrepancy. This was the first time that Newton's Theory of Gravity had been shown to be less correct than an alternative.

Since then, and so far, general relativity has been acknowledged as the theory that best explains gravity. In GR, gravitation is not viewed as a force, but rather, objects moving freely in gravitational fields travel under their own inertia in straight lines through curved space-time – defined as the shortest space-time path between two space-time events. From the perspective of the object, all motion occurs as if there were no gravitation whatsoever. It is only when observing the motion in a global sense that the curvature of space-time can be observed and the force is inferred from the object's curved path. Thus, the straight line path in space-time is seen as a curved line in space, and it is called the *ballistic trajectory* of the object. For example, a basketball thrown from the ground moves in a parabola, as it is in a uniform gravitational field. Its space-time trajectory (when the extra ct dimension is added) is almost a straight line, slightly curved (with the radius of curvature of the order of few light-years). The time derivative of the changing momentum of the object is what we label as "gravitational force".

Electromagnetic

The electrostatic force was first described in 1784 by Coulomb as a force that existed intrinsically between two charges. The properties of the electrostatic force were that it varied as an inverse square law directed in the radial direction, was both attractive and repulsive (there was intrinsic polarity), was independent of the mass of the charged objects, and followed the superposition principle. Coulomb's law unifies all these observations into one succinct statement.

Subsequent mathematicians and physicists found the construct of the *electric field* to be useful for determining the electrostatic force on an electric charge at any point in space. The electric field was based on using a hypothetical "test charge" anywhere in space and then using Coulomb's Law to determine the electrostatic force. Thus the electric field anywhere in space is defined as

$$\vec{E} = \frac{\vec{F}}{q}$$

where q is the magnitude of the hypothetical test charge.

Meanwhile, the Lorentz force of magnetism was discovered to exist between two electric currents. It has the same mathematical character as Coulomb's Law with the proviso that like currents attract and unlike currents repel. Similar to the electric field, the magnetic field can be used to determine the magnetic force on an electric current at any point in space. In this case, the magnitude of the magnetic field was determined to be

$$B = \frac{F}{I\ell}$$

where I is the magnitude of the hypothetical test current and ℓ is the length of hypothetical wire through which the test current flows. The magnetic field exerts a force on all magnets including, for example, those used in compasses. The fact that the Earth's magnetic field is aligned closely with the orientation of the Earth's axis causes compass magnets to become oriented because of the magnetic force pulling on the needle.

Through combining the definition of electric current as the time rate of change of electric charge, a rule of vector multiplication called Lorentz's Law describes the force on a charge moving in a magnetic field. The connection between electricity and magnetism allows for the description of a unified *electromagnetic force* that acts on a charge. This force can be written as a sum of the electrostatic force (due to the electric field) and the magnetic force (due to the magnetic field). Fully stated, this is the law:

$$\vec{F} = q(\vec{E} + \vec{v} \times \vec{B})$$

where $\vec{F}$ is the electromagnetic force, q is the magnitude of the charge of the particle, $\vec{E}$ is the electric field, $\vec{v}$ is the velocity of the particle that is crossed with the magnetic field ($\vec{B}$).

The origin of electric and magnetic fields would not be fully explained until 1864 when James Clerk Maxwell unified a number of earlier theories into a set of 20 scalar equations, which were later reformulated into 4 vector equations by Oliver Heaviside and Josiah Willard Gibbs. These "Maxwell Equations" fully described the sources of the fields as being stationary and moving charges, and the interactions of the fields themselves. This led Maxwell to discover that electric and magnetic fields could be "self-generating" through a wave that traveled at a speed that he calculated to be the speed of light. This insight united the nascent fields of electromagnetic theory with optics and led directly to a complete description of the electromagnetic spectrum.

However, attempting to reconcile electromagnetic theory with two observations, the photoelectric effect, and the nonexistence of the ultraviolet catastrophe, proved troublesome. Through the work of leading theoretical physicists, a new theory of electromagnetism was developed using quantum mechanics. This final modification to electromagnetic theory ultimately led to quantum electrodynamics (or QED), which fully describes all electromagnetic phenomena as being mediated by wave–particles known as photons. In QED, photons are the fundamental exchange particle, which described all interactions relating to electromagnetism including the electromagnetic force.

It is a common misconception to ascribe the stiffness and rigidity of solid matter to the repulsion of like charges under the influence of the electromagnetic force. However, these characteristics actually result from the Pauli exclusion principle. Since electrons are fermions, they cannot occupy the same quantum mechanical state as other electrons. When the electrons in a material are densely packed together, there are not enough lower energy quantum mechanical states for them all, so some of them must be in higher energy states. This means that it takes energy to pack them together. While this effect is manifested macroscopically as a structural force, it is technically only the result of the existence of a finite set of electron states.

Strong Nuclear

There are two "nuclear forces", which today are usually described as interactions that take place in quantum theories of particle physics. The strong nuclear force is the force responsible for the structural integrity of atomic nuclei while the weak nuclear force is responsible for the decay of certain nucleons into leptons and other types of hadrons.

The strong force is today understood to represent the interactions between quarks and gluons as detailed by the theory of quantum chromodynamics (QCD). The strong force is the fundamental force mediated by gluons, acting upon quarks, antiquarks, and the gluons themselves. The (aptly named) strong interaction is the "strongest" of the four fundamental forces.

The strong force only acts *directly* upon elementary particles. However, a residual of the force is observed between hadrons (the best known example being the force that acts between nucleons in atomic nuclei) as the nuclear force. Here the strong force acts indirectly, transmitted as gluons, which form part of the virtual pi and rho mesons, which classically transmit the nuclear force. The failure of many searches for free quarks has shown that the elementary particles affected are not directly observable. This phenomenon is called color confinement.

Weak Nuclear

The weak force is due to the exchange of the heavy W and Z bosons. Its most familiar effect is beta decay (of neutrons in atomic nuclei) and the associated radioactivity. The word "weak" derives from the fact that the field strength is some 10^{13} times less than that of the strong force. Still, it is stronger than gravity over short distances. A consistent electroweak theory has also been developed, which shows that electromagnetic forces and the weak force are indistinguishable at a temperatures in excess of approximately 10^{15} kelvins. Such temperatures have been probed in modern particle accelerators and show the conditions of the universe in the early moments of the Big Bang.

Non-fundamental Forces

Some forces are consequences of the fundamental ones. In such situations, idealized models can be utilized to gain physical insight.

Normal Force

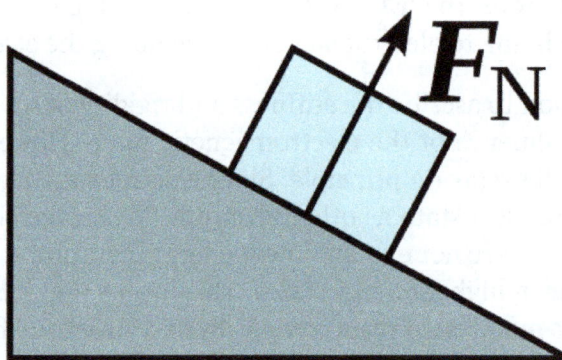

F_N represents the normal force exerted on the object.

The normal force is due to repulsive forces of interaction between atoms at close contact. When their electron clouds overlap, Pauli repulsion (due to fermionic nature of electrons) follows resulting in the force that acts in a direction normal to the surface interface between two objects. The normal force, for example, is responsible for the structural integrity of tables and floors as well as being the force that responds whenever an external force pushes on a solid object. An example of the normal force in action is the impact force on an object crashing into an immobile surface.

Friction

Friction is a surface force that opposes relative motion. The frictional force is directly related to the normal force that acts to keep two solid objects separated at the point of contact. There are two broad classifications of frictional forces: static friction and kinetic friction.

The static friction force (F_{sf}) will exactly oppose forces applied to an object parallel to a surface contact up to the limit specified by the coefficient of static friction (μ_{sf}) multiplied by the normal force (F_N). In other words, the magnitude of the static friction force satisfies the inequality:

$$0 \leq F_{sf} \leq \mu_{sf} F_N.$$

The kinetic friction force (F_{kf}) is independent of both the forces applied and the movement of the object. Thus, the magnitude of the force equals:

$$F_{kf} = \mu_{kf} F_N,$$

where μ_{kf} is the coefficient of kinetic friction. For most surface interfaces, the coefficient of kinetic friction is less than the coefficient of static friction.

Tension

Tension forces can be modeled using ideal strings that are massless, frictionless, unbreakable, and unstretchable. They can be combined with ideal pulleys, which allow ideal strings to switch physical direction. Ideal strings transmit tension forces instantaneously in action-reaction pairs so that if two objects are connected by an ideal string, any force directed along the string by the first object is accompanied by a force directed along the string in the opposite direction by the second object. By connecting the same string multiple times to the same object through the use of a set-up that uses movable pulleys, the tension force on a load can be multiplied. For every string that acts on a load, another factor of the tension force in the string acts on the load. However, even though such machines allow for an increase in force, there is a corresponding increase in the length of string that must be displaced in order to move the load. These tandem effects result ultimately in the conservation of mechanical energy since the work done on the load is the same no matter how complicated the machine.

Elastic Force

An elastic force acts to return a spring to its natural length. An ideal spring is taken to be massless, frictionless, unbreakable, and infinitely stretchable. Such springs exert forces that push when contracted, or pull when extended, in proportion to the displacement of the spring from its equilibrium position. This linear relationship was described by Robert Hooke in 1676, for whom Hooke's law is named. If $\ddot{A}x$ is the displacement, the force exerted by an ideal spring equals:

$$\vec{F} = -k\Delta\vec{x}$$

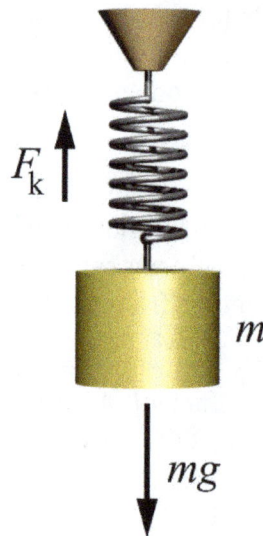

F_k is the force that responds to the load on the spring

where k is the spring constant (or force constant), which is particular to the spring. The minus sign accounts for the tendency of the force to act in opposition to the applied load.

Continuum Mechanics

Newton's laws and Newtonian mechanics in general were first developed to describe how forces affect idealized point particles rather than three-dimensional objects. However, in real life, matter has extended structure and forces that act on one part of an object might affect other parts of an object. For situations where lattice holding together the atoms in an object is able to flow, contract, expand, or otherwise change shape, the theories of continuum mechanics describe the way forces affect the material. For example, in extended fluids, differences in pressure result in forces being directed along the pressure gradients as follows:

$$\frac{\vec{F}}{V} = -\vec{\nabla}P$$

where V is the volume of the object in the fluid and P is the scalar function that describes the pressure at all locations in space. Pressure gradients and differentials result in the buoyant force for fluids suspended in gravitational fields, winds in atmospheric science, and the lift associated with aerodynamics and flight.

A specific instance of such a force that is associated with dynamic pressure is fluid resistance: a body force that resists the motion of an object through a fluid due to viscosity. For so-called "Stokes' drag" the force is approximately proportional to the velocity, but opposite in direction:

$$\vec{F}_d = -b\vec{v}$$

where:

> b is a constant that depends on the properties of the fluid and the dimensions of the object (usually the cross-sectional area), and

$\vec{v}$ is the velocity of the object.

More formally, forces in continuum mechanics are fully described by a stress–tensor with terms that are roughly defined as

$$\sigma = \frac{F}{A}$$

where A is the relevant cross-sectional area for the volume for which the stress-tensor is being calculated. This formalism includes pressure terms associated with forces that act normal to the cross-sectional area (the matrix diagonals of the tensor) as well as shear terms associated with forces that act parallel to the cross-sectional area (the off-diagonal elements). The stress tensor accounts for forces that cause all strains (deformations) including also tensile stresses and compressions.

Fictitious Forces

There are forces that are frame dependent, meaning that they appear due to the adoption of non-Newtonian (that is, non-inertial) reference frames. Such forces include the centrifugal force and the Coriolis force. These forces are considered fictitious because they do not exist in frames of reference that are not accelerating. Because these forces are not genuine they are also referred to as "pseudo forces".

In general relativity, gravity becomes a fictitious force that arises in situations where spacetime deviates from a flat geometry. As an extension, Kaluza–Klein theory and string theory ascribe electromagnetism and the other fundamental forces respectively to the curvature of differently scaled dimensions, which would ultimately imply that all forces are fictitious.

Rotations and Torque

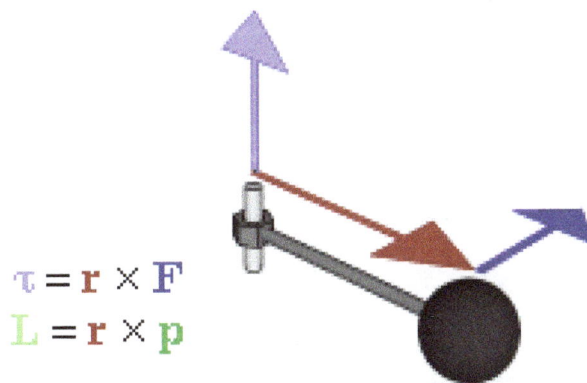

$$\tau = r \times F$$
$$L = r \times p$$

Relationship between force (F), torque (τ), and momentum vectors (p and L) in a rotating system.

Forces that cause extended objects to rotate are associated with torques. Mathematically, the torque of a force $\vec{F}$ is defined relative to an arbitrary reference point as the cross-product:

$$\vec{\tau} = \vec{r} \times \vec{F}$$

where

$\vec{r}$ is the position vector of the force application point relative to the reference point.

Torque is the rotation equivalent of force in the same way that angle is the rotational equivalent for position, angular velocity for velocity, and angular momentum for momentum. As a consequence of Newton's First Law of Motion, there exists rotational inertia that ensures that all bodies maintain their angular momentum unless acted upon by an unbalanced torque. Likewise, Newton's Second Law of Motion can be used to derive an analogous equation for the instantaneous angular acceleration of the rigid body:

$$\vec{\tau} = I\vec{\alpha}$$

where

I is the moment of inertia of the body

$\vec{\alpha}$ is the angular acceleration of the body.

This provides a definition for the moment of inertia, which is the rotational equivalent for mass. In more advanced treatments of mechanics, where the rotation over a time interval is described, the moment of inertia must be substituted by the tensor that, when properly analyzed, fully determines the characteristics of rotations including precession and nutation.

Equivalently, the differential form of Newton's Second Law provides an alternative definition of torque:

$$\vec{\tau} = \frac{d\vec{L}}{dt}, \text{ where } \vec{L} \text{ is the angular momentum of the particle.}$$

Newton's Third Law of Motion requires that all objects exerting torques themselves experience equal and opposite torques, and therefore also directly implies the conservation of angular momentum for closed systems that experience rotations and revolutions through the action of internal torques.

Centripetal Force

For an object accelerating in circular motion, the unbalanced force acting on the object equals:

$$\vec{F} = -\frac{mv^2\hat{r}}{r}$$

where m is the mass of the object, v is the velocity of the object and r is the distance to the center of the circular path and $\hat{r}$ is the unit vector pointing in the radial direction outwards from the center. This means that the unbalanced centripetal force felt by any object is always directed toward the center of the curving path. Such forces act perpendicular to the velocity vector associated with the motion of an object, and therefore do not change the speed of the object (magnitude of the velocity), but only the direction of the velocity vector. The unbalanced force that accelerates an object can be resolved into a component that is perpendicular to the path, and one that is tangential to the path. This yields both the tangential force, which accelerates the object by either slowing it down or speeding it up, and the radial (centripetal) force, which changes its direction.

Kinematic Integrals

Forces can be used to define a number of physical concepts by integrating with respect to kinematic variables. For example, integrating with respect to time gives the definition of impulse:

$$\vec{I} = \int_{t_1}^{t_2} \vec{F} dt,$$

which by Newton's Second Law must be equivalent to the change in momentum (yielding the Impulse momentum theorem).

Similarly, integrating with respect to position gives a definition for the work done by a force:

$$W = \int_{\vec{x}_1}^{\vec{x}_2} \vec{F} \cdot d\vec{x},$$

which is equivalent to changes in kinetic energy (yielding the work energy theorem).

Power P is the rate of change dW/dt of the work W, as the trajectory is extended by a position change $d\vec{x}$ in a time interval dt:

$$dW = \frac{dW}{d\vec{x}} \cdot d\vec{x} = \vec{F} \cdot d\vec{x}, \qquad \text{so} \qquad P = \frac{dW}{dt} = \frac{dW}{d\vec{x}} \cdot \frac{d\vec{x}}{dt} = \vec{F} \cdot \vec{v},$$

with $\vec{v} = d\vec{x}/dt$ the velocity.

Potential Energy

Instead of a force, often the mathematically related concept of a potential energy field can be used for convenience. For instance, the gravitational force acting upon an object can be seen as the action of the gravitational field that is present at the object's location. Restating mathematically the definition of energy (via the definition of work), a potential scalar field $U(\vec{r})$ is defined as that field whose gradient is equal and opposite to the force produced at every point:

$$\vec{F} = -\vec{\nabla} U.$$

Forces can be classified as conservative or nonconservative. Conservative forces are equivalent to the gradient of a potential while nonconservative forces are not.

Conservative Forces

A conservative force that acts on a closed system has an associated mechanical work that allows energy to convert only between kinetic or potential forms. This means that for a closed system, the net mechanical energy is conserved whenever a conservative force acts on the system. The force, therefore, is related directly to the difference in potential energy between two different locations in space, and can be considered to be an artifact of the potential field in the same way that the direction and amount of a flow of water can be considered to be an artifact of the contour map of the elevation of an area.

Conservative forces include gravity, the electromagnetic force, and the spring force. Each of these forces has models that are dependent on a position often given as a radial vector emanating from spherically symmetric potentials. Examples of this follow:

For gravity:

$$\vec{F} = -\frac{Gm_1m_2\vec{r}}{r^3}$$

where G is the gravitational constant, and m_n is the mass of object n.

For electrostatic forces:

$$\vec{F} = \frac{q_1q_2\vec{r}}{4\pi\epsilon_0 r^3}$$

where ϵ_0 is electric permittivity of free space, and q_n is the electric charge of object n.

For spring forces:

$$\vec{F} = -k\vec{r}$$

where k is the spring constant.

Nonconservative Forces

For certain physical scenarios, it is impossible to model forces as being due to gradient of potentials. This is often due to macrophysical considerations that yield forces as arising from a macroscopic statistical average of microstates. For example, friction is caused by the gradients of numerous electrostatic potentials between the atoms, but manifests as a force model that is independent of any macroscale position vector. Nonconservative forces other than friction include other contact forces, tension, compression, and drag. However, for any sufficiently detailed description, all these forces are the results of conservative ones since each of these macroscopic forces are the net results of the gradients of microscopic potentials.

The connection between macroscopic nonconservative forces and microscopic conservative forces is described by detailed treatment with statistical mechanics. In macroscopic closed systems, nonconservative forces act to change the internal energies of the system, and are often associated with the transfer of heat. According to the Second law of thermodynamics, nonconservative forces necessarily result in energy transformations within closed systems from ordered to more random conditions as entropy increases.

Units of Measurement

The SI unit of force is the newton (symbol N), which is the force required to accelerate a one kilogram mass at a rate of one meter per second squared, or kg·m·s⁻². The corresponding CGS unit is the dyne, the force required to accelerate a one gram mass by one centimeter per second squared, or g·cm·s⁻². A newton is thus equal to 100,000 dynes.

The gravitational foot-pound-second English unit of force is the pound-force (lbf), defined as the force exerted by gravity on a pound-mass in the standard gravitational field of 9.80665 m·s⁻². The pound-force provides an alternative unit of mass: one slug is the mass that will accelerate by one foot per second squared when acted on by one pound-force.

An alternative unit of force in a different foot-pound-second system, the absolute fps system, is the poundal, defined as the force required to accelerate a one-pound mass at a rate of one foot per second squared. The units of slug and poundal are designed to avoid a constant of proportionality in Newton's Second Law.

The pound-force has a metric counterpart, less commonly used than the newton: the kilogram-force (kgf) (sometimes kilopond), is the force exerted by standard gravity on one kilogram of mass. The kilogram-force leads to an alternate, but rarely used unit of mass: the metric slug (sometimes mug or hyl) is that mass that accelerates at 1 m·s⁻² when subjected to a force of 1 kgf. The kilogram-force is not a part of the modern SI system, and is generally deprecated; however it still sees use for some purposes as expressing aircraft weight, jet thrust, bicycle spoke tension, torque wrench settings and engine output torque. Other arcane units of force include the sthène, which is equivalent to 1000 N, and the kip, which is equivalent to 1000 lbf.

Units of force					
• v t e	newton (SI unit)	dyne	kilogram-force, kilopond	pound-force	poundal
1 N	$\equiv 1\,\text{kg·m/s}^2$	$= 10^5\,\text{dyn}$	$\approx 0.10197\,\text{kp}$	$\approx 0.22481\,\text{lb}_F$	$\approx 7.2330\,\text{pdl}$
1 dyn	$= 10^{-5}\,\text{N}$	$\equiv 1\,\text{g·cm/s}^2$	$\approx 1.0197 \times 10^{-6}\,\text{kp}$	$\approx 2.2481 \times 10^{-6}\,\text{lb}_F$	$\approx 7.2330 \times 10^{-5}\,\text{pdl}$
1 kp	$= 9.80665\,\text{N}$	$= 980665\,\text{dyn}$	$\equiv g_n \cdot (1\,\text{kg})$	$\approx 2.2046\,\text{lb}_F$	$\approx 70.932\,\text{pdl}$
1 lb$_F$	$\approx 4.448222\,\text{N}$	$\approx 444822\,\text{dyn}$	$\approx 0.45359\,\text{kp}$	$\equiv g_n \cdot (1\,\text{lb})$	$\approx 32.174\,\text{pdl}$
1 pdl	$\approx 0.138255\,\text{N}$	$\approx 13825\,\text{dyn}$	$\approx 0.014098\,\text{kp}$	$\approx 0.031081\,\text{lb}_F$	$\equiv 1\,\text{lb·ft/s}^2$
The value of g_n as used in the official definition of the kilogram-force is used here for all gravitational units.					

Fundamental Interaction

Fundamental interactions, also known as fundamental forces, are the interactions in physical systems that do not appear to be reducible to more basic interactions. There are four conventionally accepted fundamental interactions—gravitational, electromagnetic, strong nuclear, and weak nuclear. Each one is understood as the dynamics of a *field*. The gravitational force is modelled as a continuous classical field. The other three are each modelled as discrete quantum fields, and exhibit a measurable unit or *elementary particle*.

The two nuclear interactions produce strong forces at minuscule, subatomic distances. The strong nuclear interaction is responsible for the binding of atomic nuclei. The weak nuclear interaction also acts on the nucleus, mediating radioactive decay. Electromagnetism and gravity produce significant forces at macroscopic scales where the effects can be seen directly in everyday life. Electrical and magnetic fields tend to cancel each other out when large collections of objects are considered, so over the largest distances (on the scale of planets and galaxies), gravity tends to be the dominant force.

Theoretical physicists working beyond the Standard Model seek to quantize the gravitational field toward predictions that particle physicists can experimentally confirm, thus yielding acceptance to a theory of quantum gravity (QG). (Phenomena suitable to model as a fifth force—perhaps an added gravitational effect—remain widely disputed.) Other theorists seek to unite the electroweak and strong fields within a Grand Unified Theory (GUT). While all four fundamental interactions are widely thought to align on a highly minuscule scale, particle accelerators cannot produce the massive energy levels required to experimentally probe at that Planck scale (which would experimentally confirm such theories.) Yet some theories, such as the string theory, seek both QG and GUT within one framework, unifying all four fundamental interactions along with mass generation within a theory of everything (ToE).

History

Classical Theory

In his 1687 theory, Isaac Newton postulated space as an infinite and unalterable physical structure existing before, within, and around all objects while their states and relations unfold at a constant pace everywhere, thus absolute space and time. Inferring that all objects bearing mass approach at a constant rate, but collide by impact proportional to their masses, Newton inferred that matter exhibits an attractive force. His law of universal gravitation mathematically stated it to span the entire universe instantly (despite absolute time), or, if not actually a force, to be instant interaction among all objects (despite absolute space.) As conventionally interpreted, Newton's theory of motion modelled a *central force* without a communicating medium. Thus Newton's theory violated the first principle of mechanical philosophy, as stated by Descartes, *No action at a distance*. Conversely, during the 1820s, when explaining magnetism, Michael Faraday inferred a *field* filling space and transmitting that force. Faraday conjectured that ultimately, all forces unified into one.

In the early 1870s, James Clerk Maxwell unified electricity and magnetism as effects of an electromagnetic field whose third consequence was light, travelling at constant speed in a vacuum. The electromagnetic field theory contradicted predictions of Newton's theory of motion, unless physical states of the luminiferous aether—presumed to fill all space whether within matter or in a vacuum and to manifest the electromagnetic field—aligned all phenomena and thereby held valid the Newtonian principle relativity or invariance.

The Standard Model

The Standard Model of particle physics was developed throughout the latter half of the 20th century. In the Standard Model, the electromagnetic, strong, and weak interactions associate with elementary particles, whose behaviours are modelled in quantum mechanics (QM). For predictive success with QM's probabilistic outcomes, particle physics conventionally models QM events across a field set to special relativity, altogether relativistic quantum field theory (QFT). Force particles, called gauge bosons—*force carriers* or *messenger particles* of underlying fields—interact with matter particles, called fermions. Everyday matter is atoms, composed of three fermion types: up-quarks and down-quarks constituting, as well as electrons orbiting, the atom's nucleus. Atoms interact, form molecules, and manifest further properties through electromagnetic interactions among their electrons absorbing and emitting photons, the electromagnetic field's force carrier, which if unimpeded traverse potentially infinite distance. Electromagnetism's QFT is quantum electrodynamics (QED).

The electromagnetic interaction was modelled with the weak interaction, whose force carriers are W and Z bosons, traversing the minuscule distance, in electroweak theory (EWT). Electroweak interaction would operate at such high temperatures as soon after the presumed Big Bang, but, as the early universe cooled, split into electromagnetic and weak interactions. The strong interaction, whose force carrier is the gluon, traversing minuscule distance among quarks, is modeled in quantum chromodynamics (QCD). EWT, QCD, and the Higgs mechanism, whereby the Higgs field manifests Higgs bosons that interact with some quantum particles and thereby endow those particles with mass comprise particle physics' Standard Model (SM). Predictions are usually made using calculational approximation methods, although such perturbation theory is inadequate to model some experimental observations (for instance bound states and solitons.) Still, physicists widely accept the Standard Model as science's most experimentally confirmed theory.

Beyond the Standard Model, some theorists work to unite the electroweak and strong interactions within a Grand Unified Theory (GUT). Some attempts at GUTs hypothesize "shadow" particles, such that every known matter particle associates with an undiscovered force particle, and vice versa, altogether supersymmetry (SUSY). Other theorists seek to quantize the gravitational field by the modelling behaviour of its hypothetical force carrier, the graviton and achieve quantum gravity (QG). One approach to QG is loop quantum gravity (LQG). Still other theorists seek both QG and GUT within one framework, reducing all four fundamental interactions to a Theory of Everything (ToE). The most prevalent aim at a ToE is string theory, although to model matter particles, it added SUSY to force particles—and so, strictly speaking, became superstring theory. Multiple, seemingly disparate superstring theories were unified on a backbone, M-theory. Theories beyond the Standard Model remain highly speculative, lacking great experimental support.

Overview of the fundamental interactions

The four fundamental interactions of nature					
Property/ Interaction	**Gravitation**	**Weak**	**Electromagnetic**	**Strong**	
		(Electroweak)		**Fundamental**	**Residual**
Acts on:	Mass - Energy	Flavor	Electric charge	Color charge	Atomic nuclei
Particles experiencing:	All	Quarks, leptons	Electrically charged	Quarks, Gluons	Hadrons
Particles mediating:	Not yet observed (Graviton hypothesised)	$W^+ W^- Z^o$	γ (photon)	Gluons	Mesons
Strength at the scale of quarks:	10^{-41}	10^{-4}	1	60	Not applicable to quarks
Strength at the scale of protons/neutrons:	10^{-36}	10^{-7}	1	Not applicable to hadrons	20

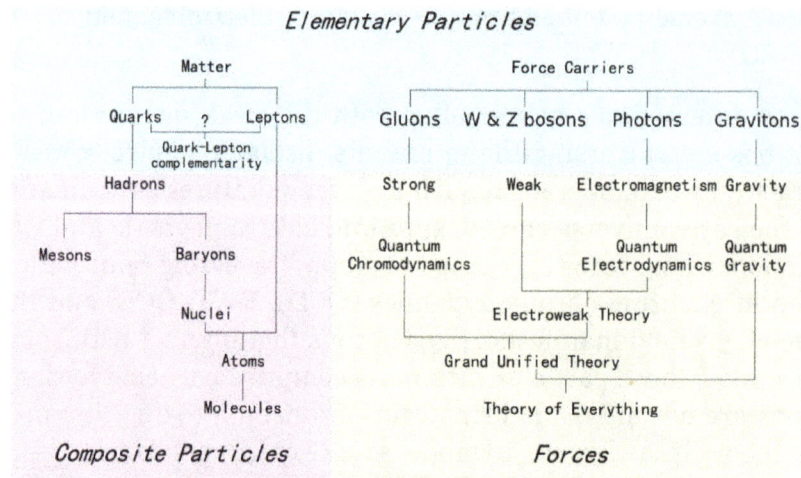

An overview of the various families of elementary and composite particles,
and the theories describing their interactions. Fermions are on the left, and Bosons are on the right.

In the conceptual model of fundamental interactions, matter consists of fermions, which carry properties called charges and spin $\pm\frac{1}{2}$ (intrinsic angular momentum $\pm\frac{\hbar}{2}$, where $\hbar$ is the reduced Planck constant). They attract or repel each other by exchanging bosons.

The interaction of any pair of fermions in perturbation theory can then be modelled thus:

Two fermions go in → *interaction* by boson exchange → Two changed fermions go out.

The exchange of bosons always carries energy and momentum between the fermions, thereby changing their speed and direction. The exchange may also transport a charge between the fermions, changing the charges of the fermions in the process (e.g., turn them from one type of fermion to another). Since bosons carry one unit of angular momentum, the fermion's spin direction will flip from $+\frac{1}{2}$ to $-\frac{1}{2}$ (or vice versa) during such an exchange (in units of the reduced Planck's constant).

Because an interaction results in fermions attracting and repelling each other, an older term for "interaction" is force.

According to the present understanding, there are four fundamental interactions or forces: gravitation, electromagnetism, the weak interaction, and the strong interaction. Their magnitude and behaviour vary greatly, as described in the table below. Modern physics attempts to explain every observed physical phenomenon by these fundamental interactions. Moreover, reducing the number of different interaction types is seen as desirable. Two cases in point are the unification of:

• Electric and magnetic force into electromagnetism;

• The electromagnetic interaction and the weak interaction into the electroweak interaction.

Both magnitude ("relative strength") and "range", as given in the table, are meaningful only within a rather complex theoretical framework. It should also be noted that the table below lists properties of a conceptual scheme that is still the subject of ongoing research.

Interaction	Current theory	Mediators	Relative strength	Long-distance behavior	Range (m)
Strong	Quantum chromodynamics (QCD)	gluons	10^{38}		10^{-15}
Electromagnetic	Quantum electrodynamics (QED)	photons	10^{36}		∞
Weak	Electroweak Theory (EWT)	W and Z bosons	10^{25}		10^{-18}
Gravitation	General Relativity (GR)	gravitons (hypothetical)	1		∞

The modern (perturbative) quantum mechanical view of the fundamental forces other than gravity is that particles of matter (fermions) do not directly interact with each other, but rather carry a charge, and exchange virtual particles (gauge bosons), which are the interaction carriers or force mediators. For example, photons mediate the interaction of electric charges, and gluons mediate the interaction of color charges.

The Interactions

Gravity

Gravitation is by far the weakest of the four interactions. The weakness of gravity can easily be demonstrated by suspending a pin using a simple magnet (such as a refrigerator magnet). The magnet is able to hold the pin against the gravitational pull of the entire Earth.

Yet gravitation is very important for macroscopic objects and over macroscopic distances for the following reasons. Gravitation:

- Is the only interaction that acts on all particles having mass, energy and/or momentum

- Has an infinite range, like electromagnetism but unlike strong and weak interaction

- Cannot be absorbed, transformed, or shielded against

- Always attracts and never repels

Even though electromagnetism is far stronger than gravitation, electrostatic attraction is not relevant for large celestial bodies, such as planets, stars, and galaxies, simply because such bodies contain equal numbers of protons and electrons and so have a net electric charge of zero. Nothing "cancels" gravity, since it is only attractive, unlike electric forces which can be attractive or repulsive. On the other hand, all objects having mass are subject to the gravitational force, which only attracts. Therefore, only gravitation matters on the large-scale structure of the universe.

The long range of gravitation makes it responsible for such large-scale phenomena as the structure of galaxies and black holes and it retards the expansion of the universe. Gravitation also explains astronomical phenomena on more modest scales, such as planetary orbits, as well as everyday experience: objects fall; heavy objects act as if they were glued to the ground, and animals can only jump so high.

Gravitation was the first interaction to be described mathematically. In ancient times, Aristotle hypothesized that objects of different masses fall at different rates. During the Scientific Revolution, Galileo Galilei experimentally determined that this was not the case — neglecting the friction due to air resistance, and buoyancy forces if an atmosphere is present (e.g. the case of a dropped air-filled balloon vs a water-filled balloon) all objects accelerate toward the Earth at the same rate. Isaac Newton's law of Universal Gravitation (1687) was a good approximation of the behaviour of gravitation. Our present-day understanding of gravitation stems from Albert Einstein's General Theory of Relativity of 1915, a more accurate (especially for cosmological masses and distances) description of gravitation in terms of the geometry of spacetime.

Merging general relativity and quantum mechanics (or quantum field theory) into a more general theory of quantum gravity is an area of active research. It is hypothesized that gravitation is mediated by a massless spin-2 particle called the graviton.

Although general relativity has been experimentally confirmed (at least for weak fields) on all but the smallest scales, there are rival theories of gravitation. Those taken seriously by the physics community all reduce to general relativity in some limit, and the focus of observational work is to establish limitations on what deviations from general relativity are possible.

Proposed extra dimensions could explain why the gravity force is so weak.

Electroweak Interaction

Electromagnetism and weak interaction appear to be very different at everyday low energies. They can be modeled using two different theories. However, above unification energy, on the order of 100 GeV, they would merge into a single electroweak force.

Electroweak theory is very important for modern cosmology, particularly on how the universe evolved. This is because shortly after the Big Bang, the temperature was approximately above 10^{15} K. Electromagnetic force and weak force were merged into a combined electroweak force.

For contributions to the unification of the weak and electromagnetic interaction between elementary particles, Abdus Salam, Sheldon Glashow and Steven Weinberg were awarded the Nobel Prize in Physics in 1979.

Electromagnetism

Electromagnetism is the force that acts between electrically charged particles. This phenomenon includes the electrostatic force acting between charged particles at rest, and the combined effect of electric and magnetic forces acting between charged particles moving relative to each other.

Electromagnetism is infinite-ranged like gravity, but vastly stronger, and therefore describes a number of macroscopic phenomena of everyday experience such as friction, rainbows, lightning, and all human-made devices using electric current, such as television, lasers, and computers. Electromagnetism fundamentally determines all macroscopic, and many atomic levels, properties of the chemical elements, including all chemical bonding.

In a four kilogram (~1 gallon) jug of water there are

$$4000 \, g H_2 O \cdot \frac{1 \, mol H_2 O}{18 \, g H_2 O} \cdot \frac{10 \, mol e^-}{1 \, mol H_2 O} \cdot \frac{96,000 \, C}{1 \, mol e^-} = 2.1 \times 10^8 C$$

of total electron charge. Thus, if we place two such jugs a meter apart, the electrons in one of the jugs repel those in the other jug with a force of

$$\frac{1}{4\pi\varepsilon_0} \frac{(2.1 \times 10^8 C)^2}{(1m)^2} = 4.1 \times 10^{26} \, N.$$

This is larger than the planet Earth would weigh if weighed on another Earth. The atomic nuclei in one jug also repel those in the other with the same force. However, these repulsive forces are canceled by the attraction of the electrons in jug A with the nuclei in jug B and the attraction of the nuclei in jug A with the electrons in jug B, resulting in no net force. Electromagnetic forces are tremendously stronger than gravity but cancel out so that for large bodies gravity dominates.

Electrical and magnetic phenomena have been observed since ancient times, but it was only in the 19th century that it was discovered that electricity and magnetism are two aspects of the same fundamental interaction. By 1864, Maxwell's equations had rigorously quantified this unified interaction. Maxwell's theory, restated using vector calculus, is the classical theory of electromagnetism, suitable for most technological purposes.

The constant speed of light in a vacuum (customarily described with the letter "c") can be derived from Maxwell's equations, which are consistent with the theory of special relativity. Einstein's 1905 theory of special relativity, however, which flows from the observation that the speed of light is constant no matter how fast the observer is moving, showed that the theoretical result implied by Maxwell's equations has profound implications far beyond electromagnetism on the very nature of time and space.

In another work that departed from classical electro-magnetism, Einstein also explained the photoelectric effect by hypothesizing that light was transmitted in quanta, which we now call photons. Starting around 1927, Paul Dirac combined quantum mechanics with the relativistic theory of electromagnetism. Further work in the 1940s, by Richard Feynman, Freeman Dyson, Julian Schwinger, and Sin-Itiro Tomonaga, completed this theory, which is now called quantum electrodynamics, the revised theory of electromagnetism. Quantum electrodynamics and quantum mechanics provide a theoretical basis for electromagnetic behavior such as quantum tunneling, in which a certain percentage of electrically charged particles move in ways that would be impossible under the classical electromagnetic theory, that is necessary for everyday electronic devices such as transistors to function.

Weak Interaction

The *weak interaction* or *weak nuclear force* is responsible for some nuclear phenomena such as beta decay. Electromagnetism and the weak force are now understood to be two aspects of a unified electroweak interaction — this discovery was the first step toward the unified theory known as the Standard Model. In the theory of the electroweak interaction, the carriers of the weak force are the massive gauge bosons called the W and Z bosons. The weak interaction is the only known interaction which does not conserve parity; it is left-right asymmetric. The weak interaction even violates CP symmetry but does conserve CPT.

Strong Interaction

The *strong interaction*, or *strong nuclear force*, is the most complicated interaction, mainly because of the way it varies with distance. At distances greater than 10 femtometers, the strong force is practically unobservable. Moreover, it holds only inside the atomic nucleus.

After the nucleus was discovered in 1908, it was clear that a new force was needed to overcome the electrostatic repulsion, a manifestation of electromagnetism, of the positively charged protons. Otherwise, the nucleus could not exist. Moreover, the force had to be strong enough to squeeze the protons into a volume that is 10^{-15} of that of the entire atom. From the short range of this force, Hideki Yukawa predicted that it was associated with a massive particle, whose mass is approximately 100 MeV.

The 1947 discovery of the pion ushered in the modern era of particle physics. Hundreds of hadrons were discovered from the 1940s to 1960s, and an extremely complicated theory of hadrons as strongly interacting particles was developed. Most notably:

- The pions were understood to be oscillations of vacuum condensates;

- Jun John Sakurai proposed the rho and omega vector bosons to be force carrying particles for approximate symmetries of isospin and hypercharge;

- Geoffrey Chew, Edward K. Burdett and Steven Frautschi grouped the heavier hadrons into families that could be understood as vibrational and rotational excitations of strings.

While each of these approaches offered deep insights, no approach led directly to a fundamental theory.

Murray Gell-Mann along with George Zweig first proposed fractionally charged quarks in 1961. Throughout the 1960s, different authors considered theories similar to the modern fundamental theory of quantum chromodynamics (QCD) as simple models for the interactions of quarks. The first to hypothesize the gluons of QCD were Moo-Young Han and Yoichiro Nambu, who introduced the quark color charge and hypothesized that it might be associated with a force-carrying field. At that time, however, it was difficult to see how such a model could permanently confine quarks. Han and Nambu also assigned each quark color an integer electrical charge, so that the quarks were fractionally charged only on average, and they did not expect the quarks in their model to be permanently confined.

In 1971, Murray Gell-Mann and Harald Fritzsch proposed that the Han/Nambu color gauge field was the correct theory of the short-distance interactions of fractionally charged quarks. A little later, David Gross, Frank Wilczek, and David Politzer discovered that this theory had the property of asymptotic freedom, allowing them to make contact with experimental evidence. They concluded that QCD was the complete theory of the strong interactions, correct at all distance scales. The discovery of asymptotic freedom led most physicists to accept QCD since it became clear that even the long-distance properties of the strong interactions could be consistent with experiment if the quarks are permanently confined.

Assuming that quarks are confined, Mikhail Shifman, Arkady Vainshtein, and Valentine Zakharov were able to compute the properties of many low-lying hadrons directly from QCD, with only a few extra parameters to describe the vacuum. In 1980, Kenneth G. Wilson published computer calculations based on the first principles of QCD, establishing, to a level of confidence tantamount to certainty, that QCD will confine quarks. Since then, QCD has been the established theory of the strong interactions.

QCD is a theory of fractionally charged quarks interacting by means of 8 photon-like particles called gluons. The gluons interact with each other, not just with the quarks, and at long distances the lines of force collimate into strings. In this way, the mathematical theory of QCD not only explains how quarks interact over short distances but also the string-like behavior, discovered by Chew and Frautschi, which they manifest over longer distances.

Beyond the Standard Model

Numerous theoretical efforts have been made to systematize the existing four fundamental interactions on the model of electroweak unification.

Grand Unified Theories (GUTs) are proposals to show that all of the fundamental interactions, other than gravity, arise from a single interaction with symmetries that break down at low energy levels. GUTs predict relationships among constants of nature that are unrelated in the SM. GUTs also predict gauge coupling unification for the relative strengths of the electromagnetic, weak, and strong forces, a prediction verified at the Large Electron–Positron Collider in 1991 for supersymmetric theories.

Theories of everything, which integrate GUTs with a quantum gravity theory face a greater barrier, because no quantum gravity theories, which include string theory, loop quantum gravity, and twistor theory, have secured wide acceptance. Some theories look for a graviton to complete the Standard Model list of force-carrying particles, while others, like loop quantum gravity, emphasize the possibility that time-space itself may have a quantum aspect to it.

Some theories beyond the Standard Model include a hypothetical fifth force, and the search for such a force is an ongoing line of experimental research in physics. In supersymmetric theories, there are particles that acquire their masses only through supersymmetry breaking effects and these particles, known as moduli can mediate new forces. Another reason to look for new forces is the recent discovery that the expansion of the universe is accelerating (also known as dark energy), giving rise to a need to explain a nonzero cosmological constant, and possibly to other modifications of general relativity. Fifth forces have also been suggested to explain phenomena such as CP violations, dark matter, and dark flow.

In December 2015, two observations in the ATLAS and CMS detectors at the Large Hadron Collider hinted at the existence of a new particle six times heavier than the Higgs Boson. However, after obtaining more experimental data, the anomaly appeared not be significant.

Friction

Friction is the force resisting the relative motion of solid surfaces, fluid layers, and material elements sliding against each other. There are several types of friction:

- Dry friction resists relative lateral motion of two solid surfaces in contact. Dry friction is subdivided into *static friction* ("stiction") between non-moving surfaces, and *kinetic friction* between moving surfaces.

- Fluid friction describes the friction between layers of a viscous fluid that are moving relative to each other.

- Lubricated friction is a case of fluid friction where a lubricant fluid separates two solid surfaces.

- Skin friction is a component of drag, the force resisting the motion of a fluid across the surface of a body.

- Internal friction is the force resisting motion between the elements making up a solid material while it undergoes deformation.

When surfaces in contact move relative to each other, the friction between the two surfaces converts kinetic energy into thermal energy (that is, it converts work to heat). This property can have dramatic consequences, as illustrated by the use of friction created by rubbing pieces of wood together to start a fire. Kinetic energy is converted to thermal energy whenever motion with friction occurs, for example when a viscous fluid is stirred. Another important consequence of many types of friction can be wear, which may lead to performance degradation and/or damage to components. Friction is a component of the science of tribology.

Friction is not itself a fundamental force. Dry friction arises from a combination of inter-surface adhesion, surface roughness, surface deformation, and surface contamination. The complexity of these interactions makes the calculation of friction from first principles impractical and necessitates the use of empirical methods for analysis and the development of theory.

Friction is a non-conservative force - work done against friction is path dependent. In the presence of friction, some energy is always lost in the form of heat. Thus mechanical energy is not conserved.

History

The Greeks, including Aristotle, Vitruvius, and Pliny the Elder, were interested in the cause and mitigation of friction. They were aware of differences between static and kinetic friction with Themistius stating in 350 A.D. that "it is easier to further the motion of a moving body than to move a body at rest".

The classic laws of sliding friction were discovered by Leonardo da Vinci in 1493, a pioneer in tribology, but the laws documented in his notebooks, were not published and remained unknown. These laws were rediscovered by Guillaume Amontons in 1699. Amontons presented the nature of friction in terms of surface irregularities and the force required to raise the weight pressing the surfaces together. This view was further elaborated by Bernard Forest de Bélidor and Leonhard Euler (1750), who derived the angle of repose of a weight on an inclined plane and first distinguished between static and kinetic friction. A different explanation was provided by John Theophilus Desaguliers (1725), who demonstrated the strong cohesion forces between lead spheres of which a small cap is cut off and which were then brought into contact with each other.

The understanding of friction was further developed by Charles-Augustin de Coulomb (1785). Coulomb investigated the influence of four main factors on friction: the nature of the materials in contact and their surface coatings; the extent of the surface area; the normal pressure (or load); and the length of time that the surfaces remained in contact (time of repose). Coulomb further considered the influence of sliding velocity, temperature and humidity, in order to decide between the different explanations on the nature of friction that had been proposed. The distinction between static and dynamic friction is made in Coulomb's friction law, although this distinction was already drawn by Johann Andreas von Segner in 1758. The effect of the time of repose was explained by Pieter van Musschenbroek (1762) by considering the surfaces of fibrous materials, with fibers meshing together, which takes a finite time in which the friction increases.

John Leslie (1766–1832) noted a weakness in the views of Amontons and Coulomb: If friction arises from a weight being drawn up the inclined plane of successive asperities, why then isn't it balanced through descending the opposite slope? Leslie was equally skeptical about the role of adhesion proposed by Desaguliers, which should on the whole have the same tendency to accelerate as to retard the motion. In Leslie's view, friction should be seen as a time-dependent process of flattening, pressing down asperities, which creates new obstacles in what were cavities before.

Arthur Jules Morin (1833) developed the concept of sliding versus rolling friction. Osborne Reynolds (1866) derived the equation of viscous flow. This completed the classic empirical model of friction (static, kinetic, and fluid) commonly used today in engineering. In 1877, Fleeming Jenkin and J. A. Ewing investigated the continuity between static and kinetic friction.

The focus of research during the 20th century has been to understand the physical mechanisms behind friction. Frank Philip Bowden and David Tabor (1950) showed that, at a microscopic level, the actual area of contact between surfaces is a very small fraction of the apparent area. This actual area of contact, caused by "asperities" (roughness) increases with pressure. The development of the atomic force microscope (ca. 1986) enabled scientists to study friction at the atomic scale, showing that, on that scale, dry friction is the product of the inter-surface shear stress and the contact area. These two discoveries explain the macroscopic proportionality between normal force and static frictional force between dry surfaces.

Laws of Dry Friction

The elementary property of sliding (kinetic) friction were discovered by experiment in the 15th to 18th centuries and were expressed as three empirical laws:

- Amontons' First Law: The force of friction is directly proportional to the applied load.

- Amontons' Second Law: The force of friction is independent of the apparent area of contact.

- Coulomb's Law of Friction: Kinetic friction is independent of the sliding velocity.

Dry Friction

Dry friction resists relative lateral motion of two solid surfaces in contact. The two regimes of dry friction are 'static friction' ("stiction") between non-moving surfaces, and *kinetic friction* (sometimes called sliding friction or dynamic friction) between moving surfaces.

Coulomb friction, named after Charles-Augustin de Coulomb, is an approximate model used to calculate the force of dry friction. It is governed by the model:

$$F_{\text{f}} \leq \mu F_{\text{n}},$$

where

- F_{f} is the force of friction exerted by each surface on the other. It is parallel to the surface, in a direction opposite to the net applied force.

- μ is the coefficient of friction, which is an empirical property of the contacting materials,

- F_{n} is the normal force exerted by each surface on the other, directed perpendicular (normal) to the surface.

The Coulomb friction F_{f} may take any value from zero up to μF_{n}, and the direction of the frictional force against a surface is opposite to the motion that surface would experience in the absence of friction. Thus, in the static case, the frictional force is exactly what it must be in order to prevent motion between the surfaces; it balances the net force tending to cause such motion. In this case, rather than providing an estimate of the actual frictional force, the Coulomb approximation provides a threshold value for this force, above which motion would commence. This maximum force is known as traction.

The force of friction is always exerted in a direction that opposes movement (for kinetic friction) or potential movement (for static friction) between the two surfaces. For example, a curling stone sliding along the ice experiences a kinetic force slowing it down. For an example of potential movement, the drive wheels of an accelerating car experience a frictional force pointing forward; if they did not, the wheels would spin, and the rubber would slide backwards along the pavement. Note that it is not the direction of movement of the vehicle they oppose, it is the direction of (potential) sliding between tire and road.

Normal Force

A block on a ramp

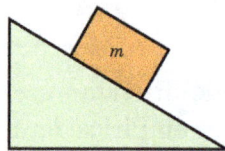

Free body diagram
of just the block

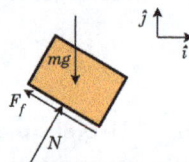

Free-body diagram for a block on a ramp. Arrows are vectors indicating directions and magnitudes of forces. N is the normal force, mg is the force of gravity, and F_f is the force of friction.

The normal force is defined as the net force compressing two parallel surfaces together; and its direction is perpendicular to the surfaces. In the simple case of a mass resting on a horizontal surface, the only component of the normal force is the force due to gravity, where $N = mg$. In this case, the magnitude of the friction force is the product of the mass of the object, the acceleration due to gravity, and the coefficient of friction. However, the coefficient of friction is not a function of mass or volume; it depends only on the material. For instance, a large aluminum block has the same coefficient of friction as a small aluminum block. However, the magnitude of the friction force itself depends on the normal force, and hence on the mass of the block.

If an object is on a level surface and the force tending to cause it to slide is horizontal, the normal force N between the object and the surface is just its weight, which is equal to its mass multiplied by the acceleration due to earth's gravity, g. If the object is on a tilted surface such as an inclined plane, the normal force is less, because less of the force of gravity is perpendicular to the face of the plane. Therefore, the normal force, and ultimately the frictional force, is determined using vector analysis, usually via a free body diagram. Depending on the situation, the calculation of the normal force may include forces other than gravity.

Coefficient of Friction

The coefficient of friction (COF), often symbolized by the Greek letter μ, is a dimensionless scalar value which describes the ratio of the force of friction between two bodies and the force pressing them together. The coefficient of friction depends on the materials used; for example, ice on steel has a low coefficient of friction, while rubber on pavement has a high coefficient of friction. Coefficients of friction range from near zero to greater than one.

For surfaces at rest relative to each other $\mu = \mu_s$, where μ_s is the *coefficient of static friction*. This is usually larger than its kinetic counterpart.

For surfaces in relative motion $\mu = \mu_k$, where μ_k is the *coefficient of kinetic friction*. The Coulomb friction is equal to F_f , and the frictional force on each surface is exerted in the direction opposite to its motion relative to the other surface.

Arthur Morin introduced the term and demonstrated the utility of the coefficient of friction. The coefficient of friction is an empirical measurement – it has to be measured experimentally, and cannot be found through calculations. Rougher surfaces tend to have higher effective values. Both static and kinetic coefficients of friction depend on the pair of surfaces in contact; for a given pair of surfaces, the coefficient of static friction is *usually* larger than that of kinetic friction; in some sets the two coefficients are equal, such as teflon-on-teflon.

Most dry materials in combination have friction coefficient values between 0.3 and 0.6. Values outside this range are rarer, but teflon, for example, can have a coefficient as low as 0.04. A value of zero would mean no friction at all, an elusive property. Rubber in contact with other surfaces can yield friction coefficients from 1 to 2. Occasionally it is maintained that μ is always < 1, but this is not true. While in most relevant applications μ < 1, a value above 1 merely implies that the force required to slide an object along the surface is greater than the normal force of the surface on the object. For example, silicone rubber or acrylic rubber-coated surfaces have a coefficient of friction that can be substantially larger than 1.

While it is often stated that the COF is a "material property," it is better categorized as a "system property." Unlike true material properties (such as conductivity, dielectric constant, yield strength), the COF for any two materials depends on system variables like temperature, velocity, atmosphere and also what are now popularly described as aging and deaging times; as well as on geometric properties of the interface between the materials. For example, a copper pin sliding against a thick copper plate can have a COF that varies from 0.6 at low speeds (metal sliding against metal) to below 0.2 at high speeds when the copper surface begins to melt due to frictional heating. The latter speed, of course, does not determine the COF uniquely; if the pin diameter is increased so that the frictional heating is removed rapidly, the temperature drops, the pin remains solid and the COF rises to that of a 'low speed' test.

Approximate Coefficients of Friction

Materials		Static Friction, μ_s		Kinetic/Sliding Friction,	
Dry and clean		Lubricated	Dry and clean	Lubricated	
Aluminium	Steel	0.61		0.47	
Alumina ceramic	Silicon Nitride ceramic				0.004 (wet)
BAM (Ceramic alloy $AlMgB_{14}$)	Titanium boride (TiB_2)	0.04–0.05	0.02		
Brass	Steel	0.35-0.51	0.19	0.44	
Cast iron	Copper	1.05		0.29	
Cast iron	Zinc	0.85		0.21	
Concrete	Rubber	1.0	0.30 (wet)	0.6-0.85	0.45-0.75 (wet)
Concrete	Wood	0.62			
Copper	Glass	0.68			
Copper	Steel	0.53		0.36	
Glass	Glass	0.9-1.0		0.4	
Human synovial fluid	Cartilage		0.01		0.003
Ice	Ice	0.02-0.09			
Polyethene	Steel	0.2	0.2		
PTFE (Teflon)	PTFE (Teflon)	0.04	0.04		0.04
Steel	Ice	0.03			
Steel	PTFE (Teflon)	0.04-0.2	0.04		0.04
Steel	Steel	0.74-0.80	0.16	0.42-0.62	
Wood	Metal	0.2–0.6	0.2 (wet)		
Wood	Wood	0.25–0.5	0.2 (wet)		

Under certain conditions some materials have very low friction coefficients. An example is (highly ordered pyrolytic) graphite which can have a friction coefficient below 0.01. This ultralow-friction regime is called superlubricity.

Static Friction

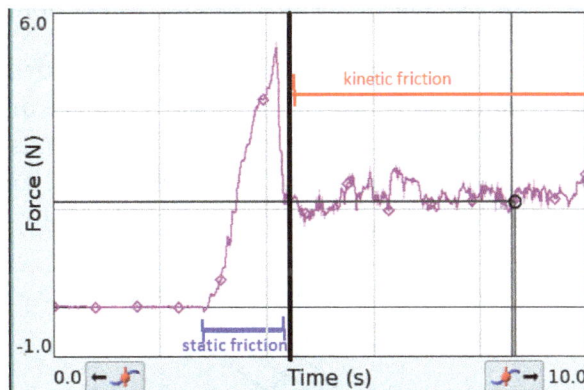

When the mass is not moving, the object experiences static friction. The friction increases as the applied force increases until the block moves. After the block moves, it experiences kinetic friction, which is less than the maximum static friction.

Static friction is friction between two or more solid objects that are not moving relative to each other. For example, static friction can prevent an object from sliding down a sloped surface. The coefficient of static friction, typically denoted as μ_s, is usually higher than the coefficient of kinetic friction.

The static friction force must be overcome by an applied force before an object can move. The maximum possible friction force between two surfaces before sliding begins is the product of the coefficient of static friction and the normal force: $F_{max} = \mu_s F_n$. When there is no sliding occurring, the friction force can have any value from zero up to F_{max}. . Any force smaller than F_{max} attempting to slide one surface over the other is opposed by a frictional force of equal magnitude and opposite direction. Any force larger than F_{max} overcomes the force of static friction and causes sliding to occur. The instant sliding occurs, static friction is no longer applicable—the friction between the two surfaces is then called kinetic friction.

An example of static friction is the force that prevents a car wheel from slipping as it rolls on the ground. Even though the wheel is in motion, the patch of the tire in contact with the ground is stationary relative to the ground, so it is static rather than kinetic friction.

The maximum value of static friction, when motion is impending, is sometimes referred to as limiting friction, although this term is not used universally.

Kinetic Friction

Kinetic friction, also known as dynamic friction or sliding friction, occurs when two objects are moving relative to each other and rub together (like a sled on the ground). The coefficient of kinetic friction is typically denoted as μ_k, and is usually less than the coefficient of static friction for the same materials. However, Richard Feynman comments that "with dry metals it is very hard to show any difference." The friction force between two surfaces after sliding begins is the product of the coefficient of kinetic friction and the normal force: $F_k = \mu_k F_n$.

New models are beginning to show how kinetic friction can be greater than static friction. Kinetic friction is now understood, in many cases, to be primarily caused by chemical bonding between the surfaces, rather than interlocking asperities; however, in many other cases roughness effects are dominant, for example in rubber to road friction. Surface roughness and contact area affect kinetic friction for micro- and nano-scale objects where surface area forces dominate inertial forces.

The origin of kinetic friction at nanoscale can be explained by thermodynamics. Upon sliding, new surface forms at the back of a sliding true contact, and existing surface disappears at the front of it. Since all surfaces involve the thermodynamic surface energy, work must be spent in creating the new surface, and energy is released as heat in removing the surface. Thus, a force is required to move the back of the contact, and frictional heat is released at the front.

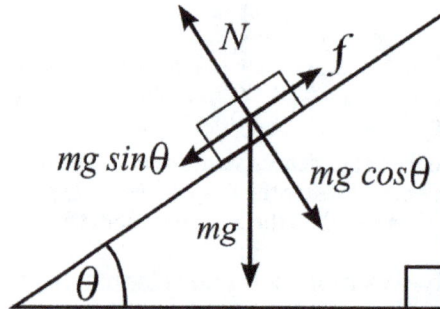

Angle of friction, θ, when block just starts to slide.

Angle of Friction

For certain applications it is more useful to define static friction in terms of the maximum angle before which one of the items will begin sliding. This is called the *angle of friction* or *friction angle*. It is defined as:

$$\tan \theta = \mu_s$$

where θ is the angle from horizontal and μ_s is the static coefficient of friction between the objects. This formula can also be used to calculate μ_s from empirical measurements of the friction angle.

Friction at the Atomic Level

Determining the forces required to move atoms past each other is a challenge in designing nano-machines. In 2008 scientists for the first time were able to move a single atom across a surface, and measure the forces required. Using ultrahigh vacuum and nearly zero temperature (5 K), a modified atomic force microscope was used to drag a cobalt atom, and a carbon monoxide molecule, across surfaces of copper and platinum.

Limitations of the Coulomb Model

The Coulomb approximation mathematically follows from the assumptions that surfaces are in atomically close contact only over a small fraction of their overall area, that this contact area is proportional to the normal force (until saturation, which takes place when all area is in atomic contact), and that the frictional force is proportional to the applied normal force, independently of the contact area. Such reasoning aside, however, the approximation is fundamentally an empirical construct. It is a rule of thumb describing the approximate outcome of an extremely complicated physical interaction. The strength of the approximation is its simplicity and versatility. Though in general the relationship between normal force and frictional force is not exactly linear (and so the frictional force is not entirely independent of the contact area of the surfaces), the Coulomb approximation is an adequate representation of friction for the analysis of many physical systems.

When the surfaces are conjoined, Coulomb friction becomes a very poor approximation (for example, adhesive tape resists sliding even when there is no normal force, or a negative normal force). In this case, the frictional force may depend strongly on the area of contact. Some drag racing tires are adhesive for this reason. However, despite the complexity of the fundamental physics behind friction, the relationships are accurate enough to be useful in many applications.

"Negative" Coefficient of Friction

As of 2012, a single study has demonstrated the potential for an *effectively negative coefficient of friction in the low-load regime,* meaning that a decrease in normal force leads to an increase in friction. This contradicts everyday experience in which an increase in normal force leads to an increase in friction. This was reported in the journal *Nature* in October 2012 and involved the friction encountered by an atomic force microscope stylus when dragged across a graphene sheet in the presence of graphene-adsorbed oxygen.

Numerical Simulation of the Coulomb Model

Despite being a simplified model of friction, the Coulomb model is useful in many numerical simulation applications such as multibody systems and granular material. Even its most simple expression encapsulates the fundamental effects of sticking and sliding which are required in many applied cases, although specific algorithms have to be designed in order to efficiently numerically integrate mechanical systems with Coulomb friction and bilateral and/or unilateral contact. Some quite nonlinear effects, such as the so-called Painlevé paradoxes, may be encountered with Coulomb friction.

Dry Friction and Instabilities

Dry friction can induce several types of instabilities in mechanical systems which display a stable behaviour in the absence of friction. These instabilities may be caused by the decrease of the friction force with an increasing velocity of sliding, by material expansion due to heat generation during friction (the thermo-elastic instabilities), or by pure dynamic effects of sliding of two elastic materials (the Adams-Martins instabilities). The latter were originally discovered in 1995 by George G. Adams and João Arménio Correia Martins for smooth surfaces and were later found in periodic rough surfaces. In particular, friction-related dynamical instabilities are thought to be responsible for brake squeal and the 'song' of a glass harp, phenomena which involve stick and slip, modelled as a drop of friction coefficient with velocity.

A practically important case is the self-oscillation of the strings of bowed instruments such as the violin, cello, hurdy-gurdy, erhu etc.

A connection between dry friction and flutter instability in a simple mechanical system has been discovered, watch the movie for more details.

Frictional instabilities can lead to the formation of new self-organized patterns (or "secondary structures") at the sliding interface, such as in-situ formed tribofilms which are utilized for the reduction of friction and wear in so-called self-lubricating materials.

Fluid Friction

Fluid friction occurs between fluid layers that are moving relative to each other. This internal resistance to flow is named *viscosity*. In everyday terms, the viscosity of a fluid is described as its "thickness". Thus, water is "thin", having a lower viscosity, while honey is "thick", having a higher viscosity. The less viscous the fluid, the greater its ease of deformation or movement.

All real fluids (except superfluids) offer some resistance to shearing and therefore are viscous. For teaching and explanatory purposes it is helpful to use the concept of an inviscid fluid or an ideal fluid which offers no resistance to shearing and so is not viscous.

Lubricated Friction

Lubricated friction is a case of fluid friction where a fluid separates two solid surfaces. Lubrication is a technique employed to reduce wear of one or both surfaces in close proximity moving relative to each another by interposing a substance called a lubricant between the surfaces.

In most cases the applied load is carried by pressure generated within the fluid due to the frictional viscous resistance to motion of the lubricating fluid between the surfaces. Adequate lubrication allows smooth continuous operation of equipment, with only mild wear, and without excessive stresses or seizures at bearings. When lubrication breaks down, metal or other components can rub destructively over each other, causing heat and possibly damage or failure.

Skin Friction

Skin friction arises from the interaction between the fluid and the skin of the body, and is directly related to the area of the surface of the body that is in contact with the fluid. Skin friction follows the drag equation and rises with the square of the velocity.

Skin friction is caused by viscous drag in the boundary layer around the object. There are two ways to decrease skin friction: the first is to shape the moving body so that smooth flow is possible, like an airfoil. The second method is to decrease the length and cross-section of the moving object as much as is practicable.

Internal Friction

Internal friction is the force resisting motion between the elements making up a solid material while it undergoes deformation.

Plastic deformation in solids is an irreversible change in the internal molecular structure of an object. This change may be due to either (or both) an applied force or a change in temperature. The change of an object's shape is called strain. The force causing it is called stress.

Elastic deformation in solids is reversible change in the internal molecular structure of an object. Stress does not necessarily cause permanent change. As deformation occurs, internal forces oppose the applied force. If the applied stress is not too large these opposing forces may completely resist the applied force, allowing the object to assume a new equilibrium state and to return to its original shape when the force is removed. This is known as elastic deformation or elasticity.

Radiation Friction

As a consequence of light pressure, Einstein in 1909 predicted the existence of "radiation friction" which would oppose the movement of matter. He wrote, "radiation will exert pressure on both sides of the plate. The forces of pressure exerted on the two sides are equal if the plate is at rest. However, if it is in motion, more radiation will be reflected on the surface that is ahead during the motion (front surface) than on the back surface. The backwardacting force of pressure exerted on the front surface is thus larger than the force of pressure acting on the back. Hence, as the resultant of the two forces, there remains a force that counteracts the motion of the plate and that increases with the velocity of the plate. We will call this resultant 'radiation friction' in brief."

Other Types of Friction

Rolling Resistance

Rolling resistance is the force that resists the rolling of a wheel or other circular object along a surface caused by deformations in the object and/or surface. Generally the force of rolling resistance is less than that associated with kinetic friction. Typical values for the coefficient of rolling resistance are 0.001. One of the most common examples of rolling resistance is the movement of motor vehicle tires on a road, a process which generates heat and sound as by-products.

Braking Friction

Any wheel equipped with a brake is capable of generating a large retarding force, usually for the purpose of slowing and stopping a vehicle or piece of rotating machinery. Braking friction differs from rolling friction because the coefficient of friction for rolling friction is small whereas the coefficient of friction for braking friction is designed to be large by choice of materials for brake pads.

Triboelectric Effect

Rubbing dissimilar materials against one another can cause a build-up of electrostatic charge, which can be hazardous if flammable gases or vapours are present. When the static build-up discharges, explosions can be caused by ignition of the flammable mixture.

Belt Friction

Belt friction is a physical property observed from the forces acting on a belt wrapped around a pulley, when one end is being pulled. The resulting tension, which acts on both ends of the belt, can be modeled by the belt friction equation.

In practice, the theoretical tension acting on the belt or rope calculated by the belt friction equation can be compared to the maximum tension the belt can support. This helps a designer of such a rig to know how many times the belt or rope must be wrapped around the pulley to prevent it from slipping. Mountain climbers and sailing crews demonstrate a standard knowledge of belt friction when accomplishing basic tasks.

Reducing Friction

Devices

Devices such as wheels, ball bearings, roller bearings, and air cushion or other types of fluid bearings can change sliding friction into a much smaller type of rolling friction.

Many thermoplastic materials such as nylon, HDPE and PTFE are commonly used in low friction bearings. They are especially useful because the coefficient of friction falls with increasing imposed load. For improved wear resistance, very high molecular weight grades are usually specified for heavy duty or critical bearings.

Lubricants

A common way to reduce friction is by using a lubricant, such as oil, water, or grease, which is placed between the two surfaces, often dramatically lessening the coefficient of friction. The science of friction and lubrication is called tribology. Lubricant technology is when lubricants are mixed with the application of science, especially to industrial or commercial objectives.

Superlubricity, a recently discovered effect, has been observed in graphite: it is the substantial decrease of friction between two sliding objects, approaching zero levels. A very small amount of frictional energy would still be dissipated.

Lubricants to overcome friction need not always be thin, turbulent fluids or powdery solids such as graphite and talc; acoustic lubrication actually uses sound as a lubricant.

Another way to reduce friction between two parts is to superimpose micro-scale vibration to one of the parts. This can be sinusoidal vibration as used in ultrasound-assisted cutting or vibration noise, known as dither.

Energy of Friction

According to the law of conservation of energy, no energy is destroyed due to friction, though it may be lost to the system of concern. Energy is transformed from other forms into thermal energy. A sliding hockey puck comes to rest because friction converts its kinetic energy into heat which raises the thermal energy of the puck and the ice surface. Since heat quickly dissipates, many early philosophers, including Aristotle, wrongly concluded that moving objects lose energy without a driving force.

When an object is pushed along a surface along a path C, the energy converted to heat is given by a line integral, in accordance with the definition of work

$$E_{th} = \int_C \mathbf{F}_{fric}(\mathbf{x}) \cdot d\mathbf{x} = \int_C \mu_k \, \mathbf{F}_n(\mathbf{x}) \cdot d\mathbf{x},$$

where

> $\mathbf{F}_{fric}$ is the friction force,
>
> $\mathbf{F}_n$ is the vector obtained by multiplying the magnitude of the normal force by a unit vector pointing *against* the object's motion,
>
> μ_k is the coefficient of kinetic friction, which is inside the integral because it may vary from location to location (e.g. if the material changes along the path),
>
> $\mathbf{x}$ is the position of the object.

Energy lost to a system as a result of friction is a classic example of thermodynamic irreversibility.

Work of Friction

In the reference frame of the interface between two surfaces, static friction does *no* work, because there is never displacement between the surfaces. In the same reference frame, kinetic friction is always in the direction opposite the motion, and does *negative* work. However, friction can do *positive* work in certain frames of reference. One can see this by placing a heavy box on a rug, then pulling on the rug quickly. In this case, the box slides backwards relative to the rug, but moves forward relative to the frame of reference in which the floor is stationary. Thus, the kinetic friction between the box and rug accelerates the box in the same direction that the box moves, doing *positive* work.

The work done by friction can translate into deformation, wear, and heat that can affect the contact surface properties (even the coefficient of friction between the surfaces). This can be beneficial as in polishing. The work of friction is used to mix and join materials such as in the process of friction welding. Excessive erosion or wear of mating sliding surfaces occurs when work due to frictional forces rise to unacceptable levels. Harder corrosion particles caught between mating surfaces in relative motion (fretting) exacerbates wear of frictional forces. Bearing seizure or failure may result from excessive wear due to work of friction. As surfaces are worn by work due to friction, fit and surface finish of an object may degrade until it no longer functions properly.

Impulse (Physics)

In classical mechanics, impulse (symbolized by J or Imp) is the integral of a force, F, over the time interval, t, for which it acts. Since force is a vector quantity, impulse is also a vector in the same direction. Impulse applied to an object produces an equivalent vector change in its linear momentum, also in the same direction. The SI unit of impulse is the newton second (N·s), and the dimensionally equivalent unit of momentum is the kilogram meter per second (kg·m/s). The corresponding English engineering units are the pound-second (lbf·s) and the slug-foot per second (slug·ft/s).

A resultant force causes acceleration and a change in the velocity of the body for as long as it acts. A resultant force applied over a longer time therefore produces a bigger change in linear momentum than the same force applied briefly: the change in momentum is equal to the product of the average force and duration. Conversely, a small force applied for a long time produces the same change in momentum—the same impulse—as a larger force applied briefly.

$$J = F_{\text{average}}(t_2 - t_1)$$

The impulse is the integral of the resultant force (F) with respect to time:

$$J = \int F \, dt$$

Mathematical Derivation in the Case of an Object of Constant Mass

The impulse delivered by the *sad* ball is mv_o, where v_o is the speed upon impact.
To the extent that it bounces back with speed v_o, the happy ball delivers an impulse of $m\Delta v = 2mv_o$.

Impulse J produced from time t_1 to t_2 is defined to be

$$\mathbf{J} = \int_{t_1}^{t_2} \mathbf{F} \, dt$$

where F is the resultant force applied from t_1 to t_2.

From Newton's second law, force is related to momentum p by

$$\mathbf{F} = \frac{d\mathbf{p}}{dt}$$

Therefore,

$$\mathbf{J} = \int_{t_1}^{t_2} \frac{d\mathbf{p}}{dt} dt$$

$$= \int_{p_1}^{p_2} d\mathbf{p}$$

$$= \mathbf{p}_2 - \mathbf{p}_1 = \Delta\mathbf{p}$$

where Δp is the change in linear momentum from time t_1 to t_2. This is often called the impulse-momentum theorem.

As a result, an impulse may also be regarded as the change in momentum of an object to which a resultant force is applied. The impulse may be expressed in a simpler form when the mass is constant:

$$\mathbf{J} = \int_{t_1}^{t_2} \mathbf{F}\,dt = \Delta p = mv_2 - mv_1$$

where

F is the resultant force applied,

t_1 and t_2 are times when the impulse begins and ends, respectively,

m is the mass of the object,

v_2 is the final velocity of the object at the end of the time interval, and

v_1 is the initial velocity of the object when the time interval begins.

Impulse has the same units and dimensions (MLT^{-1}) as momentum. In the International System of Units, these are kg·m/s = N·s. In English engineering units, they are slug·ft/s = lbf·s.

A large force applied for a very short duration, such as a golf shot, is often described as the club giving the ball an *impulse*.

The term "impulse" is also used to refer to a fast-acting force or impact. This type of impulse is often *idealized* so that the change in momentum produced by the force happens with no change in time. This sort of change is a step change, and is not physically possible. However, this is a useful model for computing the effects of ideal collisions (such as in game physics engines).

Variable Mass

The application of Newton's second law for variable mass allows impulse and momentum to be used as analysis tools for jet- or rocket-propelled vehicles. In the case of rockets, the impulse imparted can be normalized by unit of propellant expended, to create a performance parameter, specific impulse. This fact can be used to derive the Tsiolkovsky rocket equation, which relates the vehicle's propulsive change in velocity to the engine's specific impulse (or nozzle exhaust velocity) and the vehicle's propellant-mass ratio.

Work (Physics)

In physics, a force is said to do work if, when acting there is a displacement of the point of application in the direction of the force. For example, when a ball is held above the ground and then dropped, the work done on the ball as it falls is equal to the weight of the ball (a force) multiplied by the distance to the ground (a displacement).

The term *work* was introduced in 1826 by the French mathematician Gaspard-Gustave Coriolis as "weight *lifted* through a height", which is based on the use of early steam engines to lift buckets of water out of flooded ore mines. The SI unit of work is the joule (J).

Units

The SI unit of work is the joule (J), which is defined as the work expended by a force of one newton through a distance of one metre.

The dimensionally equivalent newton-metre (N·m) is sometimes used as the measuring unit for work, but this can be confused with the unit newton-metre, which is the measurement unit of torque. Usage of N·m is discouraged by the SI authority, since it can lead to confusion as to whether the quantity expressed in newton metres is a torque measurement, or a measurement of energy.

Non-SI units of work include the erg, the foot-pound, the foot-poundal, the kilowatt hour, the litre-atmosphere, and the horsepower-hour. Due to work having the same physical dimension as heat, occasionally measurement units typically reserved for heat or energy content, such as therm, BTU and Calorie, are utilized as a measuring unit.

Work and Energy

The work done by a constant force of magnitude F on a point that moves a displacement (not distance) s in the direction of the force is the product

$$W = Fs.$$

For example, if a force of 10 newtons ($F = 10$ N) acts along a point that travels 2 meters ($s = 2$ m), then it does the work $W = (10 \text{ N})(2 \text{ m}) = 20 \text{ N m} = 20$ J. This is approximately the work done lifting a 1 kg weight from ground level to over a person's head against the force of gravity. Notice that the work is doubled either by lifting twice the weight the same distance or by lifting the same weight twice the distance.

Work is closely related to energy. The work-energy principle states that an increase in the kinetic energy of a rigid body is caused by an equal amount of positive work done on the body by the resultant force acting on that body. Conversely, a decrease in kinetic energy is caused by an equal amount of negative work done by the resultant force.

From Newton's second law, it can be shown that work on a free (no fields), rigid (no internal degrees of freedom) body, is equal to the change in kinetic energy of the velocity and rotation of that body,

$$W = \Delta KE.$$

The work of forces generated by a potential function is known as potential energy and the forces are said to be conservative. Therefore, work on an object that is merely displaced in a conservative force field, without change in velocity or rotation, is equal to *minus* the change of potential energy of the object,

$$W = -\Delta PE.$$

These formulas demonstrate that work is the energy associated with the action of a force, so work subsequently possesses the physical dimensions, and units, of energy. The work/energy principles discussed here are identical to Electric work/energy principles.

Constraint Forces

Constraint forces determine the movement of components in a system, constraining the object within a boundary (in the case of a slope plus gravity, the object is *stuck to* the slope, when attached to a taut string it cannot move in an outwards direction to make the string any 'tauter'). Constraint forces ensure the velocity in the direction of the constraint is zero, which means the constraint forces do not perform work on the system.

If the system doesn't change in time, they eliminate all movement in the direction of the constraint, thus constraint forces do not perform work on the system, as the velocity of that object is constrained to be 0 parallel to this force, due to this force. This only applies for a single particle system. For example, in an Atwood machine, the rope does work on each body, but keeping always the net virtual work null. There are, however, cases where this is not true.

For example, the centripetal force exerted *inwards* by a string on a ball in uniform circular motion *sideways* constrains the ball to circular motion restricting its movement away from the center of the circle. This force does zero work because it is perpendicular to the velocity of the ball.

Another example is a book on a table. If external forces are applied to the book so that it slides on the table, then the force exerted by the table constrains the book from moving downwards. The force exerted by the table supports the book and is perpendicular to its movement which means that this constraint force does not perform work.

The magnetic force on a charged particle is $F = qv \times B$, where q is the charge, v is the velocity of the particle, and B is the magnetic field. The result of a cross product is always perpendicular to both of the original vectors, so $F \perp v$. The dot product of two perpendicular vectors is always zero, so the work $W = F \cdot v = 0$, and the magnetic force does not do work. It can change the direction of motion but never change the speed.

Mathematical calculation

For moving objects, the quantity of work/time (power) is calculated. Thus, at any instant, the rate of the work done by a force (measured in joules/second, or watts) is the scalar product of the force (a vector), and the velocity vector of the point of application. This scalar product of force and velocity is classified as instantaneous power. Just as velocities may be integrated over time to obtain a total distance, by the fundamental theorem of calculus, the total work along a path is similarly the time-integral of instantaneous power applied along the trajectory of the point of application.

Work is the result of a force on a point that moves through a distance. As the point moves, it follows a curve X, with a velocity v, at each instant. The small amount of work δW that occurs over an instant of time dt is calculated as

$$\delta W = \mathbf{F} \cdot d\mathbf{s} = \mathbf{F} \cdot \mathbf{v}dt$$

where the $\mathbf{F} \cdot \mathbf{v}$ is the power over the instant dt. The sum of these small amounts of work over the trajectory of the point yields the work,

$$W = \int_{t_1}^{t_2} \mathbf{F} \cdot \mathbf{v}dt = \int_{t_1}^{t_2} \mathbf{F} \cdot \frac{d\mathbf{s}}{dt} dt = \int_C \mathbf{F} \cdot d\mathbf{s},$$

where C is the trajectory from $x(t_1)$ to $x(t_2)$. This integral is computed along the trajectory of the particle, and is therefore said to be *path dependent*.

If the force is always directed along this line, and the magnitude of the force is F, then this integral simplifies to

$$W = \int_C F \, ds$$

where s is distance along the line. If F is constant, in addition to being directed along the line, then the integral simplifies further to

$$W = \int_C F \, ds = F \int_C ds = Fs$$

where s is the distance travelled by the point along the line.

This calculation can be generalized for a constant force that is not directed along the line, followed by the particle. In this case the dot product $\mathbf{F} \cdot d\mathbf{s} = F \cos \theta \, ds$, where θ is the angle between the force vector and the direction of movement, that is

$$W = \int_C \mathbf{F} \cdot d\mathbf{s} = Fs \cos \theta.$$

In the notable case of a force applied to a body always at an angle of 90° from the velocity vector (as when a body moves in a circle under a central force), no work is done at all, since the cosine of 90 degrees is zero. Thus, no work can be performed by gravity on a planet with a circular orbit (this is ideal, as all orbits are slightly elliptical). Also, no work is done on a body moving circularly at a constant speed while constrained by mechanical force, such as moving at constant speed in a frictionless ideal centrifuge.

Work Done by a Variable Force

Calculating the work as "force times straight path segment" would only apply in the most simple of circumstances, as noted above. If force is changing, or if the body is moving along a curved path, possibly rotating and not necessarily rigid, then only the path of the application point of the force is relevant for the work done, and only the component of the force parallel to the application point velocity is doing work (positive work when in the same direction, and negative when in the opposite direction of the velocity). This component of force can be described by the scalar quantity called *scalar tangential component* ($F \cos \theta$,, where θ is the angle between the force and the velocity). And then the most general definition of work can be formulated as follows:

> *Work of a force is the line integral of its scalar tangential component along the path of its application point.*

If the force varies (e.g. compressing a spring) we need to use calculus to find the work done. If the force is given by F(x) (a function of x) then the work done by the force along the x-axis from a to b is:

$$W = \int_a^b \mathbf{F(s)} \cdot d\mathbf{s}$$

Torque and Rotation

A force couple results from equal and opposite forces, acting on two different points of a rigid body. The sum (resultant) of these forces may cancel, but their effect on the body is the couple or torque T. The work of the torque is calculated as

$$dW = \mathbf{T} \cdot \vec{\omega} dt,$$

where the T $\cdot$ ω is the power over the instant δt. The sum of these small amounts of work over the trajectory of the rigid body yields the work,

$$W = \int_{t_1}^{t_2} \mathbf{T} \cdot \vec{\omega} dt.$$

This integral is computed along the trajectory of the rigid body with an angular velocity ω that varies with time, and is therefore said to be *path dependent*.

If the angular velocity vector maintains a constant direction, then it takes the form,

$$\vec{\omega} = \dot{\phi} \mathbf{S},$$

where φ is the angle of rotation about the constant unit vector S. In this case, the work of the torque becomes,

$$W = \int_{t_1}^{t_2} \mathbf{T} \cdot \vec{\omega} dt = \int_{t_1}^{t_2} \mathbf{T} \cdot \mathbf{S} \frac{d\phi}{dt} dt = \int_C \mathbf{T} \cdot \mathbf{S} d\phi,$$

where C is the trajectory from $\varphi(t_1)$ to $\varphi(t_2)$. This integral depends on the rotational trajectory $\varphi(t)$, and is therefore path-dependent.

If the torque T is aligned with the angular velocity vector so that,

$$\mathbf{T} = \tau \mathbf{S},$$

and both the torque and angular velocity are constant, then the work takes the form,

$$W = \int_{t_1}^{t_2} \tau \dot{\phi} dt = \tau(\phi_2 - \phi_1).$$

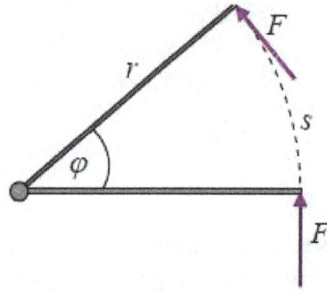

A force of constant magnitude and perpendicular to the lever arm

This result can be understood more simply by considering the torque as arising from a force of constant magnitude F, being applied perpendicularly to a lever arm at a distance r, as shown in the figure. This force will act through the distance along the circular arc $s = r\varphi$, so the work done is

$$W = Fs = Fr\phi.$$

Introduce the torque $\tau = Fr$, to obtain

$$W = Fr\phi = \tau\phi,$$

as presented above.

Notice that only the component of torque in the direction of the angular velocity vector contributes to the work.

Work and Potential Energy

The scalar product of a force F and the velocity v of its point of application defines the power input to a system at an instant of time. Integration of this power over the trajectory of the point of application, $C = x(t)$, defines the work input to the system by the force.

Path Dependence

Therefore, the work done by a force F on an object that travels along a curve C is given by the line integral:

$$W = \int_C \mathbf{F} \cdot d\mathbf{x} = \int_{t_1}^{t_2} \mathbf{F} \cdot \mathbf{v} dt,$$

where $dx(t)$ defines the trajectory C and v is the velocity along this trajectory. In general this integral requires the path along which the velocity is defined, so the evaluation of work is said to be path dependent.

The time derivative of the integral for work yields the instantaneous power,

$$\frac{dW}{dt} = P(t) = \mathbf{F} \cdot \mathbf{v}.$$

Path Independence

If the work for an applied force is independent of the path, then the work done by the force is, by the gradient theorem, the potential function evaluated at the start and end of the trajectory of the point of application. Such a force is said to be conservative. This means that there is a potential function $U(x)$, that can be evaluated at the two points $x(t_1)$ and $x(t_2)$ to obtain the work over any trajectory between these two points. It is tradition to define this function with a negative sign so that positive work is a reduction in the potential, that is

$$W = \int_C \mathbf{F} \cdot d\mathbf{x} = \int_{\mathbf{x}(t_1)}^{\mathbf{x}(t_2)} \mathbf{F} \cdot d\mathbf{x} = U(\mathbf{x}(t_1)) - U(\mathbf{x}(t_2)).$$

The function $U(x)$ is called the potential energy associated with the applied force. Examples of forces that have potential energies are gravity and spring forces.

In this case, the gradient of work yields

$$\nabla W = -\nabla U = -\left(\frac{\partial U}{\partial x}, \frac{\partial U}{\partial y}, \frac{\partial U}{\partial z}\right) = \mathbf{F},$$

and the force F is said to be "derivable from a potential."

Because the potential U defines a force F at every point x in space, the set of forces is called a force field. The power applied to a body by a force field is obtained from the gradient of the work, or potential, in the direction of the velocity V of the body, that is

$$P(t) = -\nabla U \cdot \mathbf{v} = \mathbf{F} \cdot \mathbf{v}.$$

Work by Gravity

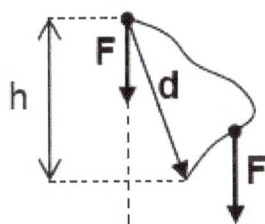

Gravity $F = mg$ does work $W = mgh$ along any descending path

In the absence of other forces, gravity results in a constant downward acceleration of every freely moving object. Near Earth's surface the acceleration due to gravity is $g = 9.8$ m·s^{-2} and the gravitational force on an object of mass m is $F_g = mg$. It is convenient to imagine this gravitational force concentrated at the center of mass of the object.

If an object is displaced upwards or downwards a vertical distance $y_2 - y_1$, the work W done on the object by its weight mg is:

$$W = F_g(y_2 - y_1) = F_g \Delta y = -mg\Delta y$$

where F_g is weight (pounds in imperial units, and newtons in SI units), and Δy is the change in height y. Notice that the work done by gravity depends only on the vertical movement of the object. The presence of friction does not affect the work done on the object by its weight.

Work by Gravity in Space

The force of gravity exerted by a mass M on another mass m is given by

$$\mathbf{F} = -\frac{GMm}{r^3}\mathbf{r},$$

where r is the position vector from M to m.

Let the mass m move at the velocity v then the work of gravity on this mass as it moves from position $r(t_1)$ to $r(t_2)$ is given by

$$W = -\int_{\mathbf{r}(t_1)}^{\mathbf{r}(t_2)} \frac{GMm}{r^3}\mathbf{r} \cdot d\mathbf{r} = -\int_{t_1}^{t_2} \frac{GMm}{r^3}\mathbf{r} \cdot \mathbf{v}\,dt.$$

Notice that the position and velocity of the mass m are given by

$$\mathbf{r} = r\mathbf{e}_r, \qquad \mathbf{v} = \dot{r}\mathbf{e}_r + r\dot{\theta}\mathbf{e}_t,$$

where e_r and e_t are the radial and tangential unit vectors directed relative to the vector from M to m. Use this to simplify the formula for work of gravity to,

$$W = -\int_{t_1}^{t_2} \frac{GmM}{r^3}(r\mathbf{e}_r) \cdot (\dot{r}\mathbf{e}_r + r\dot{\theta}\mathbf{e}_t)\,dt = -\int_{t_1}^{t_2} \frac{GmM}{r^3}r\dot{r}\,dt = \frac{GMm}{r(t_2)} - \frac{GMm}{r(t_1)}.$$

This calculation uses the fact that

$$\frac{d}{dt}r^{-1} = -r^{-2}\dot{r} = -\frac{\dot{r}}{r^2}.$$

The function

$$U = -\frac{GMm}{r},$$

is the gravitational potential function, also known as gravitational potential energy. The negative sign follows the convention that work is gained from a loss of potential energy.

Work by a Spring

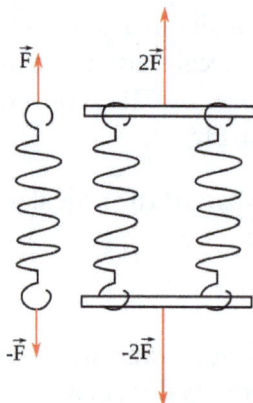

Forces in springs assembled in parallel

Consider a spring that exerts a horizontal force F = (−kx, 0, 0) that is proportional to its deflection in the x direction independent of how a body moves. The work of this spring on a body moving along the space curve X(t) = ($x(t)$, $y(t)$, $z(t)$), is calculated using its velocity, v = (v_x, v_y, v_z), to obtain

$$W = \int_0^t \mathbf{F} \cdot \mathbf{v}\,dt = -\int_0^t kx v_x\,dt = -\frac{1}{2}kx^2.$$

For convenience, consider contact with the spring occurs at t = 0, then the integral of the product of the distance x and the x-velocity, xv_x, is $(1/2)x^2$.

Work by a Gas

$$W = \int_a^b P\,dV$$

Where P is pressure, V is volume, and a and b are initial and final volumes.

Work–energy Principle

The principle of work and kinetic energy (also known as the work–energy principle) states that *the work done by all forces acting on a particle (the work of the resultant force) equals the change in the kinetic energy of the particle.* That is, the work W done by the resultant force on a particle equals the change in the particle's kinetic energy E_k,

$$W = \Delta E_k = \tfrac{1}{2}mv_2^2 - \tfrac{1}{2}mv_1^2,$$

where v_1 and v_2 are the speeds of the particle before and after the work is done and m is its mass.

The derivation of the *work–energy principle* begins with Newton's second law and the resultant force on a particle which includes forces applied to the particle and constraint forces imposed on its movement. Computation of the scalar product of the forces with the velocity of the particle evaluates the instantaneous power added to the system.

Constraints define the direction of movement of the particle by ensuring there is no component of velocity in the direction of the constraint force. This also means the constraint forces do not add to the instantaneous power. The time integral of this scalar equation yields work from the instantaneous power, and kinetic energy from the scalar product of velocity and acceleration. The fact the work–energy principle eliminates the constraint forces underlies Lagrangian mechanics.

This section focuses on the work–energy principle as it applies to particle dynamics. In more general systems work can change the potential energy of a mechanical device, the heat energy in a thermal system, or the electrical energy in an electrical device. Work transfers energy from one place to another or one form to another.

Derivation for a Particle Moving Along a Straight Line

In the case the resultant force F is constant in both magnitude and direction, and parallel to the velocity of the particle, the particle is moving with constant acceleration a along a straight line. The relation between the net force and the acceleration is given by the equation $F = ma$ (Newton's second law), and the particle displacement s can be expressed by the equation

$$s = \frac{v_2^2 - v_1^2}{2a}$$

which follows from $v_2^2 = v_1^2 + 2as$

The work of the net force is calculated as the product of its magnitude and the particle displacement. Substituting the above equations, one obtains:

$$W = Fs = mas = ma\left(\frac{v_2^2 - v_1^2}{2a}\right) = \frac{mv_2^2}{2} - \frac{mv_1^2}{2} = \Delta E_k$$

Other derivation:

$$W = Fs = mas = m\left(\frac{v_2^2 - v_1^2}{2s}\right)s$$

$$W = m\left(\frac{0_2^2 - v_1^2}{2s}\right)s$$

$$W = -\frac{1}{2}mv^2$$

Vertical displacement derivation

$$W = F \times S = mg \times h$$

In the general case of rectilinear motion, when the net force F is not constant in magnitude, but is constant in direction, and parallel to the velocity of the particle, the work must be integrated along the path of the particle:

$$W = \int_{t_1}^{t_2} \mathbf{F} \cdot \mathbf{v}dt = \int_{t_1}^{t_2} Fvdt = \int_{t_1}^{t_2} mavdt = m\int_{t_1}^{t_2} v\frac{dv}{dt}dt = m\int_{v_1}^{v_2} vdv = \tfrac{1}{2}m(v_2^2 - v_1^2).a$$

General Derivation of the Work–energy Theorem for a Particle

For any net force acting on a particle moving along any curvilinear path, it can be demonstrated that its work equals the change in the kinetic energy of the particle by a simple derivation analogous to the equation above. Some authors call this result *work–energy principle*, but it is more widely known as the work–energy theorem:

$$W = \int_{t_1}^{t_2} \mathbf{F} \cdot \mathbf{v}dt = m\int_{t_1}^{t_2} \mathbf{a} \cdot \mathbf{v}dt = \frac{m}{2}\int_{t_1}^{t_2} \frac{dv^2}{dt}dt = \frac{m}{2}\int_{v_1^2}^{v_2^2} dv^2 = \frac{mv_2^2}{2} - \frac{mv_1^2}{2} = \Delta E_k$$

The identity $\mathbf{a} \cdot \mathbf{v} = \frac{1}{2} \frac{dv^2}{dt}$ requires some algebra. From the identity $v^2 = \mathbf{v} \cdot \mathbf{v}$ and definition $\mathbf{a} = \frac{d\mathbf{v}}{dt}$ it follows

$$\frac{dv^2}{dt} = \frac{d(\mathbf{v} \cdot \mathbf{v})}{dt} = \frac{d\mathbf{v}}{dt} \cdot \mathbf{v} + \mathbf{v} \cdot \frac{d\mathbf{v}}{dt} = 2\frac{d\mathbf{v}}{dt} \cdot \mathbf{v} = 2\mathbf{a} \cdot \mathbf{v} \ .$$

The remaining part of the above derivation is just simple calculus, same as in the preceding rectilinear case.

Derivation for a Particle in Constrained Movement

In particle dynamics, a formula equating work applied to a system to its change in kinetic energy is obtained as a first integral of Newton's second law of motion. It is useful to notice that the resultant force used in Newton's laws can be separated into forces that are applied to the particle and forces imposed by constraints on the movement of the particle. Remarkably, the work of a constraint force is zero, therefore only the work of the applied forces need be considered in the work–energy principle.

To see this, consider a particle P that follows the trajectory X(t) with a force F acting on it. Isolate the particle from its environment to expose constraint forces R, then Newton's Law takes the form

$$\mathbf{F} + \mathbf{R} = m\,\ddot{\mathbf{X}},$$

where m is the mass of the particle.

Vector Formulation

Note that n dots above a vector indicates its nth time derivative. The scalar product of each side of Newton's law with the velocity vector yields

$$\mathbf{F} \cdot \dot{\mathbf{X}} = m\,\ddot{\mathbf{X}} \cdot \dot{\mathbf{X}},$$

because the constraint forces are perpendicular to the particle velocity. Integrate this equation along its trajectory from the point $X(t_1)$ to the point $X(t_2)$ to obtain

$$\int_{t_1}^{t_2} \mathbf{F} \cdot \dot{\mathbf{X}}\,dt = m \int_{t_1}^{t_2} \ddot{\mathbf{X}} \cdot \dot{\mathbf{X}}\,dt.$$

The left side of this equation is the work of the applied force as it acts on the particle along the trajectory from time t_1 to time t_2. This can also be written as

$$W = \int_{t_1}^{t_2} \mathbf{F} \cdot \dot{\mathbf{X}}\,dt = \int_{X(t_1)}^{X(t_2)} \mathbf{F} \cdot d\mathbf{X}.$$

This integral is computed along the trajectory X(t) of the particle and is therefore path dependent.

The right side of the first integral of Newton's equations can be simplified using the following identity

$$\frac{1}{2}\frac{d}{dt}(\dot{\mathbf{X}}\cdot\dot{\mathbf{X}}) = \ddot{\mathbf{X}}\cdot\dot{\mathbf{X}},$$

Now it is integrated explicitly to obtain the change in kinetic energy,

$$\Delta K = m\int_{t_1}^{t_2}\ddot{\mathbf{X}}\cdot\dot{\mathbf{X}}\,dt = \frac{m}{2}\int_{t_1}^{t_2}\frac{d}{dt}(\dot{\mathbf{X}}\cdot\dot{\mathbf{X}})dt = \frac{m}{2}\dot{\mathbf{X}}\cdot\dot{\mathbf{X}}(t_2) - \frac{m}{2}\dot{\mathbf{X}}\cdot\dot{\mathbf{X}}(t_1) = \frac{1}{2}m\Delta v^2,$$

where the kinetic energy of the particle is defined by the scalar quantity,

$$K = \frac{m}{2}\dot{\mathbf{X}}\cdot\dot{\mathbf{X}} = \frac{1}{2}mv^2$$

Tangential and Normal Components

It is useful to resolve the velocity and acceleration vectors into tangential and normal components along the trajectory X(t), such that

$$\dot{\mathbf{X}} = v\mathbf{T}, \quad \text{and} \quad \ddot{\mathbf{X}} = \dot{v}\mathbf{T} + v^2\kappa\mathbf{N}.$$

where

$$v = |\dot{\mathbf{X}}| = \sqrt{\dot{\mathbf{X}}\cdot\dot{\mathbf{X}}}.$$

Then, the scalar product of velocity with acceleration in Newton's second law takes the form

$$\Delta K = m\int_{t_1}^{t_2}\dot{v}v\,dt = \frac{m}{2}\int_{t_1}^{t_2}\frac{d}{dt}v^2\,dt = \frac{m}{2}v^2(t_2) - \frac{m}{2}v^2(t_1),$$

where the kinetic energy of the particle is defined by the scalar quantity,

$$K = \frac{m}{2}v^2 = \frac{m}{2}\dot{\mathbf{X}}\cdot\dot{\mathbf{X}}.$$

The result is the work–energy principle for particle dynamics,

$$W = \Delta K.$$

This derivation can be generalized to arbitrary rigid body systems.

Moving in a Straight Line (Skid to a Stop)

Consider the case of a vehicle moving along a straight horizontal trajectory under the action of a driving force and gravity that sum to F. The constraint forces between the vehicle and the road define R, and we have

$$\mathbf{F} + \mathbf{R} = m\ddot{\mathbf{X}}.$$

For convenience let the trajectory be along the X-axis, so X = $(d, 0)$ and the velocity is V = $(v, 0)$, then R · V = 0, and F · V = $F_x v$, where F_x is the component of F along the X-axis, so

$$F_x v = m\dot{v}v.$$

Integration of both sides yields

$$\int_{t_1}^{t_2} F_x v\, dt = \frac{m}{2}v^2(t_2) - \frac{m}{2}v^2(t_1).$$

If F_x is constant along the trajectory, then the integral of velocity is distance, so

$$F_x(d(t_2) - d(t_1)) = \frac{m}{2}v^2(t_2) - \frac{m}{2}v^2(t_1).$$

As an example consider a car skidding to a stop, where k is the coefficient of friction and W is the weight of the car. Then the force along the trajectory is $F_x = -kW$. The velocity v of the car can be determined from the length s of the skid using the work–energy principle,

$$kWs = \frac{W}{2g}v^2, \quad \text{or} \quad v = \sqrt{2ksg}.$$

Notice that this formula uses the fact that the mass of the vehicle is $m = W/g$.

Lotus type 119B gravity racer at Lotus 60th celebration.

Gravity racing championship in Campos Novos, Santa Catarina, Brazil, 8 September 2010.

Coasting Down a Mountain Road (Gravity Racing)

Consider the case of a vehicle that starts at rest and coasts down a mountain road, the work-energy principle helps compute the minimum distance that the vehicle travels to reach a velocity V, of say 60 mph (88 fps). Rolling resistance and air drag will slow the vehicle down so the actual distance will be greater than if these forces are neglected.

Let the trajectory of the vehicle following the road be $X(t)$ which is a curve in three-dimensional space. The force acting on the vehicle that pushes it down the road is the constant force of gravity $F = (0, 0, W)$, while the force of the road on the vehicle is the constraint force R. Newton's second law yields,

$$\mathbf{F} + \mathbf{R} = m\ddot{\mathbf{X}}.$$

The scalar product of this equation with the velocity, $V = (v_x, v_y, v_z)$, yields

$$Wv_z = m\dot{V}V,$$

where V is the magnitude of V. The constraint forces between the vehicle and the road cancel from this equation because $R \cdot V = 0$, which means they do no work. Integrate both sides to obtain

$$\int_{t_1}^{t_2} Wv_z dt = \frac{m}{2}V^2(t_2) - \frac{m}{2}V^2(t_1).$$

The weight force W is constant along the trajectory and the integral of the vertical velocity is the vertical distance, therefore,

$$W\Delta z = \frac{m}{2}V^2.$$

Recall that $V(t_1)=0$. Notice that this result does not depend on the shape of the road followed by the vehicle.

In order to determine the distance along the road assume the downgrade is 6%, which is a steep road. This means the altitude decreases 6 feet for every 100 feet traveled—for angles this small the sin and tan functions are approximately equal. Therefore, the distance s in feet down a 6% grade to reach the velocity V is at least

$$s = \frac{\Delta z}{0.06} = 8.3\frac{V^2}{g}, \quad \text{or} \quad s = 8.3\frac{88^2}{32.2} \approx 2000\text{ft.}$$

This formula uses the fact that the weight of the vehicle is $W = mg$.

Work of Forces Acting on a Rigid Body

The work of forces acting at various points on a single rigid body can be calculated from the work of a resultant force and torque. To see this, let the forces F_1, F_2 ... F_n act on the points X_1, X_2 ... X_n in a rigid body.

The trajectories of X_i, $i = 1, ..., n$ are defined by the movement of the rigid body. This movement is given by the set of rotations $[A(t)]$ and the trajectory $d(t)$ of a reference point in the body. Let the coordinates x_i $i = 1, ..., n$ define these points in the moving rigid body's reference frame M, so that the trajectories traced in the fixed frame F are given by

$$\mathbf{X}_i(t) = [A(t)]\mathbf{x}_i + \mathbf{d}(t) \quad i = 1,\ldots,n.$$

The velocity of the points X_i along their trajectories are

$$\mathbf{V}_i = \vec{\omega} \times (\mathbf{X}_i - \mathbf{d}) + \dot{\mathbf{d}},$$

where ω is the angular velocity vector obtained from the skew symmetric matrix

$$[\Omega] = \dot{A}A^{\mathrm{T}},$$

known as the angular velocity matrix.

The small amount of work by the forces over the small displacements δr_i can be determined by approximating the displacement by $\delta r = v\delta t$ so

$$\delta W = \mathbf{F}_1 \cdot \mathbf{V}_1 \delta t + \mathbf{F}_2 \cdot \mathbf{V}_2 \delta t + \ldots + \mathbf{F}_n \cdot \mathbf{V}_n \delta t$$

or

$$\delta W = \sum_{i=1}^{n} \mathbf{F}_i \cdot (\vec{\omega} \times (\mathbf{X}_i - \mathbf{d}) + \dot{\mathbf{d}})\delta t.$$

This formula can be rewritten to obtain

$$\delta W = (\sum_{i=1}^{n} \mathbf{F}_i) \cdot \dot{\mathbf{d}}\,\delta t + (\sum_{i=1}^{n} (\mathbf{X}_i - \mathbf{d}) \times \mathbf{F}_i) \cdot \vec{\omega}\delta t = (\mathbf{F} \cdot \dot{\mathbf{d}} + \mathbf{T} \cdot \vec{\omega})\delta t,$$

where F and T are the resultant force and torque applied at the reference point d of the moving frame M in the rigid body.

Power (Physics)

In physics, power is the rate of doing work. It is the amount of energy consumed per unit time. Having no direction, it is a scalar quantity. In the SI system, the unit of power is the joule per second (J/s), known as the watt in honour of James Watt, the eighteenth-century developer of the steam engine. Another common and traditional measure is horsepower (comparing to the power of a horse). Being the rate of work, the equation for power can be written:

$$P(t) = \frac{W}{t}$$

The integral of power over time defines the work performed. Because this integral depends on the trajectory of the point of application of the force and torque, this calculation of work is said to be path dependent.

As a physical concept, power requires both a change in the physical universe and a specified time in which the change occurs. This is distinct from the concept of work, which is only measured in terms of a net change in the state of the physical universe. The same amount of work is done when carrying a load up a flight of stairs whether the person carrying it walks or runs, but more power is needed for running because the work is done in a shorter amount of time.

The output power of an electric motor is the product of the torque that the motor generates and the angular velocity of its output shaft. The power involved in moving a vehicle is the product of the traction force of the wheels and the velocity of the vehicle. The rate at which a light bulb converts electrical energy into light and heat is measured in watts—the higher the wattage, the more power, or equivalently the more electrical energy is used per unit time.

Units

The dimension of power is energy divided by time. The SI unit of power is the watt (W), which is equal to one joule per second. Other units of power include ergs per second (erg/s), horsepower (hp), metric horsepower (Pferdestärke (PS) or cheval vapeur (CV)), and foot-pounds per minute. One horsepower is equivalent to 33,000 foot-pounds per minute, or the power required to lift 550 pounds by one foot in one second, and is equivalent to about 746 watts. Other units include dBm, a relative logarithmic measure with 1 milliwatt as reference; food calories per hour (often referred to as kilocalories per hour); Btu per hour (Btu/h); and tons of refrigeration (12,000 Btu/h).

Equations for Power

Power, as a function of time, is the rate at which work is done, so can be expressed by this equation:

$$P(t) = \frac{W}{t}$$

Because work is a force applied over a distance, this can be rewritten as:

$$P(t) = \frac{W}{t} = \frac{\mathbf{F} \cdot \mathbf{d}}{t}$$

And with distance per unit time being a velocity, power can likewise be understood as:

$$P(t) = \mathbf{F} \cdot \mathbf{v}$$

Knowing from Newton's 2nd Law that force is mass times acceleration, the expression for power can also be written as:

$$P(t) = \mathbf{ma} \cdot \mathbf{v}$$

Power will change over time as velocity changes due to acceleration. Knowing that acceleration is the time rate of change of velocity, this can then be written:

$$P(t) = m\mathbf{v} \cdot \frac{d\mathbf{v}}{dt}$$

Comparing with the equation for kinetic energy:

$$E_k = \tfrac{1}{2}mv^2$$

It can be seen from the previous equation that power is mass times a velocity term times another velocity term divided by time. This shows how power is an amount of energy consumed per unit time.

Average Power

As a simple example, burning a kilogram of coal releases much more energy than does detonating a kilogram of TNT, but because the TNT reaction releases energy much more quickly, it delivers far more power than the coal. If ΔW is the amount of work performed during a period of time of duration Δt, the average power P_{avg} over that period is given by the formula

$$P_{avg} = \frac{\Delta W}{\Delta t}.$$

It is the average amount of work done or energy converted per unit of time. The average power is often simply called "power" when the context makes it clear.

The instantaneous power is then the limiting value of the average power as the time interval Δt approaches zero.

$$P = \lim_{\Delta t \to 0} P_{avg} = \lim_{\Delta t \to 0} \frac{\Delta W}{\Delta t} = \frac{dW}{dt}.$$

In the case of constant power P, the amount of work performed during a period of duration T is given by:

$$W = Pt.$$

In the context of energy conversion, it is more customary to use the symbol E rather than W.

Mechanical Power

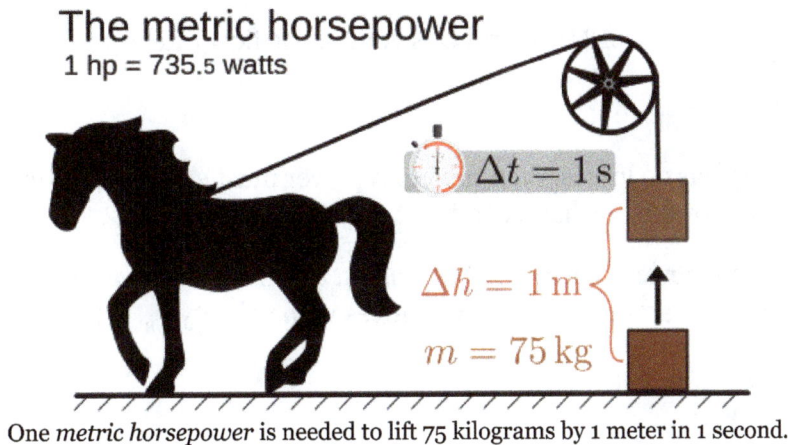

The metric horsepower
1 hp = 735.5 watts

$\Delta t = 1\,\text{s}$

$\Delta h = 1\,\text{m}$

$m = 75\,\text{kg}$

One *metric horsepower* is needed to lift 75 kilograms by 1 meter in 1 second.

Power in mechanical systems is the combination of forces and movement. In particular, power is the product of a force on an object and the object's velocity, or the product of a torque on a shaft and the shaft's angular velocity.

Mechanical power is also described as the time derivative of work. In mechanics, the work done by a force F on an object that travels along a curve C is given by the line integral:

$$W_C = \int_C \mathbf{F} \cdot \mathbf{v} \, dt = \int_C \mathbf{F} \cdot d\mathbf{x},$$

where x defines the path C and v is the velocity along this path.

If the force F is derivable from a potential (conservative), then applying the gradient theorem (and remembering that force is the negative of the gradient of the potential energy) yields:

$$W_C = U(B) - U(A),$$

where A and B are the beginning and end of the path along which the work was done.

The power at any point along the curve C is the time derivative

$$P(t) = \frac{dW}{dt} = \mathbf{F} \cdot \mathbf{v} = -\frac{dU}{dt}.$$

In one dimension, this can be simplified to:

$$P(t) = F \cdot v.$$

In rotational systems, power is the product of the torque τ and angular velocity ω,

$$P(t) = \tau \cdot \omega,$$

where ω measured in radians per second. The represents scalar product.

In fluid power systems such as hydraulic actuators, power is given by

$$P(t) = pQ,$$

where p is pressure in pascals, or N/m² and Q is volumetric flow rate in m³/s in SI units.

Mechanical Advantage

If a mechanical system has no losses, then the input power must equal the output power. This provides a simple formula for the mechanical advantage of the system.

Let the input power to a device be a force F_A acting on a point that moves with velocity v_A and the output power be a force F_B acts on a point that moves with velocity v_B. If there are no losses in the system, then

$$P = F_B v_B = F_A v_A,$$

and the mechanical advantage of the system (output force per input force) is given by

$$\text{MA} = \frac{F_B}{F_A} = \frac{v_A}{v_B}.$$

The similar relationship is obtained for rotating systems, where T_A and ω_A are the torque and angular velocity of the input and T_B and ω_B are the torque and angular velocity of the output. If there are no losses in the system, then

$$P = T_A\omega_A = T_B\omega_B,$$

which yields the mechanical advantage

$$\text{MA} = \frac{T_B}{T_A} = \frac{\omega_A}{\omega_B}.$$

These relations are important because they define the maximum performance of a device in terms of velocity ratios determined by its physical dimensions.

Electrical Power

Ansel Adams photograph of electrical wires of the Boulder Dam Power Units, 1941–1942

The instantaneous electrical power P delivered to a component is given by

$$P(t) = I(t) \cdot V(t)$$

where

      $P(t)$ is the instantaneous power, measured in watts (joules per second)

      $V(t)$ is the potential difference (or voltage drop) across the component, measured in volts

      $I(t)$ is the current through it, measured in amperes

If the component is a resistor with time-invariant voltage to current ratio, then:

$$P = I \cdot V = I^2 \cdot R = \frac{V^2}{R}$$

where

$$R = \frac{V}{I}$$

is the resistance, measured in ohms.

Peak Power and Duty Cycle

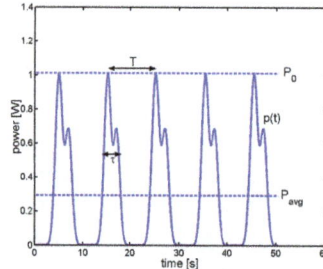

In a train of identical pulses, the instantaneous power is a periodic function of time. The ratio of the pulse duration to the period is equal to the ratio of the average power to the peak power. It is also called the duty cycle

In the case of a periodic signal $s(t)$ of period T, like a train of identical pulses, the instantaneous power $p(t) = |s(t)|^2$ is also a periodic function of period T. The *peak power* is simply defined by:

$$P_0 = \max[p(t)].$$

The peak power is not always readily measurable, however, and the measurement of the average power P_{avg} is more commonly performed by an instrument. If one defines the energy per pulse as:

$$\epsilon_{pulse} = \int_0^T p(t)dt$$

then the average power is:

$$P_{avg} = \frac{1}{T}\int_0^T p(t)dt = \frac{\epsilon_{pulse}}{T}.$$

One may define the pulse length τ such that $P_0\tau = \epsilon_{pulse}$ so so that the ratios

$$\frac{P_{avg}}{P_0} = \frac{\tau}{T}$$

are equal. These ratios are called the *duty cycle* of the pulse train.

Energy

In physics, energy is a property of objects which can be transferred to other objects or converted into different forms. The "ability of a system to perform work" is a common description, but it is misleading because energy is not necessarily available to do work. For instance, in SI units, energy is measured in joules, and one joule is defined "mechanically", being the energy transferred to an object by the mechanical work of moving it a distance of 1 metre against a force of 1 newton.

However, there are many other definitions of energy, depending on the context, such as thermal energy, radiant energy, electromagnetic, nuclear, etc., where definitions are derived that are the most convenient.

The Sun is the source of energy for life on Earth. It releases mainly thermal energy, heat energy and radiant (light) energy.

Common energy forms include the kinetic energy of a moving object, the potential energy stored by an object's position in a force field (gravitational, electric or magnetic), the elastic energy stored by stretching solid objects, the chemical energy released when a fuel burns, the radiant energy carried by light, and the thermal energy due to an object's temperature. All of the many forms of energy are convertible to other kinds of energy. In Newtonian physics, there is a universal law of conservation of energy which says that energy can be neither created nor be destroyed; however, it can change from one form to another.

For "closed systems" with no external source or sink of energy, the first law of thermodynamics states that a system's energy is constant unless energy is transferred in or out by mechanical work or heat, and that no energy is lost in transfer. This means that it is impossible to create or destroy energy. While heat can always be fully converted into work in a reversible isothermal expansion of an ideal gas, for cyclic processes of practical interest in heat engines the second law of thermodynamics states that the system doing work always loses some energy as waste heat. This creates a limit to the amount of heat energy that can do work in a cyclic process, a limit called the available energy. Mechanical and other forms of energy can be transformed in the other direction into thermal energy without such limitations. The total energy of a system can be calculated by adding up all forms of energy in the system.

Examples of energy transformation include generating electric energy from heat energy via a steam turbine, or lifting an object against gravity using electrical energy driving a crane motor. Lifting against gravity performs mechanical work on the object and stores gravitational potential energy in the object. If the object falls to the ground, gravity does mechanical work on the object which transforms the potential energy in the gravitational field to the kinetic energy released as heat on impact with the ground. Our Sun transforms nuclear potential energy to other forms of energy; its total mass does not decrease due to that in itself (since it still contains the same total energy even if in different forms), but its mass does decrease when the energy escapes out to its surroundings, largely as radiant energy.

Mass and energy are closely related. According to the theory of mass–energy equivalence, any object that has mass when stationary in a frame of reference (called rest mass) also has an equivalent amount of energy whose form is called rest energy in that frame, and any additional energy acquired by the object above that rest energy will increase an object's mass. For example, with a sensitive enough scale, one could measure an increase in mass after heating an object.

Living organisms require available energy to stay alive, such as the energy humans get from food. Civilisation gets the energy it needs from energy resources such as fossil fuels, nuclear fuel, or renewable energy. The processes of Earth's climate and ecosystem are driven by the radiant energy Earth receives from the sun and the geothermal energy contained within the earth. In biology, energy can be thought of as what's needed to keep entropy low.

Forms

In a typical lightning strike, 500 megajoules of electric potential energy is converted into the same amount of energy in other forms, mostly light energy, sound energy and thermal energy.

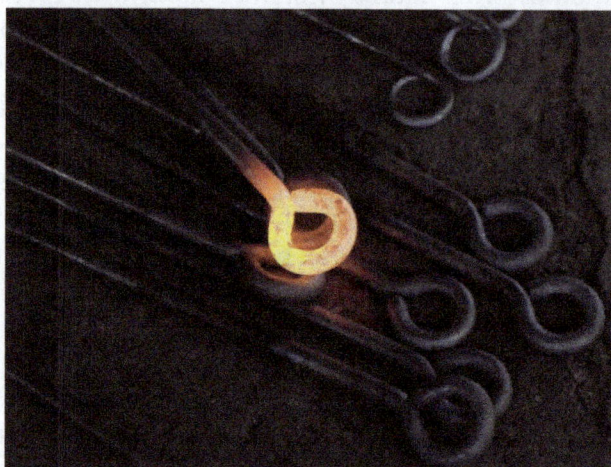

Thermal energy is energy of microscopic constituents of matter, which may include both kinetic and potential energy.

The total energy of a system can be subdivided and classified in various ways. For example, classical mechanics distinguishes between kinetic energy, which is determined by an object's movement through space, and potential energy, which is a function of the position of an object within a field. It may also be convenient to distinguish gravitational energy, thermal energy, several types of nuclear energy (which utilize potentials from the nuclear force and the weak force), electric energy (from the electric field), and magnetic energy (from the magnetic field), among others. Many of these classifications overlap; for instance, thermal energy usually consists partly of kinetic and partly of potential energy.

Some types of energy are a varying mix of both potential and kinetic energy. An example is mechanical energy which is the sum of (usually macroscopic) kinetic and potential energy in a system. Elastic energy in materials is also dependent upon electrical potential energy (among atoms and molecules), as is chemical energy, which is stored and released from a reservoir of electrical potential energy between electrons, and the molecules or atomic nuclei that attract them. The list is also not necessarily complete. Whenever physical scientists discover that a certain phenomenon appears to violate the law of energy conservation, new forms are typically added that account for the discrepancy.

Heat and work are special cases in that they are not properties of systems, but are instead properties of *processes* that transfer energy. In general we cannot measure how much heat or work are present in an object, but rather only how much energy is transferred among objects in certain ways during the occurrence of a given process. Heat and work are measured as positive or negative depending on which side of the transfer we view them from.

Potential energies are often measured as positive or negative depending on whether they are greater or less than the energy of a specified base state or configuration such as two interacting bodies being infinitely far apart. Wave energies (such as radiant or sound energy), kinetic energy, and rest energy are each greater than or equal to zero because they are measured in comparison to a base state of zero energy: "no wave", "no motion", and "no inertia", respectively.

The distinctions between different kinds of energy is not always clear-cut. As Richard Feynman points out:

These notions of potential and kinetic energy depend on a notion of length scale. For example, one can speak of *macroscopic* potential and kinetic energy, which do not include thermal potential and kinetic energy. Also what is called chemical potential energy is a macroscopic notion, and closer examination shows that it is really the sum of the potential *and kinetic* energy on the atomic and subatomic scale. Similar remarks apply to nuclear "potential" energy and most other forms of energy. This dependence on length scale is non-problematic if the various length scales are decoupled, as is often the case ... but confusion can arise when different length scales are coupled, for instance when friction converts macroscopic work into microscopic thermal energy.

Some examples of different kinds of energy:

Forms of energy	
Type of energy	**Description**
Kinetic	(≥ 0), that of the motion of a body
Potential	A category comprising many forms in this list
Mechanical	The sum of (usually macroscopic) kinetic and potential energies
Mechanical wave	(≥ 0), a form of mechanical energy propagated by a material's oscillations
Chemical	that contained in molecules

Electric	that from electric fields
Magnetic	that from magnetic fields
Radiant	(≥ 0), that of electromagnetic radiation, including light
Chromodynamic	that of binding quarks to form hadrons
Nuclear	that of binding nucleons to form the atomic nucleus
Ionization	that of binding an electron to its atom or molecule
Elastic	that of deformation of a material (or its container) exhibiting a restorative force
Gravitational	that from gravitational fields
Rest	(≥ 0) that equivalent to an object's rest mass
Thermal	A microscopic, disordered equivalent of mechanical energy
Heat	an amount of thermal energy being transferred (in a given process) in the direction of decreasing temperature
Mechanical work	an amount of energy being transferred in a given process due to displacement in the direction of an applied force

History

Thomas Young – the first to use the term "energy" in the modern sense.

The word *energy* derives from the Ancient Greek, which possibly appears for the first time in the work of Aristotle in the 4th century BC. In contrast to the modern definition, energeia was a qualitative philosophical concept, broad enough to include ideas such as happiness and pleasure.

In the late 17th century, Gottfried Leibniz proposed the idea of the Latin: *vis viva*, or living force, which defined as the product of the mass of an object and its velocity squared; he believed that total *vis viva* was conserved. To account for slowing due to friction, Leibniz theorized that thermal energy consisted of the random motion of the constituent parts of matter, a view shared by Isaac Newton, although it would be more than a century until this was generally accepted. The modern analog of this property, kinetic energy, differs from *vis viva* only by a factor of two.

In 1807, Thomas Young was possibly the first to use the term "energy" instead of *vis viva*, in its modern sense. Gustave-Gaspard Coriolis described "kinetic energy" in 1829 in its modern sense, and in 1853, William Rankine coined the term "potential energy". The law of conservation of energy was also first postulated in the early 19th century, and applies to any isolated system. It was argued for some years whether heat was a physical substance, dubbed the caloric, or merely a physical quantity, such as momentum. In 1845 James Prescott Joule discovered the link between mechanical work and the generation of heat.

These developments led to the theory of conservation of energy, formalized largely by William Thomson (Lord Kelvin) as the field of thermodynamics. Thermodynamics aided the rapid development of explanations of chemical processes by Rudolf Clausius, Josiah Willard Gibbs, and Walther Nernst. It also led to a mathematical formulation of the concept of entropy by Clausius and to the introduction of laws of radiant energy by Jožef Stefan. According to Noether's theorem, the conservation of energy is a consequence of the fact that the laws of physics do not change over time. Thus, since 1918, theorists have understood that the law of conservation of energy is the direct mathematical consequence of the translational symmetry of the quantity conjugate to energy, namely time.

Units of Measure

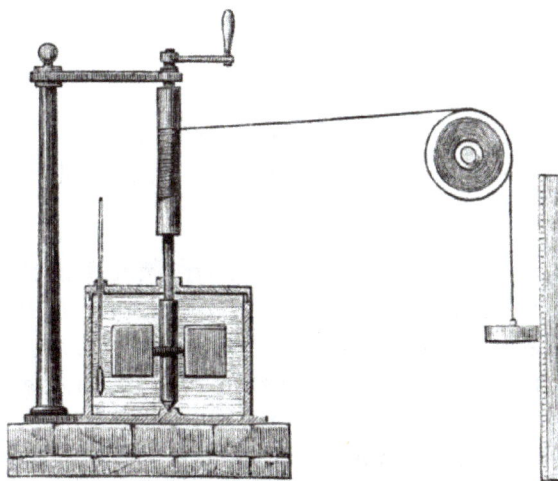

Joule's apparatus for measuring the mechanical equivalent of heat.
A descending weight attached to a string causes a paddle immersed in water to rotate.

In 1843 James Prescott Joule independently discovered the mechanical equivalent in a series of experiments. The most famous of them used the "Joule apparatus": a descending weight, attached to a string, caused rotation of a paddle immersed in water, practically insulated from heat transfer. It showed that the gravitational potential energy lost by the weight in descending was equal to the internal energy gained by the water through friction with the paddle.

In the International System of Units (SI), the unit of energy is the joule, named after James Prescott Joule. It is a derived unit. It is equal to the energy expended (or work done) in applying a force of one newton through a distance of one metre. However energy is also expressed in many other units not part of the SI, such as ergs, calories, British Thermal Units, kilowatt-hours and kilocalories, which require a conversion factor when expressed in SI units.

The SI unit of energy rate (energy per unit time) is the watt, which is a joule per second. Thus, one joule is one watt-second, and 3600 joules equal one watt-hour. The CGS energy unit is the erg and the imperial and US customary unit is the foot pound. Other energy units such as the electronvolt, food calorie or thermodynamic kcal (based on the temperature change of water in a heating process), and BTU are used in specific areas of science and commerce.

Scientific Use

Classical Mechanics

In classical mechanics, energy is a conceptually and mathematically useful property, as it is a conserved quantity. Several formulations of mechanics have been developed using energy as a core concept.

Work, a form of energy, is force times distance.

$$W = \int_C \mathbf{F} \cdot d\mathbf{s}$$

This says that the work (W) is equal to the line integral of the force F along a path C. Work and thus energy is frame dependent. For example, consider a ball being hit by a bat. In the center-of-mass reference frame, the bat does no work on the ball. But, in the reference frame of the person swinging the bat, considerable work is done on the ball.

The total energy of a system is sometimes called the Hamiltonian, after William Rowan Hamilton. The classical equations of motion can be written in terms of the Hamiltonian, even for highly complex or abstract systems. These classical equations have remarkably direct analogs in nonrelativistic quantum mechanics.

Another energy-related concept is called the Lagrangian, after Joseph-Louis Lagrange. This formalism is as fundamental as the Hamiltonian, and both can be used to derive the equations of motion or be derived from them. It was invented in the context of classical mechanics, but is generally useful in modern physics. The Lagrangian is defined as the kinetic energy *minus* the potential energy. Usually, the Lagrange formalism is mathematically more convenient than the Hamiltonian for non-conservative systems (such as systems with friction).

Noether's theorem (1918) states that any differentiable symmetry of the action of a physical system has a corresponding conservation law. Noether's theorem has become a fundamental tool of modern theoretical physics and the calculus of variations. A generalisation of the seminal formulations on constants of motion in Lagrangian and Hamiltonian mechanics (1788 and 1833, respectively), it does not apply to systems that cannot be modeled with a Lagrangian; for example, dissipative systems with continuous symmetries need not have a corresponding conservation law.

Chemistry

In the context of chemistry, energy is an attribute of a substance as a consequence of its atomic, molecular or aggregate structure. Since a chemical transformation is accompanied by a change in one or more of these kinds of structure, it is invariably accompanied by an increase or decrease of energy of the substances involved. Some energy is transferred between the surroundings and the reactants of the reaction in the form of heat or light; thus the products of a reaction may have more or less energy than the reactants. A reaction is said to be exergonic if the final state is lower on the energy scale than the initial state; in the case of endergonic reactions the situation is the reverse. Chemical reactions are invariably not possible unless the reactants surmount an energy barrier known as the activation energy. The *speed* of a chemical reaction (at given temperature T) is related to the activation energy E, by the Boltzmann's population factor $e^{-E/kT}$ – that is the probability of molecule to have energy greater than or equal to E at the given temperature T. This exponential dependence of a reaction rate on temperature is known as the Arrhenius equation. The activation energy necessary for a chemical reaction can be in the form of thermal energy.

Biology

Energy and human life

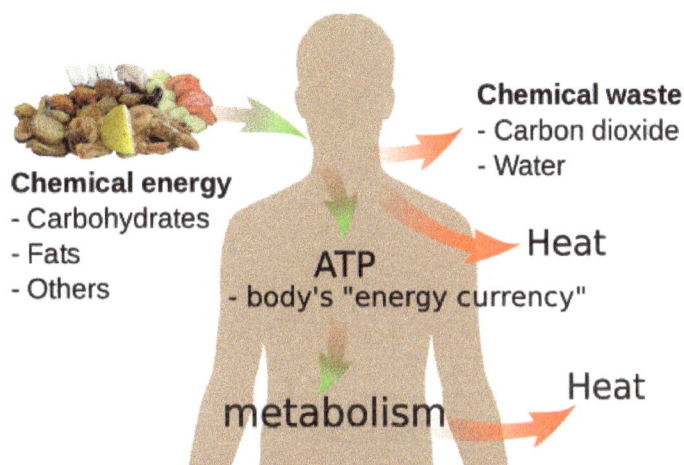

Basic overview of energy and human life.

In biology, energy is an attribute of all biological systems from the biosphere to the smallest living organism. Within an organism it is responsible for growth and development of a biological cell or an organelle of a biological organism. Energy is thus often said to be stored by cells in the structures of molecules of substances such as carbohydrates (including sugars), lipids, and proteins, which release energy when reacted with oxygen in respiration. In human terms, the human equivalent (H-e) (Human energy conversion) indicates, for a given amount of energy expenditure, the relative quantity of energy needed for human metabolism, assuming an average human energy expenditure of 12,500 kJ per day and a basal metabolic rate of 80 watts. For example, if our bodies run (on average) at 80 watts, then a light bulb running at 100 watts is running at 1.25 human equivalents (100 ÷ 80) i.e. 1.25 H-e. For a difficult task of only a few seconds' duration, a person can put out thousands of watts, many times the 746 watts in one official horsepower. For tasks lasting a few minutes, a fit human can generate perhaps 1,000 watts. For an activity that must be sustained for an hour, output drops to around 300; for an activity kept up all day, 150 watts is about the maximum. The human equivalent assists understanding of energy flows in physical and biological systems by expressing energy units in human terms: it provides a "feel" for the use of a given amount of energy.

Sunlight is also captured by plants as *chemical potential energy* in photosynthesis, when carbon dioxide and water (two low-energy compounds) are converted into the high-energy compounds carbohydrates, lipids, and proteins. Plants also release oxygen during photosynthesis, which is utilized by living organisms as an electron acceptor, to release the energy of carbohydrates, lipids, and proteins. Release of the energy stored during photosynthesis as heat or light may be triggered suddenly by a spark, in a forest fire, or it may be made available more slowly for animal or human metabolism, when these molecules are ingested, and catabolism is triggered by enzyme action.

Any living organism relies on an external source of energy—radiation from the Sun in the case of green plants, chemical energy in some form in the case of animals—to be able to grow and reproduce. The daily 1500–2000 Calories (6–8 MJ) recommended for a human adult are taken as a combination of oxygen and food molecules, the latter mostly carbohydrates and fats, of which glucose ($C_6H_{12}O_6$) and stearin ($C_{57}H_{110}O_6$) are convenient examples. The food molecules are oxidised to carbon dioxide and water in the mitochondria

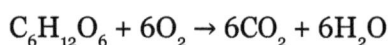

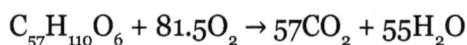

$$C_6H_{12}O_6 + 6O_2 \rightarrow 6CO_2 + 6H_2O$$

$$C_{57}H_{110}O_6 + 81.5O_2 \rightarrow 57CO_2 + 55H_2O$$

and some of the energy is used to convert ADP into ATP.

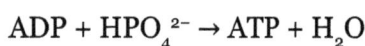

$$ADP + HPO_4{}^{2-} \rightarrow ATP + H_2O$$

The rest of the chemical energy in O_2 and the carbohydrate or fat is converted into heat: the ATP is used as a sort of "energy currency", and some of the chemical energy it contains is used for other metabolism when ATP reacts with OH groups and eventually splits into ADP and phosphate (at each stage of a metabolic pathway, some chemical energy is converted into heat). Only a tiny fraction of the original chemical energy is used for work:

gain in kinetic energy of a sprinter during a 100 m race: 4 kJ

gain in gravitational potential energy of a 150 kg weight lifted through 2 metres: 3 kJ

Daily food intake of a normal adult: 6–8 MJ

It would appear that living organisms are remarkably inefficient (in the physical sense) in their use of the energy they receive (chemical energy or radiation), and it is true that most real machines manage higher efficiencies. In growing organisms the energy that is converted to heat serves a vital purpose, as it allows the organism tissue to be highly ordered with regard to the molecules it is built from. The second law of thermodynamics states that energy (and matter) tends to become more evenly spread out across the universe: to concentrate energy (or matter) in one specific place, it is necessary to spread out a greater amount of energy (as heat) across the remainder of the universe ("the surroundings"). Simpler organisms can achieve higher energy efficiencies than more complex ones, but the complex organisms can occupy ecological niches that are not available to their simpler brethren. The conversion of a portion of the chemical energy to heat at each step in a metabolic pathway is the physical reason behind the pyramid of biomass observed in ecology: to take just the first step in the food chain, of the estimated 124.7 Pg/a of carbon that is fixed by photosynthesis, 64.3 Pg/a (52%) are used for the metabolism of green plants, i.e. reconverted into carbon dioxide and heat.

Earth Sciences

In geology, continental drift, mountain ranges, volcanoes, and earthquakes are phenomena that can be explained in terms of energy transformations in the Earth's interior, while meteorological phenomena like wind, rain, hail, snow, lightning, tornadoes and hurricanes are all a result of energy transformations brought about by solar energy on the atmosphere of the planet Earth.

Sunlight may be stored as gravitational potential energy after it strikes the Earth, as (for example) water evaporates from oceans and is deposited upon mountains (where, after being released at a hydroelectric dam, it can be used to drive turbines or generators to produce electricity). Sunlight also drives many weather phenomena, save those generated by volcanic events. An example of a solar-mediated weather event is a hurricane, which occurs when large unstable areas of warm ocean, heated over months, give up some of their thermal energy suddenly to power a few days of violent air movement.

In a slower process, radioactive decay of atoms in the core of the Earth releases heat. This thermal energy drives plate tectonics and may lift mountains, via orogenesis. This slow lifting represents a kind of gravitational potential energy storage of the thermal energy, which may be later released to active kinetic energy in landslides, after a triggering event. Earthquakes also release stored elastic potential energy in rocks, a store that has been produced ultimately from the same radioactive heat sources. Thus, according to present understanding, familiar events such as landslides and earthquakes release energy that has been stored as potential energy in the Earth's gravitational field or elastic strain (mechanical potential energy) in rocks. Prior to this, they represent release of energy that has been stored in heavy atoms since the collapse of long-destroyed supernova stars created these atoms.

Cosmology

In cosmology and astronomy the phenomena of stars, nova, supernova, quasars and gamma-ray bursts are the universe's highest-output energy transformations of matter. All stellar phenomena (including solar activity) are driven by various kinds of energy transformations. Energy in such transformations is either from gravitational collapse of matter (usually molecular hydrogen) into various classes of astronomical objects (stars, black holes, etc.), or from nuclear fusion (of lighter elements, primarily hydrogen). The nuclear fusion of hydrogen in the Sun also releases another store of potential energy which was created at the time of the Big Bang. At that time, according to theory, space expanded and the universe cooled too rapidly for hydrogen to completely fuse into heavier elements. This meant that hydrogen represents a store of potential energy that can be released by fusion. Such a fusion process is triggered by heat and pressure generated from gravitational collapse of hydrogen clouds when they produce stars, and some of the fusion energy is then transformed into sunlight.

Quantum Mechanics

In quantum mechanics, energy is defined in terms of the energy operator as a time derivative of the wave function. The Schrödinger equation equates the energy operator to the full energy of a particle or a system. Its results can be considered as a definition of measurement of energy in quantum mechanics. The Schrödinger equation describes the space- and time-dependence of a slowly changing (non-relativistic) wave function of quantum systems. The solution of this equation for a bound system is discrete (a set of permitted states, each characterized by an energy level) which results in the concept of quanta. In the solution of the Schrödinger equation for any oscillator (vibrator) and for electromagnetic waves in a vacuum, the resulting energy states are related to the frequency by Planck's relation: $E = h\nu$ (where h is Planck's constant and ν the frequency). In the case of an electromagnetic wave these energy states are called quanta of light or photons.

Relativity

When calculating kinetic energy (work to accelerate a mass from zero speed to some finite speed) relativistically – using Lorentz transformations instead of Newtonian mechanics – Einstein discovered an unexpected by-product of these calculations to be an energy term which does not vanish at zero speed. He called it rest mass energy: energy which every mass must possess even when being at rest. The amount of energy is directly proportional to the mass of body:

$$E = mc^2,$$

where

> m is the mass,

> c is the speed of light in vacuum,

> E is the rest mass energy.

For example, consider electron–positron annihilation, in which the rest mass of individual particles is destroyed, but the inertia equivalent of the system of the two particles (its invariant mass) remains (since all energy is associated with mass), and this inertia and invariant mass is carried off by photons which individually are massless, but as a system retain their mass. This is a reversible process – the inverse process is called pair creation – in which the rest mass of particles is created from energy of two (or more) annihilating photons. In this system the matter (electrons and positrons) is destroyed and changed to non-matter energy (the photons). However, the total system mass and energy do not change during this interaction.

In general relativity, the stress–energy tensor serves as the source term for the gravitational field, in rough analogy to the way mass serves as the source term in the non-relativistic Newtonian approximation.

It is not uncommon to hear that energy is "equivalent" to mass. It would be more accurate to state that every energy has an inertia and gravity equivalent, and because mass is a form of energy, then mass too has inertia and gravity associated with it.

In classical physics, energy is a scalar quantity, the canonical conjugate to time. In special relativity energy is also a scalar (although not a Lorentz scalar but a time component of the energy–momentum 4-vector). In other words, energy is invariant with respect to rotations of space, but not invariant with respect to rotations of space-time (= boosts).

Transformation

A turbo generator transforms the energy of pressurised steam into electrical energy

Energy may be transformed between different forms at various efficiencies. Items that transform between these forms are called transducers. Examples of transducers include a battery, from chemical energy to electric energy; a dam: gravitational potential energy to kinetic energy of moving water (and the blades of a turbine) and ultimately to electric energy through an electric generator; or a heat engine, from heat to work.

There are strict limits to how efficiently heat can be converted into work in a cyclic process, e.g. in a heat engine, as described by Carnot's theorem and the second law of thermodynamics. However, some energy transformations can be quite efficient. The direction of transformations in energy (what kind of energy is transformed to what other kind) is often determined by entropy (equal energy spread among all available degrees of freedom) considerations. In practice all energy transformations are permitted on a small scale, but certain larger transformations are not permitted because it is statistically unlikely that energy or matter will randomly move into more concentrated forms or smaller spaces.

Energy transformations in the universe over time are characterized by various kinds of potential energy that has been available since the Big Bang later being "released" (transformed to more active types of energy such as kinetic or radiant energy) when a triggering mechanism is available. Familiar examples of such processes include nuclear decay, in which energy is released that was originally "stored" in heavy isotopes (such as uranium and thorium), by nucleosynthesis, a process ultimately using the gravitational potential energy released from the gravitational collapse of supernovae, to store energy in the creation of these heavy elements before they were incorporated into the solar system and the Earth. This energy is triggered and released in nuclear fission bombs or in civil nuclear power generation. Similarly, in the case of a chemical explosion, chemical potential energy is transformed to kinetic energy and thermal energy in a very short time. Yet another example is that of a pendulum. At its highest points the kinetic energy is zero and the gravitational potential energy is at maximum. At its lowest point the kinetic energy is at maximum and is equal to the decrease of potential energy. If one (unrealistically) assumes that there is no friction or other losses, the conversion of energy between these processes would be perfect, and the pendulum would continue swinging forever.

Energy is also transferred from potential energy (E_p) to kinetic energy (E_k) and then back to potential energy constantly. This is referred to as conservation of energy. In this closed system, energy cannot be created or destroyed; therefore, the initial energy and the final energy will be equal to each other. This can be demonstrated by the following:

$$E_{pi} + E_{ki} = E_{pF} + E_{kF} \qquad\qquad (4)$$

The equation can then be simplified further since $E_p = mgh$ (mass times acceleration due to gravity times the height) and $E_k = \frac{1}{2}mv^2$ (half mass times velocity squared). Then the total amount of energy can be found by adding .

Conservation of Energy and Mass in Transformation

Energy gives rise to weight when it is trapped in a system with zero momentum, where it can be weighed. It is also equivalent to mass, and this mass is always associated with it. Mass is also equivalent to a certain amount of energy, and likewise always appears associated with it, as described in mass-energy equivalence. The formula $E = mc^2$, derived by Albert Einstein (1905) quantifies the relationship between rest-mass and rest-energy within the concept of special relativity. In different theoretical frameworks, similar formulas were derived by J. J. Thomson (1881), Henri Poincaré (1900), Friedrich Hasenöhrl (1904) and others.

Matter may be converted to energy (and vice versa), but mass cannot ever be destroyed; rather, mass/energy equivalence remains a constant for both the matter and the energy, during any process when they are converted into each other. However, since c^2 is extremely large relative to ordinary human scales, the conversion of ordinary amount of matter (for example, 1 kg) to other forms of energy (such as heat, light, and other radiation) can liberate tremendous amounts of energy (~ 9×10^{16} joules = 21 megatons of TNT), as can be seen in nuclear reactors and nuclear weapons. Conversely, the mass equivalent of a unit of energy is minuscule, which is why a loss of energy (loss of mass) from most systems is difficult to measure by weight, unless the energy loss is very large. Examples of energy transformation into matter (i.e., kinetic energy into particles with rest mass) are found in high-energy nuclear physics.

Reversible and Non-reversible Transformations

Thermodynamics divides energy transformation into two kinds: reversible processes and irreversible processes. An irreversible process is one in which energy is dissipated (spread) into empty energy states available in a volume, from which it cannot be recovered into more concentrated forms (fewer quantum states), without degradation of even more energy. A reversible process is one in which this sort of dissipation does not happen. For example, conversion of energy from one type of potential field to another, is reversible, as in the pendulum system described above. In processes where heat is generated, quantum states of lower energy, present as possible excitations in fields between atoms, act as a reservoir for part of the energy, from which it cannot be recovered, in order to be converted with 100% efficiency into other forms of energy. In this case, the energy must partly stay as heat, and cannot be completely recovered as usable energy, except at the price of an increase in some other kind of heat-like increase in disorder in quantum states, in the universe (such as an expansion of matter, or a randomisation in a crystal).

As the universe evolves in time, more and more of its energy becomes trapped in irreversible states (i.e., as heat or other kinds of increases in disorder). This has been referred to as the inevitable thermodynamic heat death of the universe. In this heat death the energy of the universe does not change, but the fraction of energy which is available to do work through a heat engine, or be transformed to other usable forms of energy (through the use of generators attached to heat engines), grows less and less.

Conservation of Energy

According to conservation of energy, energy can neither be created (produced) nor destroyed by itself. It can only be transformed. The total inflow of energy into a system must equal the total outflow of energy from the system, plus the change in the energy contained within the system. Energy is subject to a strict global conservation law; that is, whenever one measures (or calculates) the total energy of a system of particles whose interactions do not depend explicitly on time, it is found that the total energy of the system always remains constant.

Richard Feynman said during a 1961 lecture:

There is a fact, or if you wish, a *law*, governing all natural phenomena that are known to date. There is no known exception to this law—it is exact so far as we know. The law is called the *conservation of energy*. It states that there is a certain quantity, which we call energy, that does not change in manifold changes which nature undergoes. That is a most abstract idea, because it is a mathematical principle; it says that there is a numerical quantity which does not change when something happens. It is not a description of a mechanism, or anything concrete; it is just a strange fact that we can calculate some number and when we finish watching nature go through her tricks and calculate the number again, it is the same.

— The Feynman Lectures on Physics

Most kinds of energy (with gravitational energy being a notable exception) are subject to strict local conservation laws as well. In this case, energy can only be exchanged between adjacent regions of space, and all observers agree as to the volumetric density of energy in any given space. There is also a global law of conservation of energy, stating that the total energy of the universe cannot change; this is a corollary of the local law, but not vice versa.

This law is a fundamental principle of physics. As shown rigorously by Noether's theorem, the conservation of energy is a mathematical consequence of translational symmetry of time, a property of most phenomena below the cosmic scale that makes them independent of their locations on the time coordinate. Put differently, yesterday, today, and tomorrow are physically indistinguishable. This is because energy is the quantity which is canonical conjugate to time. This mathematical entanglement of energy and time also results in the uncertainty principle - it is impossible to define the exact amount of energy during any definite time interval. The uncertainty principle should not be confused with energy conservation - rather it provides mathematical limits to which energy can in principle be defined and measured.

Each of the basic forces of nature is associated with a different type of potential energy, and all types of potential energy (like all other types of energy) appears as system mass, whenever present. For example, a compressed spring will be slightly more massive than before it was compressed. Likewise, whenever energy is transferred between systems by any mechanism, an associated mass is transferred with it.

In quantum mechanics energy is expressed using the Hamiltonian operator. On any time scales, the uncertainty in the energy is by

$$\Delta E \Delta t \geq \frac{\hbar}{2}$$

which is similar in form to the Heisenberg Uncertainty Principle (but not really mathematically equivalent thereto, since H and t are not dynamically conjugate variables, neither in classical nor in quantum mechanics).

In particle physics, this inequality permits a qualitative understanding of virtual particles which carry momentum, exchange by which and with real particles, is responsible for the creation of all known fundamental forces (more accurately known as fundamental interactions). Virtual photons (which are simply lowest quantum mechanical energy state of photons) are also responsible for electrostatic interaction between electric charges (which results in Coulomb law), for spontaneous radiative decay of exited atomic and nuclear states, for the Casimir force, for van der Waals bond forces and some other observable phenomena.

Energy Transfer

Closed Systems

Energy transfer can be considered for the special case of systems which are closed to transfers of matter. The portion of the energy which is transferred by conservative forces over a distance is measured as the work the source system does on the receiving system. The portion of the energy which does not do work during the transfer is called heat. Energy can be transferred between systems in a variety of ways. Examples include the transmission of electromagnetic energy via photons, physical collisions which transfer kinetic energy, and the conductive transfer of thermal energy.

Energy is strictly conserved and is also locally conserved wherever it can be defined. In thermodynamics, for closed systems, the process of energy transfer is described by the first law:

$$\Delta E = W + Q \tag{1}$$

where E is the amount of energy transferred, W represents the work done on the system, and Q represents the heat flow into the system. As a simplification, the heat term, Q, is sometimes ignored, especially when the thermal efficiency of the transfer is high.

$$\Delta E = W \tag{2}$$

This simplified equation is the one used to define the joule, for example.

Open Systems

Beyond the constraints of closed systems, open systems can gain or lose energy in association with matter transfer (both of these process are illustrated by fueling an auto, a system which gains in energy thereby, without addition of either work or heat). Denoting this energy by E, one may write

$$\Delta E = W + Q + E. \tag{3}$$

Thermodynamics

Internal Energy

Internal energy is the sum of all microscopic forms of energy of a system. It is the energy needed to create the system. It is related to the potential energy, e.g., molecular structure, crystal structure, and other geometric aspects, as well as the motion of the particles, in form of kinetic energy. Thermodynamics is chiefly concerned with changes in internal energy and not its absolute value, which is impossible to determine with thermodynamics alone.

First Law of Thermodynamics

The first law of thermodynamics asserts that energy (but not necessarily thermodynamic free energy) is always conserved and that heat flow is a form of energy transfer. For homogeneous systems, with a well-defined temperature and pressure, a commonly used corollary of the first law is that, for a system subject only to pressure forces and heat transfer (e.g., a cylinder-full of gas) without chemical changes, the differential change in the internal energy of the system (with a *gain* in energy signified by a positive quantity) is given as

$$dE = T dS - P dV \text{ ,}$$

where the first term on the right is the heat transferred into the system, expressed in terms of temperature T and entropy S (in which entropy increases and the change dS is positive when the system is heated), and the last term on the right hand side is identified as work done on the system, where pressure is P and volume V (the negative sign results since compression of the system requires work to be done on it and so the volume change, dV, is negative when work is done on the system).

This equation is highly specific, ignoring all chemical, electrical, nuclear, and gravitational forces, effects such as advection of any form of energy other than heat and pV-work. The general formulation of the first law (i.e., conservation of energy) is valid even in situations in which the system is not homogeneous. For these cases the change in internal energy of a *closed* system is expressed in a general form by

$$dE = \delta Q + \delta W$$

where δQ is the heat supplied to the system and δW is the work applied to the system.

Equipartition of Energy

The energy of a mechanical harmonic oscillator (a mass on a spring) is alternatively kinetic and potential. At two points in the oscillation cycle it is entirely kinetic, and alternatively at two other points it is entirely potential. Over the whole cycle, or over many cycles, net energy is thus equally split between kinetic and potential. This is called equipartition principle; total energy of a system with many degrees of freedom is equally split among all available degrees of freedom.

This principle is vitally important to understanding the behaviour of a quantity closely related to energy, called entropy. Entropy is a measure of evenness of a distribution of energy between parts of a system. When an isolated system is given more degrees of freedom (i.e., given new available energy states that are the same as existing states), then total energy spreads over all available degrees equally without distinction between "new" and "old" degrees. This mathematical result is called the second law of thermodynamics.

Forms of Energy

In the context of physical science, several forms of energy have been identified. These include:

Forms of energy	
Type of energy	**Description**
Kinetic	(≥ 0), that of the motion of a body
Potential	A category comprising many forms in this list
Mechanical	The sum of (usually macroscopic) kinetic and potential energies
Mechanical wave	(≥ 0), a form of mechanical energy propagated by a material's oscillations
Chemical	that contained in molecules
Electric	that from electric fields
Magnetic	that from magnetic fields
Radiant	(≥ 0), that of electromagnetic radiation, including light
Chromodynamic	that of binding quarks to form hadrons
Nuclear	that of binding nucleons to form the atomic nucleus
Ionization	that of binding an electron to its atom or molecule
Elastic	that of deformation of a material (or its container) exhibiting a restorative force
Gravitational	that from gravitational fields
Rest	(≥ 0) that equivalent to an object's rest mass
Thermal	A microscopic, disordered equivalent of mechanical energy
Heat	an amount of thermal energy being transferred (in a given process) in the direction of decreasing temperature
Mechanical work	an amount of energy being transferred in a given process due to displacement in the direction of an applied force

Some entries in the above list constitute or comprise others in the list. The list is not necessarily complete. Whenever physical scientists discover that a certain phenomenon appears to violate the law of energy conservation, new forms are typically added that account for the discrepancy.

Heat and work are special cases in that they are not properties of systems, but are instead properties of *processes* that transfer energy. In general we cannot measure how much heat or work are present in an object, but rather only how much energy is transferred among objects in certain ways during the occurrence of a given process. Heat and work are measured as positive or negative depending on which side of the transfer we view them from.

Classical mechanics distinguishes between kinetic energy, which is determined by an object's movement through space, and potential energy, which is a function of the position of an object within a field, which may itself be related to the arrangement of other objects or particles. These include gravitational energy (which is stored in the way masses are arranged in a gravitational field), several types of nuclear energy (which utilize potentials from the nuclear force and the weak force), electric energy (from the electric field), and magnetic energy (from the magnetic field).

Other familiar types of energy are a varying mix of both potential and kinetic energy. An example is mechanical energy which is the sum of (usually macroscopic) kinetic and potential energy in a system. Elastic energy in materials is also dependent upon electrical potential energy (among atoms and molecules), as is chemical energy, which is stored and released from a reservoir of electrical potential energy between electrons, and the molecules or atomic nuclei that attract them..

Potential energies are often measured as positive or negative depending on whether they are greater or less than the energy of a specified base state or configuration such as two interacting bodies being infinitely far apart.

Wave energies (such as light or sound energy), kinetic energy, and rest energy are each greater than or equal to zero because they are measured in comparison to a base state of zero energy: "no wave", "no motion", and "no inertia", respectively.

It has been attempted to categorize *all* forms of energy as either kinetic or potential, but as Richard Feynman points out:

These notions of potential and kinetic energy depend on a notion of length scale. For example, one can speak of *macroscopic* potential and kinetic energy, which do not include thermal potential and kinetic energy. Also what is called chemical potential energy is a macroscopic notion, and closer examination shows that it is really the sum of the potential *and kinetic* energy on the atomic and subatomic scale. Similar remarks apply to nuclear "potential" energy and most other forms of energy. This dependence on length scale is non-problematic if the various length scales are decoupled, as is often the case ... but confusion can arise when different length scales are coupled, for instance when friction converts macroscopic work into microscopic thermal energy.

Also, at relativistic speeds, defining kinetic energy is problematic because the energy due to the body's motion does not simply contribute additively to the total energy as it does at classical speeds.

Energy may be transformed between different forms at various efficiencies. Items that transform between these forms are called transducers.

Mechanical Energy

Examples of the interconversion of energy	
Mechanical energy is converted	
into	**by**
Mechanical energy	Lever
Thermal energy	Brakes
Electric energy	Dynamo

Electromagnetic radiation	Synchrotron
Chemical energy	Matches
Nuclear energy	Particle accelerator

General Non-relativistic Mechanics

Mechanical energy (symbols E_M or E) manifest in many forms, but can be broadly classified into potential energy (E_p, V, U or Φ) and kinetic energy (E_k or T). The term potential energy is a very general term, because it exists in all force fields, such as gravitation, electrostatic and magnetic fields. Potential energy refers to the energy any object gain due to its position in a force field.

The relation between mechanical energy with kinetic and potential energy is simply

$$E = T + V ..$$

Lagrangian and Hamiltonian Mechanics

In more advanced topics, kinetic plus potential energy is physically the total energy of the system, but also known as the Hamiltonian of the system:

$$H = T + V,$$

used in Hamilton's equations of motion, to obtain equations describing a classical system in terms of energy rather than forces. The Hamiltonian is just a mathematical expression, rather than a *form* of energy.

Another analogous quantity of diverse applicability and efficiency is the Lagrangian of the system:

$$L = T - V ,$$

used in Lagrange's equations of motion, which serve the same purpose as Hamilton's equations.

Kinetic Energy

General Scope

Kinetic energy is the work required to accelerate an object to a given speed. In general:

$$E_k = \int \mathbf{F} \cdot d\mathbf{x} = \int \mathbf{v} \cdot d\mathbf{p}$$

Classical Mechanics

In classical mechanics, for a particle of constant mass m, in which case the force acting on it is F = ma where a is the particle's acceleration vector, the integral is:

$$E_k = \int \mathbf{F} \cdot d\mathbf{x} = m \int \frac{d\mathbf{v}}{dt} \cdot \mathbf{v} dt = m \int d\mathbf{v} \cdot \mathbf{v} = m \int \frac{1}{2} d(\mathbf{v} \cdot \mathbf{v}) = \frac{1}{2} m v^2$$

Special Relativistic Mechanics

At speeds approaching the speed of light c, this work must be calculated using Lorentz transformations, and applying mass and energy conservation, which results in

$$E_k = (\gamma - 1) mc^2,$$

where

$$\gamma = \frac{1}{\sqrt{1 - \left(\dfrac{v}{c}\right)^2}}$$

is the lorentz factor.

Here the two terms on the right hand side are identified with the total energy and the rest energy of the object, respectively. This equation reduces to the one above it, at small (compared to c) speed. The kinetic energy is zero at v=0 (when $\gamma = 1$), so that at rest, the total energy is the rest energy. So a mass at rest in some inertial reference frame has a corresponding amount of rest energy equal to:

$$E_0 = m_0 c^2$$

All masses at rest have a tremendous amount of energy, due to the proportionality factor of c^2.

Potential Energy

Potential energy is defined as the work done *against a given force* in changing the position of an object with respect to a reference position, often taken to be infinite separation. In other words, it is the work done on the object to give it that much energy. Changes in work and potential energy are related simply,

$$\Delta U = -\Delta W..$$

The name "potential" energy originally signified the idea that the energy could readily be transferred as work — at least in an idealized system. This is not completely true for any real system, but is often a reasonable first approximation in classical mechanics.

Mechanical Work

Translational Motion

If F is the force and r is the displacement, then the change in mechanical work done along the path between positions r_1 and r_2 due to the force is, in integral form:

$$\Delta W = \int_{r_1}^{r_2} \mathbf{F} \cdot d\mathbf{r} ,$$

(the dot represents the scalar product of the two vectors). The general equation above can be simplified in a number of common cases, notably when dealing with gravity or with elastic forces. If the force is conservative the equation can be written in differential form as

$$\mathbf{F} = \nabla W .$$

Rotational Motion

The rotational analogue is the work done by a torque τ, between the angles θ_1 and θ_2,

$$\Delta W = \int_{\theta_1}^{\theta_2} |\tau| d\theta.$$

Elastic Potential Energy

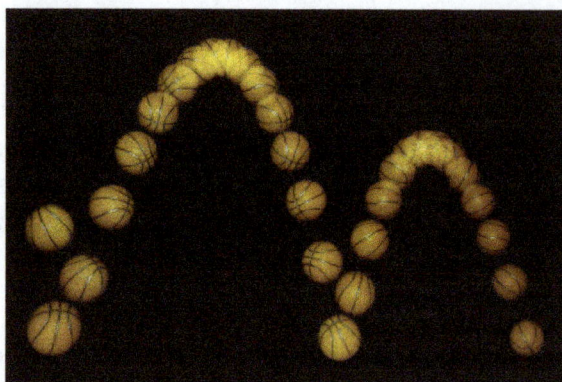

As a ball falls freely under the influence of gravity, it accelerates downward, its initial potential energy converting into kinetic energy. On impact with a hard surface the ball deforms, converting the kinetic energy into elastic potential energy. As the ball springs back, the energy converts back firstly to kinetic energy and then as the ball re-gains height into potential energy. Energy conversion to heat due to inelastic deformation and air resistance cause each successive bounce to be lower than the last.

Elastic potential energy is defined as a work needed to compress or extend a spring. The tension/compression force F in a spring or any other system which obeys Hooke's law is proportional to the extension/compression x,

$$\mathbf{F} = k\mathbf{x} ,$$

where k is the force constant of the particular spring or system. In this case the force is conservative, the calculated work becomes

$$E_{p,e} = \frac{1}{2} kx^2 .$$

If k is not constant the above equation will fail. Hooke's law is a good approximation for behaviour of chemical bonds under stable conditions, i.e. when they are not being broken or formed.

Surface Energy

If there is any kind of tension in a surface, such as a stretched sheet of rubber or material interfaces, it is possible to define surface energy.

If γ is the surface tension, and S = surface area, then the work done W to increase the area by a unit area is the surface energy:

$$dW = \gamma dS.$$

In particular, any meeting of dissimilar materials that do not mix will result in some kind of surface tension, if there is freedom for the surfaces to move then, as seen in capillary surfaces for example, the minimum energy will as usual be sought.

A minimal surface, for example, represents the smallest possible energy that a surface can have if its energy is proportional to the area of the surface. For this reason, (open) soap films of small size are minimal surfaces (small size reduces gravity effects, and openness prevents pressure from building up. Note that a bubble is a minimum energy surface but not a minimal surface by definition).

Sound Energy

Sound is a form of mechanical vibration which propagates through any mechanical medium. It is closely related to the ability of the human ear to perceive sound. The wide outer area of the ear is maximized to collect sound vibrations. It is amplified and passed through the outer ear, striking the eardrum, which transmits sounds into the inner ear. Auditory nerves fire according to the particular vibrations of the sound waves in the inner ear, which designate such things as the pitch and volume of the sound. The ear is set up in an optimal way to interpret sound energy in the form of vibrations.

Gravitational Potential Energy

The gravitational force very near the surface of a massive body (e.g. a planet) varies very little with small changes in height, h, and locally is equal mg where m is mass, and g is the gravitational acceleration (AKA field strength). At the Earth's surface $g = 9.81$ m s^{-1}. In these cases, the gravitational potential energy is given by

$$E_{p,g} \approx mgh$$

A more general expression for the potential energy due to Newtonian gravitation between two bodies of masses m_1 and m_2, is

$$E_{p,g} = -\frac{Gm_1m_2}{r},$$

where r is the separation between the two bodies and G is the gravitational constant, $6.6742(10) \times 10^{-11}$ m^3 kg^{-1} s^{-2}. In this case, the zero potential reference point is the infinite separation of the two bodies. Care must be taken that these masses are point masses or uniform spherical solids/shells. It cannot be applied directly to any objects of any shape and any mass.

In terms of the gravitational potential (Φ, U or V), the potential energy is (by definition of gravitational potential),

$$E_{p,g} = -\Phi m .$$

Thermal Energy

Examples of the interconversion of energy	
Thermal energy is converted	
into	**by**
Mechanical energy	Steam turbine
Thermal energy	Heat exchanger
Electric energy	Thermocouple
Electromagnetic radiation	Hot objects
Chemical energy	Blast furnace
Nuclear energy	Supernova

General Scope

Thermal energy (of some state of matter - gas, plasma, solid, etc.) is the energy associated with the microscopical random motion of particles constituting the media. For example, in case of monatomic gas it is just a kinetic energy of motion of atoms of gas as measured in the reference frame of the center of mass of gas. In case of molecules in the gas rotational and vibrational energy is involved. In the case of liquids and solids there is also potential energy (of interaction of atoms) involved, and so on.

Heat is defined as a transfer (flow) of thermal energy across certain boundary (for example, from a hot body to cold via the area of their contact). A practical definition for small transfers of heat is

$$\Delta q = \int_{T_1}^{T_2} C_v dT$$

where C_v is the heat capacity of the system. This definition will fail if the system undergoes a phase transition—e.g. if ice is melting to water—as in these cases the system can absorb heat without increasing its temperature. In more complex systems, it is preferable to use the concept of internal energy rather than that of thermal energy.

Despite the theoretical problems, the above definition is useful in the experimental measurement of energy changes. In a wide variety of situations, it is possible to use the energy released by a system to raise the temperature of another object, e.g. a bath of water. It is also possible to measure the amount of electric energy required to raise the temperature of the object by the same amount. The calorie was originally defined as the amount of energy required to raise the temperature of one gram of water by 1 °C (approximately 4.1855 J, although the definition later changed), and the British thermal unit was defined as the energy required to heat one pound of water by 1 °F (later fixed as 1055.06 J).

Kinetic Theory

In kinetic theory which describes the ideal gas, the thermal energy per degree of freedom is given by:

$$U = \frac{d_f}{2} k_B T$$

where d_f is the number of degrees of freedom and k_B is the Boltzmann constant. The total thermal energies would equal the total internal energy of the gas, since intermolecular potential energy is neglected in this theory. The term $k_B T$ occurs very frequently in statistical thermodynamics.

Chemical Energy

Examples of the interconversion of energy	
Chemical energy is converted	
into	**by**
Mechanical energy	Muscle
Thermal energy	Fire
Electric energy	Fuel cell
Electromagnetic radiation	Glowworms
Chemical energy	Chemical reaction

Chemical energy is the energy due to excretion of atoms in molecules and various other kinds of aggregates of matter. It may be defined as a work done by electric forces during re-arrangement of mutual positions of electric charges, electrons and protons, in the process of aggregation. So, basically it is electrostatic potential energy of electric charges. If the chemical energy of a system decreases during a chemical reaction, the difference is transferred to the surroundings in some form (often heat or light); on the other hand if the chemical energy of a system increases as a result of a chemical reaction - the difference then is supplied by the surroundings (usually again in form of heat or light). For example,

when two hydrogen atoms react to form a dihydrogen molecule, the chemical energy *decreases* by 724 zJ (the bond energy of the H–H bond);

when the electron is completely removed from a hydrogen atom, forming a hydrogen ion (in the gas phase), the chemical energy *increases* by 2.18 aJ (the ionization energy of hydrogen).

It is common to quote the changes in chemical energy for one mole of the substance in question: typical values for the change in molar chemical energy during a chemical reaction range from tens to hundreds of kilojoules per mole.

The chemical energy as defined above is also referred to by chemists as the internal energy, U: technically, this is measured by keeping the volume of the system constant. Most practical chemistry is performed at constant pressure and, if the volume changes during the reaction (e.g. a gas is given off), a correction must be applied to take account of the work done by or on the atmosphere to obtain the enthalpy, H, this correction is the work done by an expanding gas,

$$\Delta E = p\Delta V,,$$

so the enthalpy now reads;

$$\Delta H = \Delta U + p\Delta V .$$

A second correction, for the change in entropy, S, must also be performed to determine whether a chemical reaction will take place or not, giving the Gibbs free energy, G. The correction is the energy required to create order from disorder,

$$\Delta E = T\Delta S \ ,$$

so we have;

$$\Delta G = \Delta H - T\Delta S \ .$$

These corrections are sometimes negligible, but often not (especially in reactions involving gases).

Since the industrial revolution, the burning of coal, oil, natural gas or products derived from them has been a socially significant transformation of chemical energy into other forms of energy. the energy "consumption" (one should really speak of "energy transformation") of a society or country is often quoted in reference to the average energy released by the combustion of these fossil fuels:

1 tonne of coal equivalent (TCE) = 29.3076 GJ = 8,141 kilowatt hour

1 tonne of oil equivalent (TOE) = 41.868 GJ = 11,630 kilowatt hour

On the same basis, a tank-full of gasoline (45 litres, 12 gallons) is equivalent to about 1.6 GJ of chemical energy. Another chemically based unit of measurement for energy is the "tonne of TNT", taken as 4.184 GJ. Hence, burning a tonne of oil releases about ten times as much energy as the explosion of one tonne of TNT: fortunately, the energy is usually released in a slower, more controlled manner.

Simple examples of storage of chemical energy are batteries and food. When food is digested and metabolized (often with oxygen), chemical energy is released, which can in turn be transformed into heat, or by muscles into kinetic energy.

According to the Bohr theory of the atom, the chemical energy is characterized by the Rydberg constant.

$$R_y = \frac{m_e e^4}{8\varepsilon_0^2 h^2} = \frac{1}{2}\alpha^2 m_e c^2 = 13.605\ 692\ 53(30)\ \text{eV}$$

Electrostatic Energy

General Scope

Examples of the interconversion of energy	
Electric energy is converted	
into	**by**
Mechanical energy	Electric motor
Thermal energy	Resistor
Electric energy	Transformer
Electromagnetic radiation	Light-emitting diode

| Chemical energy | Electrolysis |
| Nuclear energy | Synchrotron |

The electric potential energy of given configuration of charges is defined as the work which must be done against the Coulomb force to rearrange charges from infinite separation to this configuration (or the work done by the Coulomb force separating the charges from this configuration to infinity). For two point-like charges Q_1 and Q_2 at a distance r this work, and hence electric potential energy is equal to:

$$E_{p,e} = \frac{1}{4\pi\epsilon_0} \frac{Q_1 Q_2}{r}$$

where ε_0 is the electric constant of a vacuum, $10^7/4\pi c_0^2$ or $8.854188\ldots \times 10^{-12}$ F m^{-1}. In terms of electrostatic potential (ϕ for absolute, V for difference in potential), again by definition, electrostatic potential energy is given by:

$$E_{p,e} = \phi q. .$$

If the charge is accumulated in a capacitor (of capacitance C), the reference configuration is usually selected not to be infinite separation of charges, but vice versa - charges at an extremely close proximity to each other (so there is zero net charge on each plate of a capacitor). The justification for this choice is purely practical - it is easier to measure both voltage difference and magnitude of charges on a capacitor plates not versus infinite separation of charges but rather versus discharged capacitor where charges return to close proximity to each other (electrons and ions recombine making the plates neutral). In this case the work and thus the electric potential energy becomes

$$E_{p,e} = \frac{Q^2}{2C} = \frac{1}{2}CV^2 = \frac{1}{2}VQ ,$$

(different forms obtained using the definition of capacitance).

Electric Energy

Electric Circuits

If an electric current passes through a resistor, electric energy is converted to heat; if the current passes through an electric appliance, some of the electric energy will be converted into other forms of energy (although some will always be lost as heat). The amount of electric energy due to an electric current can be expressed in a number of different ways:

$$E = VQ = VIt = Pt = \frac{V^2 t}{R} = I^2 Rt$$

where V is the electric potential difference (in volts), Q is the charge (in coulombs), I is the current (in amperes), t is the time for which the current flows (in seconds), P is the power (in watts) and R is the electric resistance (in ohms). The last of these expressions is important in the practical measurement of energy, as potential difference, resistance and time can all be measured with considerable accuracy.

Magnetic Energy

General Scope

There is no fundamental difference between magnetic energy and electric energy: the two phenomena are related by Maxwell's equations. The potential energy of a magnet of magnetic moment m in a magnetic field B is defined as the work of magnetic force (actually of magnetic torque) on re-alignment of the vector of the magnetic dipole moment, and is equal to:

$$E_{p,m} = -\mathbf{m} \cdot \mathbf{B}.$$

Electric Circuits

The energy stored in an inductor (of inductance L) carrying current I is

$$E_{p,m} = \frac{1}{2} L I^2.$$

This second expression forms the basis for superconducting magnetic energy storage.

Electromagnetic Energy

Examples of the interconversion of energy	
Electromagnetic radiation is converted	
into	by
Mechanical energy	Solar sail
Thermal energy	Solar collector
Electric energy	Solar cell
Electromagnetic radiation	Non-linear optics
Chemical energy	Photosynthesis
Nuclear energy	Mössbauer spectroscopy

Calculating work needed to create an electric or magnetic field in unit volume (say, in a capacitor or an inductor) results in the electric and magnetic fields energy densities:

$$u_e = \frac{\epsilon_0}{2} |\mathbf{E}|^2, \quad u_m = \frac{1}{2\mu_0} |\mathbf{B}|^2,$$

in SI units.

Electromagnetic radiation, such as microwaves, visible light or gamma rays, represents a flow of electromagnetic energy. Applying the above expressions to magnetic and electric components of electromagnetic field both the volumetric density and the flow of energy in EM field can be calculated. The resulting Poynting vector, which is expressed as

$$\mathbf{S} = \frac{1}{\mu} \mathbf{E} \times \mathbf{B},$$

in SI units, gives the density of the flow of energy and its direction.

The energy of electromagnetic radiation is quantized (has discrete energy levels). The energy of a photon is:

$$E = h\nu = \frac{hc}{\lambda},$$

so the spacing between energy levels is:

$$\Delta E = E_2 - E_1 = hc\left(\nu_2 - \nu_1\right) = hc\left(\frac{1}{\lambda_2} - \frac{1}{\lambda_1}\right),$$

where h is the Planck constant, $6.6260693(11)\times10^{-34}$ Js, and ν is the frequency of the radiation. This quantity of electromagnetic energy is usually called a photon. The photons which make up visible light have energies of 270–520 yJ, equivalent to 160–310 kJ/mol, the strength of weaker chemical bonds.

Nuclear Energy

Examples of the interconversion of energy	
Nuclear binding energy is converted	
into	**by**
Mechanical energy	Alpha radiation
Thermal energy	Sun
Electrical energy	Beta radiation
Electromagnetic radiation	Gamma radiation
Chemical energy	Radioactive decay
Nuclear energy	Nuclear isomerism

Nuclear potential energy, along with electric potential energy, provides the energy released from nuclear fission and nuclear fusion processes. The result of both these processes are nuclei in which the more-optimal size of the nucleus allows the nuclear force (which is opposed by the electromagnetic force) to bind nuclear particles more tightly together than before the reaction.

The Weak nuclear force (different from the strong force) provides the potential energy for certain kinds of radioactive decay, such as beta decay.

The energy released in nuclear processes is so large that the relativistic change in mass (after the energy has been removed) can be as much as several parts per thousand.

Nuclear particles (nucleons) like protons and neutrons are *not* destroyed (law of conservation of baryon number) in fission and fusion processes. A few lighter particles may be created or destroyed (example: beta minus and beta plus decay, or electron capture decay), but these minor processes are not important to the immediate energy release in fission and fusion. Rather, fission and fusion release energy when collections of baryons become more tightly bound, and it is the energy associated with a fraction of the mass of the nucleons (but not the whole particles) which appears as the heat and electromagnetic radiation generated by nuclear reactions. This heat and radiation retains the "missing" mass, but the mass is missing only because it escapes in the form of heat or light, which retain the mass and conduct it out of the system where it is not measured.

The energy from the Sun, also called solar energy, is an example of this form of energy conversion. In the Sun, the process of hydrogen fusion converts about 4 million metric tons of solar "matter" per second into light, which is radiated into space, but during this process, although protons change into neutrons, the number of total protons-plus-neutrons does not change. In this system, the radiated light itself (as a system) retains the "missing" mass, which represents 4 million tons per second of electromagnetic radiation, moving into space. Each of the helium nuclei which are formed in the process are less massive than the four protons from they were formed, but (to a good approximation), no particles are destroyed in the process of turning the Sun's nuclear potential energy into light. Instead, the four nucleons in a helium nucleus in the Sun have an average mass that is less than the protons which formed them, and this mass difference (4 million tons/second) is the mass that moves off as sunlight.

Conservation of Energy

In physics, the law of conservation of energy states that the total energy of an isolated system remains constant—it is said to be *conserved* over time. Energy can neither be created nor destroyed; rather, it transforms from one form to another. For instance, chemical energy can be converted to kinetic energy in the explosion of a stick of dynamite.

A consequence of the law of conservation of energy is that a perpetual motion machine of the first kind cannot exist. That is to say, no system without an external energy supply can deliver an unlimited amount of energy to its surroundings.

History

Gottfried Leibniz

Ancient philosophers as far back as Thales of Miletus c. 550 BCE had inklings of the conservation of some underlying substance of which everything is made. However, there is no particular reason to identify this with what we know today as "mass-energy" (for example, Thales thought it was water). Empedocles (490–430 BCE) wrote that in his universal system, composed of four roots (earth, air, water, fire), "nothing comes to be or perishes"; instead, these elements suffer continual rearrangement.

In 1605, Simon Stevinus was able to solve a number of problems in statics based on the principle that perpetual motion was impossible.

In 1638, Galileo published his analysis of several situations—including the celebrated "interrupted pendulum"—which can be described (in modern language) as conservatively converting potential energy to kinetic energy and back again. Essentially, he pointed out that the height a moving body rises is equal to the height from which it falls, and used this observation to infer the idea of inertia. The remarkable aspect of this observation is that the height that a moving body ascends to does not depend on the shape of the frictionless surface that the body is moving on.

In 1669, Christian Huygens published his laws of collision. Among the quantities he listed as being invariant before and after the collision of bodies were both the sum of their linear momentums as well as the sum of their kinetic energies. However, the difference between elastic and inelastic collision was not understood at the time. This led to the dispute among later researchers as to which of these conserved quantities was the more fundamental. In his Horologium Oscillatorium, he gave a much more explicit and clearer statement regarding the height of ascent of a moving body, and connected this idea with the impossibility of a perpetual motion. Huygens' study of the dynamics of pendulum motion was based on a single principle: that the center of gravity of heavy objects cannot lift itself.

The fact that kinetic energy is scalar, unlike linear momentum which is a vector, and hence easier to work with did not escape the attention of Gottfried Wilhelm Leibniz. It was Leibniz during 1676–1689 who first attempted a mathematical formulation of the kind of energy which is connected with *motion* (kinetic energy). Using Huygens' work on collision, Leibniz noticed that in many mechanical systems (of several masses, m_i each with velocity v_i),

$$\sum_i m_i v_i^2$$

was conserved so long as the masses did not interact. He called this quantity the *vis viva* or *living force* of the system. The principle represents an accurate statement of the approximate conservation of kinetic energy in situations where there is no friction. Many physicists at that time, such as Newton, held that the conservation of momentum, which holds even in systems with friction, as defined by the momentum:

$$\sum_i m_i v_i$$

was the conserved *vis viva*. It was later shown that both quantities are conserved simultaneously, given the proper conditions such as an elastic collision.

In 1687, Isaac Newton published his Principia, which was organized around the concept of force and momentum. However, the researchers were quick to recognize that the principles set out in the book, while fine for point masses, were not sufficient to tackle the motions of rigid and fluid bodies. Some other principles were also required.

Daniel Bernoulli

The law of conservation of vis viva was championed by the father and son duo, Johann and Daniel Bernoulli. The former enunciated the principle of virtual work as used in statics in its full generality in 1715, while the later based his Hydrodynamica, published in 1738, on this single conservation principle. Daniel's study of loss of vis viva of flowing water led him to formulate the Bernoulli's principle, which relates the loss to be proportional to the change in hydrodynamic pressure. Daniel also formulated the notion of work and efficiency for hydraulic machines; and he gave a kinetic theory of gases, and linked the kinetic energy of gas molecules with the temperature of the gas.

This focus on the vis viva by the continental physicists eventually led to the discovery of stationarity principles governing mechanics, such as the D'Alembert's principle and Lagrangian and Hamiltonian formulations of mechanics.

Émilie du Châtelet (1706 – 1749) proposed and tested the hypothesis of the conservation of total energy, as distinct from momentum. Inspired by the theories of Gottfried Leibniz, she repeated and publicized an experiment originally devised by Willem 's Gravesande in 1722 in which balls were dropped from different heights into a sheet of soft clay. Each ball's kinetic energy - as indicated by the quantity of material displaced - was shown to be proportional to the square of the velocity. The deformation of the clay was found to be directly proportional to the height the balls were dropped from, equal to the initial potential energy. Earlier workers, including Newton and Voltaire, had all believed that "energy" (so far as they understood the concept at all) was not distinct from momentum and therefore proportional to velocity. According to this understanding, the deformation of the clay should have been proportional to the square root of the height from which the balls were dropped from. In classical physics the correct formula is $E_k = \frac{1}{2}mv^2$, , where E_k is the kinetic energy of an object, m its mass and v its speed. On this basis, Châtelet proposed that energy must always have the same dimensions in any form, which is necessary to be able to relate it in different forms (kinetic, potential, heat...).

Engineers such as John Smeaton, Peter Ewart, Carl Holtzmann, Gustave-Adolphe Hirn and Marc Seguin recognized that conservation of momentum alone was not adequate for practical calculation and made use of Leibniz's principle. The principle was also championed by some chemists such as William Hyde Wollaston. Academics such as John Playfair were quick to point out that kinetic energy is clearly not conserved. This is obvious to a modern analysis based on the second law of thermodynamics, but in the 18th and 19th centuries the fate of the lost energy was still unknown.

Gradually it came to be suspected that the heat inevitably generated by motion under friction was another form of *vis viva*. In 1783, Antoine Lavoisier and Pierre-Simon Laplace reviewed the two competing theories of *vis viva* and caloric theory. Count Rumford's 1798 observations of heat generation during the boring of cannons added more weight to the view that mechanical motion could be converted into heat, and (as importantly) that the conversion was quantitative and could be predicted (allowing for a universal conversion constant between kinetic energy and heat). *Vis viva* then started to be known as *energy*, after the term was first used in that sense by Thomas Young in 1807.

Gaspard-Gustave Coriolis

The recalibration of *vis viva* to

$$\frac{1}{2}\sum_i m_i v_i^2$$

which can be understood as converting kinetic energy to work, was largely the result of Gaspard-Gustave Coriolis and Jean-Victor Poncelet over the period 1819–1839. The former called the quantity *quantité de travail* (quantity of work) and the latter, *travail mécanique* (mechanical work), and both championed its use in engineering calculation.

In a paper *Über die Natur der Wärme*(German "On the Nature of Heat/Warmth"), published in the *Zeitschrift für Physik* in 1837, Karl Friedrich Mohr gave one of the earliest general statements of the doctrine of the conservation of energy in the words: "besides the 54 known chemical elements there is in the physical world one agent only, and this is called *Kraft* [energy or work]. It may appear, according to circumstances, as motion, chemical affinity, cohesion, electricity, light and magnetism; and from any one of these forms it can be transformed into any of the others."

Mechanical Equivalent of Heat

A key stage in the development of the modern conservation principle was the demonstration of the *mechanical equivalent of heat*. The caloric theory maintained that heat could neither be created nor destroyed, whereas conservation of energy entails the contrary principle that heat and mechanical work are interchangeable.

In the middle of the eighteenth century Mikhail Lomonosov, a Russian scientist, postulated his corpusculo-kinetic theory of heat, which rejected the idea of a caloric. Through the results of empirical studies, Lomonosov came to the conclusion that heat was not transferred through the particles of the caloric fluid.

In 1798, Count Rumford (Benjamin Thompson) performed measurements of the frictional heat generated in boring cannons, and developed the idea that heat is a form of kinetic energy; his measurements refuted caloric theory, but were imprecise enough to leave room for doubt.

James Prescott Joule

The mechanical equivalence principle was first stated in its modern form by the German surgeon Julius Robert von Mayer in 1842. Mayer reached his conclusion on a voyage to the Dutch East Indies, where he found that his patients' blood was a deeper red because they were consuming less oxygen, and therefore less energy, to maintain their body temperature in the hotter climate. He discovered that heat and mechanical work were both forms of energy and in 1845, after improving his knowledge of physics, he published a monograph that stated a quantitative relationship between them.

Meanwhile, in 1843, James Prescott Joule independently discovered the mechanical equivalent in a series of experiments. In the most famous, now called the "Joule apparatus", a descending weight attached to a string caused a paddle immersed in water to rotate. He showed that the gravitational potential energy lost by the weight in descending was equal to the internal energy gained by the water through friction with the paddle.

Over the period 1840–1843, similar work was carried out by engineer Ludwig A. Colding though it was little known outside his native Denmark.

Both Joule's and Mayer's work suffered from resistance and neglect but it was Joule's that eventually drew the wider recognition.

In 1844, William Robert Grove postulated a relationship between mechanics, heat, light, electricity and magnetism by treating them all as manifestations of a single "force" (*energy* in modern terms). In 1874, Grove published his theories in his book *The Correlation of Physical Forces*. In 1847, drawing on the earlier work of Joule, Sadi Carnot and Émile Clapeyron, Hermann von Helmholtz arrived at conclusions similar to Grove's and published his theories in his book *Über die Erhaltung der Kraft* (*On the Conservation of Force*, 1847). The general modern acceptance of the principle stems from this publication.

In 1850, William Rankine first used the phrase *the law of the conservation of energy* for the principle.

In 1877, Peter Guthrie Tait claimed that the principle originated with Sir Isaac Newton, based on a creative reading of propositions 40 and 41 of the *Philosophiae Naturalis Principia Mathematica*. This is now regarded as an example of Whig history.

Mass−energy Equivalence

Matter is composed of such things as atoms, electrons, neutrons, and protons. It has *intrinsic* or *rest* mass. In the limited range of recognized experience of the nineteenth century it was found that such rest mass is conserved. Einstein's 1905 theory of special relativity showed that it corresponds to an equivalent amount of *rest energy*. This means that it can be converted to or from equivalent amounts of *other* (non-material) forms of energy, for example kinetic energy, potential energy, and electromagnetic radiant energy. When this happens, as recognized in twentieth century experience, rest mass is not conserved, unlike the *total* mass or *total* energy. All forms of energy contribute to the total mass and total energy.

For example, an electron and a positron each have rest mass. They can perish together, converting their combined rest energy into photons having electromagnetic radiant energy, but no rest mass. If this occurs within an isolated system that does not release the photons or their energy into the external surroundings, then neither the total *mass* nor the total *energy* of the system will change. The produced electromagnetic radiant energy contributes just as much to the inertia (and to any weight) of the system as did the rest mass of the electron and positron before their demise. Likewise, non-material forms of energy can perish into matter, which has rest mass.

Thus, conservation of energy (*total,* including material or *rest* energy), and conservation of mass (*total,* not just *rest*), each still holds as an (equivalent) law. In the nineteenth century these had appeared as two seemingly-distinct laws.

Conservation of Energy in Beta Decay

The discovery in 1911 that electrons emitted in beta decay have a continuous rather than a discrete spectrum appeared to contradict conservation of energy, under the then-current assumption that beta decay is the simple emission of an electron from a nucleus. This problem was eventually resolved in 1933 by Enrico Fermi who proposed the correct description of beta-decay as the emission of both an electron and an antineutrino, which carries away the apparently missing energy.

First Law of Thermodynamics

For a closed thermodynamic system, the first law of thermodynamics may be stated as:

$$\delta Q = dU + \delta W \text{ , or equivalently, } dU = \delta Q - \delta W$$

where δQ is the quantity of energy added to the system by a heating process, δW is the quantity of energy lost by the system due to work done by the system on its surroundings and dU is the change in the internal energy of the system.

The δ's before the heat and work terms are used to indicate that they describe an increment of energy which is to be interpreted somewhat differently than the dU increment of internal energy. Work and heat refer to kinds of process which add or subtract energy to or from a system, while the internal energy U is a property of a particular state of the system when it is in unchanging thermodynamic equilibrium. Thus the term "heat energy" for δQ means "that amount of energy added as the result of heating" rather than referring to a particular form of energy. Likewise, the term "work energy" for δW means "that amount of energy lost as the result of work". Thus one can state the amount of internal energy possessed by a thermodynamic system that one knows is presently in a given state, but one cannot tell, just from knowledge of the given present state, how much energy has in the past flowed into or out of the system as a result of its being heated or cooled, nor as the result of work being performed on or by the system.

Entropy is a function of the state of a system which tells of limitations of the possibility of conversion of heat into work.

For a simple compressible system, the work performed by the system may be written:

$$\delta W = PdV,$$

where P is the pressure and dV is a small change in the volume of the system, each of which are system variables. In the fictive case in which the process is idealized and infinitely slow, so as to be called *quasi-static,* and regarded as reversible, the heat being transferred from a source with temperature infinitesimally above the system temperature, then the heat energy may be written

$$\delta Q = TdS,$$

where T is the temperature and dS is a small change in the entropy of the system. Temperature and entropy are variables of state of a system.

If an open system (in which mass may be exchanged with the environment) has several walls such that the mass transfer is through rigid walls separate from the heat and work transfers, then the first law may be written:

$$dU = \delta Q - \delta W + u'dM,$$

where dM is the added mass and u' is the internal energy per unit mass of the added mass, measured in the surroundings before the process.

Noether's Theorem

The conservation of energy is a common feature in many physical theories. From a mathematical point of view it is understood as a consequence of Noether's theorem, developed by Emmy Noether in 1915 and first published in 1918. The theorem states every continuous symmetry of a physical theory has an associated conserved quantity; if the theory's symmetry is time invariance then the conserved quantity is called "energy". The energy conservation law is a consequence of the shift symmetry of time; energy conservation is implied by the empirical fact that the laws of physics do not change with time itself. Philosophically this can be stated as "nothing depends on time per se". In other words, if the physical system is invariant under the continuous symmetry of time translation then its energy (which is canonical conjugate quantity to time) is conserved. Conversely, systems which are not invariant under shifts in time (an example, systems with time dependent potential energy) do not exhibit conservation of energy – unless we consider them to exchange energy with another, external system so that the theory of the enlarged system becomes time invariant again. Since any time-varying system can be embedded within a larger time-invariant system (with the exception of the universe), conservation can always be recovered by a suitable re-definition of what energy is and extending the scope of your system. Conservation of energy for finite systems is valid in such physical theories as special relativity and quantum theory (including QED) in the flat space-time.

Relativity

With the discovery of special relativity by Henri Poincaré and Albert Einstein, energy was proposed to be one component of an energy-momentum 4-vector. Each of the four components (one of energy and three of momentum) of this vector is separately conserved across time, in any closed system, as seen from any given inertial reference frame. Also conserved is the vector length (Minkowski norm), which is the rest mass for single particles, and the invariant mass for systems of particles.

The relativistic energy of a single massive particle contains a term related to its rest mass in addition to its kinetic energy of motion. In the limit of zero kinetic energy (or equivalently in the rest frame) of a massive particle; or else in the center of momentum frame for objects or systems which retain kinetic energy, the total energy of particle or object (including internal kinetic energy in systems) is related to its rest mass or its invariant mass via the famous equation $E = mc^2$.

Thus, the rule of *conservation of energy* over time in special relativity continues to hold, so long as the reference frame of the observer is unchanged. This applies to the total energy of systems, although different observers disagree as to the energy value. Also conserved, and invariant to all observers, is the invariant mass, which is the minimal system mass and energy that can be seen by any observer, and which is defined by the energy–momentum relation.

In general relativity conservation of energy-momentum is expressed with the aid of a stress-energy-momentum pseudotensor. The theory of general relativity leaves open the question of whether there is a conservation of energy for the entire universe.

Quantum Theory

In quantum mechanics, energy of a quantum system is described by a self-adjoint (or Hermitian) operator called the Hamiltonian, which acts on the Hilbert space (or a space of wave functions) of the system. If the Hamiltonian is a time-independent operator, emergence probability of the measurement result does not change in time over the evolution of the system. Thus the expectation value of energy is also time independent. The local energy conservation in quantum field theory is ensured by the quantum Noether's theorem for energy-momentum tensor operator. Note that due to the lack of the (universal) time operator in quantum theory, the uncertainty relations for time and energy are not fundamental in contrast to the position-momentum uncertainty principle, and merely holds in specific cases. Energy at each fixed time can in principle be exactly measured without any trade-off in precision forced by the time-energy uncertainty relations. Thus the conservation of energy in time is a well defined concept even in quantum mechanics.

References

- Lang, Helen S. (1998). The order of nature in Aristotle's physics : place and the elements (1. publ. ed.). Cambridge: Cambridge Univ. Press. ISBN 9780521624534.

- Hetherington, Norriss S. (1993). Cosmology: Historical, Literary, Philosophical, Religious, and Scientific Perspectives. Garland Reference Library of the Humanities. p. 100. ISBN 0-8153-1085-4.

- Howland, R. A. (2006). Intermediate dynamics a linear algebraic approach (Online-Ausg. ed.). New York: Springer. pp. 255–256. ISBN 9780387280592.

- Jammer, Max (1999). Concepts of force : a study in the foundations of dynamics (Facsim. ed.). Mineola, N.Y.: Dover Publications. pp. 220–222. ISBN 9780486406893.

- Scharf, Toralf (2007). Polarized light in liquid crystals and polymers. John Wiley and Sons. p. 19. ISBN 0-471-74064-0., Chapter 2, p. 19

- Wandmacher, Cornelius; Johnson, Arnold (1995). Metric Units in Engineering. ASCE Publications. p. 15. ISBN 0-7844-0070-9.

- Beer, Ferdinand P.; Johnston, E. Russel, Jr. (1996). Vector Mechanics for Engineers (Sixth ed.). McGraw-Hill. p. 397. ISBN 0-07-297688-8.

- Bhavikatti, S. S.; K. G. Rajashekarappa (1994). Engineering Mechanics. New Age International. p. 112. ISBN 978-81-224-0617-7. Retrieved 2007-10-21.

- Bigoni, D. Nonlinear Solid Mechanics: Bifurcation Theory and Material Instability. Cambridge University Press, 2012. ISBN 9781107025417.

- Andrew Pytel; Jaan Kiusalaas (2010). Engineering Mechanics: Dynamics – SI Version, Volume 2 (3rd ed.). Cengage Learning,. p. 654. ISBN 9780495295631.

- Hugh D. Young & Roger A. Freedman (2008). University Physics (12th ed.). Addison-Wesley. p. 329. ISBN 978-0-321-50130-1.

- Smith, Crosbie (1998). The Science of Energy – a Cultural History of Energy Physics in Victorian Britain. The University of Chicago Press. ISBN 0-226-76420-6.

- Lofts, G; O'Keeffe D; et al. (2004). "11 — Mechanical Interactions". Jacaranda Physics 1 (2 ed.). Milton, Queensland, Australia: John Willey & Sons Australia Ltd. p. 286. ISBN 0-7016-3777-3.

- Arianrhod, Robyn (2012). Seduced by logic : Émilie du Châtelet, Mary Somerville, and the Newtonian revolution (US ed.). New York: Oxford University Press. ISBN 978-0-19-993161-3.

- Persson, B. N. J. (2000). Sliding friction: physical principles and applications. Springer. ISBN 978-3-540-67192-3. Retrieved 2016-01-23.

- Nosonovsky, Michael (2013). Friction-Induced Vibrations and Self-Organization: Mechanics and Non-Equilibrium Thermodynamics of Sliding Contact. CRC Press. p. 333. ISBN 978-1466504011.

- Hadden, Richard W. (1994). On the shoulders of merchants: exchange and the mathematical conception of nature in early modern Europe. SUNY Press. p. 13. ISBN 0-7914-2011-6., Chapter 1, p. 13

Fundamental States of Matter

Matter is observed on a daily basis in the form of solid, liquid, gas and plasma. The other states of matter exist only in extreme conditions. The topics discussed in the section are of great importance to broaden the existing on fundamental states of matter.

Matter

In the classical physics observed in everyday life, matter can be defined as any substance that has mass and takes up space, or as any substance made up of atoms, thus excluding other energy phenomena or waves such as light or sound. Observable physical objects are said to be composed of matter. More generally, in (modern) physics, matter is not a fundamental concept because a universal definition of it is elusive: elementary constituents may not take up space per se, and individually-massless particles may be composed to form objects that have mass (even when at rest).

All the everyday objects that we can bump into, touch or squeeze are composed of atoms. This ordinary atomic matter is in turn made up of interacting subatomic particles—usually a nucleus of protons and neutrons, and a cloud of orbiting electrons. Typically, science considers these composite particles matter because they have both rest mass and volume. By contrast, massless particles, such as photons, are not considered matter, because they have neither rest mass nor volume. However, not all particles with rest mass have a classical volume, since fundamental particles such as quarks and leptons (sometimes equated with matter) are considered "point particles" with no effective size or volume. Nevertheless, quarks and leptons together make up "ordinary matter", and their interactions contribute to the effective volume of the composite particles that make up ordinary matter.

Matter exists in *states* (or *phases*): the classical solid, liquid, and gas; as well as the more exotic plasma, Bose–Einstein condensates, fermionic condensates, and quark–gluon plasma.

For much of the history of the natural sciences people have contemplated the exact nature of matter. The idea that matter was built of discrete building blocks, the so-called *particulate theory of matter*, was first put forward by the Greek philosophers Leucippus (~490 BC) and Democritus (~470–380 BC).

Comparison with Mass

Matter should not be confused with mass, as the two are not quite the same in modern physics. For example, mass is a conserved quantity, which means that its value is unchanging through time, within closed systems. However, matter is *not* conserved in such systems, although this is not obvious in ordinary conditions on Earth, where matter is approximately conserved. Still, special relativity shows that matter may disappear by conversion into energy, even inside closed systems, and it can also be created from energy, within such systems. However, because *mass* (like energy) can neither be created nor destroyed, the quantity of mass and the quantity of energy remain the same during a transformation of matter (which represents a certain amount of energy) into non-material (i.e., non-matter) energy. This is also true in the reverse transformation of energy into matter.

Different fields of science use the term matter in different, and sometimes incompatible, ways. Some of these ways are based on loose historical meanings, from a time when there was no reason to distinguish mass and matter. As such, there is no single universally agreed scientific meaning of the word "matter". Scientifically, the term "mass" is well-defined, but "matter" is not. Sometimes in the field of physics "matter" is simply equated with particles that exhibit rest mass (i.e., that cannot travel at the speed of light), such as quarks and leptons. However, in both physics and chemistry, matter exhibits both wave-like and particle-like properties, the so-called wave–particle duality.

Definition

Based on Mass, Volume, and Space

The common definition of matter is *anything that has mass and volume (occupies space)*. For example, a car would be said to be made of matter, as it has mass and volume (occupies space).

The observation that matter occupies space goes back to antiquity. However, an explanation for why matter occupies space is recent, and is argued to be a result of the phenomenon described in the Pauli exclusion principle. Two particular examples where the exclusion principle clearly relates matter to the occupation of space are white dwarf stars and neutron stars.

Based on Atoms

A definition of "matter" based on its physical and chemical structure is: *matter is made up of atoms*. As an example, deoxyribonucleic acid molecules (DNA) are matter under this definition because they are made of atoms. This definition can extend to include charged atoms and molecules, so as to include plasmas (gases of ions) and electrolytes (ionic solutions), which are not obviously included in the atoms definition. Alternatively, one can adopt the *protons, neutrons, and electrons* definition.

Based on Protons, Neutrons and Electrons

A definition of "matter" more fine-scale than the atoms and molecules definition is: *matter is made up of what atoms and molecules are made of*, meaning anything made of positively charged protons, neutral neutrons, and negatively charged electrons. This definition goes beyond atoms and molecules, however, to include substances made from these building blocks that are *not* simply atoms or molecules, for example white dwarf matter—typically, carbon and oxygen nuclei in a sea of degenerate electrons. At a microscopic level, the constituent "particles" of matter such as protons, neutrons, and electrons obey the laws of quantum mechanics and exhibit wave–particle duality. At an even deeper level, protons and neutrons are made up of quarks and the force fields (gluons) that bind them together.

Based on Quarks and Leptons

As seen in the above discussion, many early definitions of what can be called *ordinary matter* were based upon its structure or *building blocks*. On the scale of elementary particles, a definition that follows this tradition can be stated as: *ordinary matter is everything that is composed of elementary fermions, namely quarks and leptons*. The connection between these formulations follows.

Leptons (the most famous being the electron), and quarks (of which baryons, such as protons and neutrons, are made) combine to form atoms, which in turn form molecules. Because atoms and molecules are said to be matter, it is natural to phrase the definition as: *ordinary matter is anything that is made of the same things that atoms and molecules are made of.* (However, notice that one also can make from these building blocks matter that is *not* atoms or molecules.) Then, because electrons are leptons, and protons, and neutrons are made of quarks, this definition in turn leads to the definition of matter as being *quarks and leptons*, which are the two types of elementary fermions. Carithers and Grannis state: *Ordinary matter is composed entirely of first-generation particles, namely the [up] and [down] quarks, plus the electron and its neutrino.* (Higher generations particles quickly decay into first-generation particles, and thus are not commonly encountered.)

This definition of ordinary matter is more subtle than it first appears. All the particles that make up ordinary matter (leptons and quarks) are elementary fermions, while all the force carriers are elementary bosons. The W and Z bosons that mediate the weak force are not made of quarks or leptons, and so are not ordinary matter, even if they have mass. In other words, mass is not something that is exclusive to ordinary matter.

The quark–lepton definition of ordinary matter, however, identifies not only the elementary building blocks of matter, but also includes composites made from the constituents (atoms and molecules, for example). Such composites contain an interaction energy that holds the constituents together, and may constitute the bulk of the mass of the composite. As an example, to a great extent, the mass of an atom is simply the sum of the masses of its constituent protons, neutrons and electrons. However, digging deeper, the protons and neutrons are made up of quarks bound together by gluon fields and these gluons fields contribute significantly to the mass of hadrons. In other words, most of what composes the "mass" of ordinary matter is due to the binding energy of quarks within protons and neutrons. For example, the sum of the mass of the three quarks in a nucleon is approximately 12.5 MeV/c^2, which is low compared to the mass of a nucleon (approximately 938 MeV/c^2). The bottom line is that most of the mass of everyday objects comes from the interaction energy of its elementary components.

The Standard Model groups matter particles into three generations, where each generation consists of two quarks and two leptons. The first generation is the *up* and *down* quarks, the *electron* and the *electron neutrino*; the second includes the *charm* and *strange* quarks, the *muon* and the *muon neutrino*; the third generation consists of the *top* and *bottom* quarks and the *tau* and *tau neutrino*. The most natural explanation for this would be that quarks and leptons of higher generations are excited states of the first generations. If this turns out to be the case, it would imply that quarks and leptons are composite particles, rather than elementary particles.

Based on Theories of Relativity

In the context of relativity, mass is not an additive quantity, in the sense that one can add the rest masses of particles in a system to get the total rest mass of the system. Thus, in relativity usually a more general view is that it is not the sum of rest masses, but the energy–momentum tensor that quantifies the amount of matter. This tensor gives the rest mass for the entire system. "Matter" therefore is sometimes considered as anything that contributes to the energy–momentum of a system, that is, anything that is not purely gravity. This view is commonly held in fields that deal with general relativity such as cosmology. In this view, light and other massless particles and fields are part of matter.

The reason for this is that in this definition, electromagnetic radiation (such as light) as well as the energy of electromagnetic fields contributes to the mass of systems, and therefore appears to add matter to them. For example, light radiation (or thermal radiation) trapped inside a box would contribute to the mass of the box, as would any kind of energy inside the box, including the kinetic energy of particles held by the box. Nevertheless, isolated individual particles of light (photons) and the isolated kinetic energy of massive particles, are normally not considered to be *matter*.

A difference between matter and mass therefore may seem to arise when single particles are examined. In such cases, the mass of single photons is zero. For particles with rest mass, such as leptons and quarks, isolation of the particle in a frame where it is not moving, removes its kinetic energy.

A source of definition difficulty in relativity arises from two definitions of mass in common use, one of which is formally equivalent to total energy (and is thus observer dependent), and the other of which is referred to as rest mass or invariant mass and is independent of the observer. Only "rest mass" is loosely equated with matter (since it can be weighed). Invariant mass is usually applied in physics to unbound systems of particles. However, energies which contribute to the "invariant mass" may be weighed also in special circumstances, such as when a system that has invariant mass is confined and has no net momentum (as in the box example above). Thus, a photon with no mass may (confusingly) still add mass to a system in which it is trapped. The same is true of the kinetic energy of particles, which by definition is not part of their rest mass, but which does add rest mass to systems in which these particles reside (an example is the mass added by the motion of gas molecules of a bottle of gas, or by the thermal energy of any hot object).

Since such mass (kinetic energies of particles, the energy of trapped electromagnetic radiation and stored potential energy of repulsive fields) is measured as part of the mass of ordinary *matter* in complex systems, the "matter" status of "massless particles" and fields of force becomes unclear in such systems. These problems contribute to the lack of a rigorous definition of matter in science, although mass is easier to define as the total stress–energy above (this is also what is weighed on a scale, and what is the source of gravity).

Structure

In particle physics, fermions are particles that obey Fermi–Dirac statistics. Fermions can be elementary, like the electron—or composite, like the proton and neutron. In the Standard Model, there are two types of elementary fermions: quarks and leptons, which are discussed next.

Quarks

Quarks are particles of spin-$\frac{1}{2}$, implying that they are fermions. They carry an electric charge of $-\frac{1}{3}$ e (down-type quarks) or $+\frac{2}{3}$ e (up-type quarks). For comparison, an electron has a charge of -1 e. They also carry colour charge, which is the equivalent of the electric charge for the strong interaction. Quarks also undergo radioactive decay, meaning that they are subject to the weak interaction. Quarks are massive particles, and therefore are also subject to gravity.

Quark properties							
name	symbol	spin	electric charge (e)	mass (MeV/c²)	mass comparable to	antiparticle	antiparticle symbol
up-type quarks							
up	u	$\frac{1}{2}$	$+\frac{2}{3}$	1.5 to 3.3	~ 5 electrons	antiup	u
charm	c	$\frac{1}{2}$	$+\frac{2}{3}$	1160 to 1340	~1 proton	anticharm	c
top	t	$\frac{1}{2}$	$+\frac{2}{3}$	169,100 to 173,300	~180 protons or ~1 tungsten atom	antitop	t
down-type quarks							
down	d	$\frac{1}{2}$	$-\frac{1}{3}$	3.5 to 6.0	~10 electrons	antidown	d
strange	s	$\frac{1}{2}$	$-\frac{1}{3}$	70 to 130	~ 200 electrons	antistrange	s
bottom	b	$\frac{1}{2}$	$-\frac{1}{3}$	4130 to 4370	~ 5 protons	antibottom	b

Quark structure of a proton: 2 up quarks and 1 down quark.

Baryonic Matter

Baryons are strongly interacting fermions, and so are subject to Fermi–Dirac statistics. Amongst the baryons are the protons and neutrons, which occur in atomic nuclei, but many other unstable baryons exist as well. The term baryon usually refers to triquarks—particles made of three quarks.

"Exotic" baryons made of four quarks and one antiquark are known as the pentaquarks, but their existence is not generally accepted.

Baryonic matter is the part of the universe that is made of baryons (including all atoms). This part of the universe does not include dark energy, dark matter, black holes or various forms of degenerate matter, such as compose white dwarf stars and neutron stars. Microwave light seen by Wilkinson Microwave Anisotropy Probe (WMAP), suggests that only about 4.6% of that part of the universe within range of the best telescopes (that is, matter that may be visible because light could reach us from it), is made of baryonic matter. About 23% is dark matter, and about 72% is dark energy.

As a matter of fact, the great majority of ordinary matter in the universe is unseen, since visible stars and gas inside galaxies and clusters account for less than 10 per cent of the ordinary matter contribution to the mass-energy density of the universe.

A comparison between the white dwarf IK Pegasi B (center), its A-class companion IK Pegasi A (left) and the Sun (right). This white dwarf has a surface temperature of 35,500 K.

Degenerate Matter

In physics, degenerate matter refers to the ground state of a gas of fermions at a temperature near absolute zero. The Pauli exclusion principle requires that only two fermions can occupy a quantum state, one spin-up and the other spin-down. Hence, at zero temperature, the fermions fill up sufficient levels to accommodate all the available fermions—and in the case of many fermions, the maximum kinetic energy (called the *Fermi energy*) and the pressure of the gas becomes very large, and depends on the number of fermions rather than the temperature, unlike normal states of matter.

Degenerate matter is thought to occur during the evolution of heavy stars. The demonstration by Subrahmanyan Chandrasekhar that white dwarf stars have a maximum allowed mass because of the exclusion principle caused a revolution in the theory of star evolution.

Degenerate matter includes the part of the universe that is made up of neutron stars and white dwarfs.

Strange Matter

Strange matter is a particular form of quark matter, usually thought of as a *liquid* of up, down, and strange quarks. It is contrasted with nuclear matter, which is a liquid of neutrons and protons (which themselves are built out of up and down quarks), and with non-strange quark matter, which is a quark liquid that contains only up and down quarks. At high enough density, strange matter is expected to be color superconducting. Strange matter is hypothesized to occur in the core of neutron stars, or, more speculatively, as isolated droplets that may vary in size from femtometers (strangelets) to kilometers (quark stars).

Two Meanings of the Term "Strange Matter"

In particle physics and astrophysics, the term is used in two ways, one broader and the other more specific.

1. The broader meaning is just quark matter that contains three flavors of quarks: up, down, and strange. In this definition, there is a critical pressure and an associated critical density, and when nuclear matter (made of protons and neutrons) is compressed beyond this density, the protons and neutrons dissociate into quarks, yielding quark matter (probably strange matter).

2. The narrower meaning is quark matter that is *more stable than nuclear matter*. The idea that this could happen is the "strange matter hypothesis" of Bodmer and Witten. In this definition, the critical pressure is zero: the true ground state of matter is *always* quark matter. The nuclei that we see in the matter around us, which are droplets of nuclear matter, are actually metastable, and given enough time (or the right external stimulus) would decay into droplets of strange matter, i.e. strangelets.

Leptons

Leptons are particles of spin-$\frac{1}{2}$, meaning that they are fermions. They carry an electric charge of -1 e (charged leptons) or 0 e (neutrinos). Unlike quarks, leptons do not carry colour charge, meaning that they do not experience the strong interaction. Leptons also undergo radioactive decay, meaning that they are subject to the weak interaction. Leptons are massive particles, therefore are subject to gravity.

name	symbol	spin	electric charge (e)	mass (MeV/c²)	mass comparable to	antiparticle	antiparticle symbol
Lepton properties							
charged leptons							
electron	e−	$\frac{1}{2}$	−1	0.5110	1 electron	antielectron	e+
muon	μ−	$\frac{1}{2}$	−1	105.7	~ 200 electrons	antimuon	μ+
tau	τ−	$\frac{1}{2}$	−1	1,777	~ 2 protons	antitau	τ+
neutrinos							
electron neutrino	ν e	$\frac{1}{2}$	0	< 0.000460	< $\frac{1}{1000}$ electron	electron antineutrino	ν e

| muon neutrino | ν μ | $\frac{1}{2}$ | 0 | < 0.19 | < $\frac{1}{2}$ electron | muon antineutrino | $\overline{\nu}$ μ |
| tau neutrino | ν τ | $\frac{1}{2}$ | 0 | < 18.2 | < 40 electrons | tau antineutrino | $\overline{\nu}$ τ |

Phases

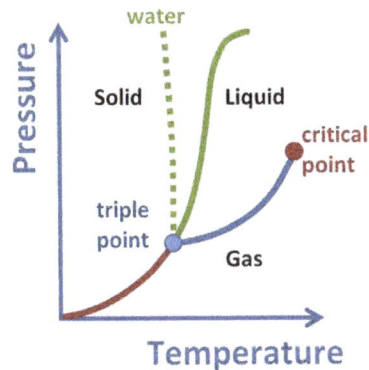

Phase diagram for a typical substance at a fixed volume. Vertical axis is *Pressure*, horizontal axis is *Temperature*. The green line marks the freezing point (above the green line is *solid*, below it is *liquid*) and the blue line the boiling point (above it is *liquid* and below it is *gas*). So, for example, at higher *T*, a higher *P* is necessary to maintain the substance in liquid phase. At the triple point the three phases; liquid, gas and solid; can coexist. Above the critical point there is no detectable difference between the phases. The dotted line shows the anomalous behavior of water: ice melts at constant temperature with increasing pressure.

In bulk, matter can exist in several different forms, or states of aggregation, known as *phases*, depending on ambient pressure, temperature and volume. A phase is a form of matter that has a relatively uniform chemical composition and physical properties (such as density, specific heat, refractive index, and so forth). These phases include the three familiar ones (solids, liquids, and gases), as well as more exotic states of matter (such as plasmas, superfluids, supersolids, Bose–Einstein condensates, ...). A *fluid* may be a liquid, gas or plasma. There are also paramagnetic and ferromagnetic phases of magnetic materials. As conditions change, matter may change from one phase into another. These phenomena are called phase transitions, and are studied in the field of thermodynamics. In nanomaterials, the vastly increased ratio of surface area to volume results in matter that can exhibit properties entirely different from those of bulk material, and not well described by any bulk phase.

Phases are sometimes called *states of matter*, but this term can lead to confusion with thermodynamic states. For example, two gases maintained at different pressures are in different *thermodynamic states* (different pressures), but in the same *phase* (both are gases).

Antimatter

In particle physics and quantum chemistry, antimatter is matter that is composed of the antiparticles of those that constitute ordinary matter. If a particle and its antiparticle come into contact with each other, the two annihilate; that is, they may both be converted into other particles with equal energy in accordance with Einstein's equation $E = mc^2$. These new particles may be high-energy photons (gamma rays) or other particle–antiparticle pairs. The resulting particles are endowed with an amount of kinetic energy equal to the difference between the rest mass of the products of the annihilation and the rest mass of the original particle–antiparticle pair, which is often quite large.

Antimatter is not found naturally on Earth, except very briefly and in vanishingly small quantities (as the result of radioactive decay, lightning or cosmic rays). This is because antimatter that came to exist on Earth outside the confines of a suitable physics laboratory would almost instantly meet the ordinary matter that Earth is made of, and be annihilated. Antiparticles and some stable antimatter (such as antihydrogen) can be made in tiny amounts, but not in enough quantity to do more than test a few of its theoretical properties.

There is considerable speculation both in science and science fiction as to why the observable universe is apparently almost entirely matter, and whether other places are almost entirely antimatter instead. In the early universe, it is thought that matter and antimatter were equally represented, and the disappearance of antimatter requires an asymmetry in physical laws called the charge parity (or CP symmetry) violation. CP symmetry violation can be obtained from the Standard Model, but at this time the apparent asymmetry of matter and antimatter in the visible universe is one of the great unsolved problems in physics. Possible processes by which it came about are explored in more detail under baryogenesis.

Other Types

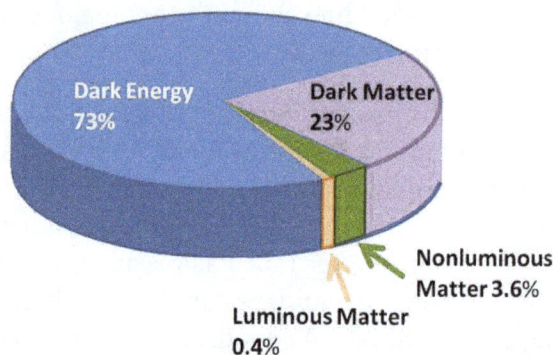

Pie chart showing the fractions of energy in the universe contributed by different sources. *Ordinary matter* is divided into *luminous matter* (the stars and luminous gases and 0.005% radiation) and *nonluminous matter* (intergalactic gas and about 0.1% neutrinos and 0.04% supermassive black holes). Ordinary matter is uncommon. Modeled after Ostriker and Steinhardt.

Ordinary matter, in the quarks and leptons definition, constitutes about 4% of the energy of the observable universe. The remaining energy is theorized to be due to exotic forms, of which 23% is dark matter and 73% is dark energy.

Galaxy rotation curve for the Milky Way. Vertical axis is speed of rotation about the galactic center. Horizontal axis is distance from the galactic center. The sun is marked with a yellow ball. The observed curve of speed of rotation is blue. The predicted curve based upon stellar mass and gas in the Milky Way is red. The difference is due to dark matter or perhaps a modification of the law of gravity. Scatter in observations is indicated roughly by gray bars.

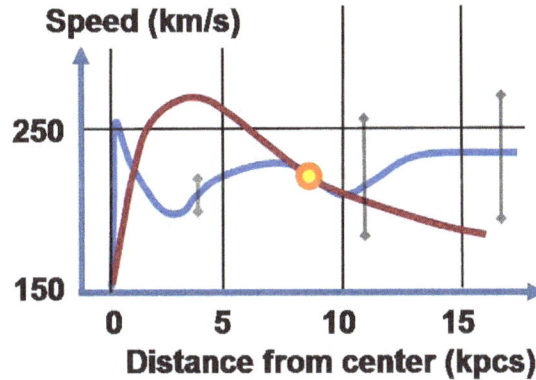

Dark Matter

In astrophysics and cosmology, dark matter is matter of unknown composition that does not emit or reflect enough electromagnetic radiation to be observed directly, but whose presence can be inferred from gravitational effects on visible matter. Observational evidence of the early universe and the big bang theory require that this matter have energy and mass, but is not composed of either elementary fermions (as above) OR gauge bosons. The commonly accepted view is that most of the dark matter is non-baryonic in nature. As such, it is composed of particles as yet unobserved in the laboratory. Perhaps they are supersymmetric particles, which are not Standard Model particles, but relics formed at very high energies in the early phase of the universe and still floating about.

Dark Energy

In cosmology, dark energy is the name given to the antigravitating influence that is accelerating the rate of expansion of the universe. It is known not to be composed of known particles like protons, neutrons or electrons, nor of the particles of dark matter, because these all gravitate.

Fully 70% of the matter density in the universe appears to be in the form of dark energy. Twenty-six percent is dark matter. Only 4% is ordinary matter. So less than 1 part in 20 is made out of matter we have observed experimentally or described in the standard model of particle physics. Of the other 96%, apart from the properties just mentioned, we know absolutely nothing.

— Lee Smolin: The Trouble with Physics

Exotic Matter

Exotic matter is a hypothetical concept of particle physics. It covers any material that violates one or more classical conditions or is not made of known baryonic particles. Such materials would possess qualities like negative mass or being repelled rather than attracted by gravity.

Historical Development

Antiquity (c. 610 BC–c. 322 BC)

The pre-Socratics were among the first recorded speculators about the underlying nature of the visible world. Thales (c. 624 BC–c. 546 BC) regarded water as the fundamental material of the world. Anaximander (c. 610 BC–c. 546 BC) posited that the basic material was wholly characterless or limitless: the Infinite (*apeiron*). Anaximenes (flourished 585 BC, d. 528 BC) posited that the basic stuff was *pneuma* or air. Heraclitus (c. 535–c. 475 BC) seems to say the basic element is fire, though perhaps he means that all is change. Empedocles (c. 490–430 BC) spoke of four elements of which everything was made: earth, water, air, and fire. Meanwhile, Parmenides argued that change does not exist, and Democritus argued that everything is composed of minuscule, inert bodies of all shapes called atoms, a philosophy called atomism. All of these notions had deep philosophical problems.

Aristotle (384 BC – 322 BC) was the first to put the conception on a sound philosophical basis, which he did in his natural philosophy, especially in *Physics* book I. He adopted as reasonable suppositions the four Empedoclean elements, but added a fifth, aether. Nevertheless, these elements are not basic in Aristotle's mind. Rather they, like everything else in the visible world, are composed of the basic *principles* matter and form.

For my definition of matter is just this—the primary substratum of each thing, from which it comes to be without qualification, and which persists in the result.

—Aristotle, Physics I:9a32

The word Aristotle uses for matter, (*hyle* or *hule*), can be literally translated as wood or timber, that is, "raw material" for building. Indeed, Aristotle's conception of matter is intrinsically linked to something being made or composed. In other words, in contrast to the early modern conception of matter as simply occupying space, matter for Aristotle is definitionally linked to process or change: matter is what underlies a change of substance. For example, a horse eats grass: the horse changes the grass into itself; the grass as such does not persist in the horse, but some aspect of it— its matter—does. The matter is not specifically described (e.g., as atoms), but consists of whatever persists in the change of substance from grass to horse. Matter in this understanding does not exist independently (i.e., as a substance), but exists interdependently (i.e., as a "principle") with form and only insofar as it underlies change. It can be helpful to conceive of the relationship of matter and form as very similar to that between parts and whole. For Aristotle, matter as such can only *receive* actuality from form; it has no activity or actuality in itself, similar to the way that parts as such only have their existence *in* a whole (otherwise they would be independent wholes).

Seventeenth and Eighteenth Centuries

René Descartes (1596–1650) originated the modern conception of matter. He was primarily a geometer. Instead of, like Aristotle, deducing the existence of matter from the physical reality of change, Descartes arbitrarily postulated matter to be an abstract, mathematical substance that occupies space:

So, extension in length, breadth, and depth, constitutes the nature of bodily substance; and thought constitutes the nature of thinking substance. And everything else attributable to body presupposes extension, and is only a mode of extended

—René Descartes, Principles of Philosophy

For Descartes, matter has only the property of extension, so its only activity aside from locomotion is to exclude other bodies: this is the mechanical philosophy. Descartes makes an absolute distinction between mind, which he defines as unextended, thinking substance, and matter, which he defines as unthinking, extended substance. They are independent things. In contrast, Aristotle defines matter and the formal/forming principle as complementary *principles* that together compose one independent thing (substance). In short, Aristotle defines matter (roughly speaking) as what things are actually made of (with a *potential* independent existence), but Descartes elevates matter to an actual independent thing in itself.

The continuity and difference between Descartes' and Aristotle's conceptions is noteworthy. In both conceptions, matter is passive or inert. In the respective conceptions matter has different relationships to intelligence. For Aristotle, matter and intelligence (form) exist together in an interdependent relationship, whereas for Descartes, matter and intelligence (mind) are definitionally opposed, independent substances.

Descartes' justification for restricting the inherent qualities of matter to extension is its permanence, but his real criterion is not permanence (which equally applied to color and resistance), but his desire to use geometry to explain all material properties. Like Descartes, Hobbes, Boyle, and Locke argued that the inherent properties of bodies were limited to extension, and that so-called secondary qualities, like color, were only products of human perception.

Isaac Newton (1643–1727) inherited Descartes' mechanical conception of matter. In the third of his "Rules of Reasoning in Philosophy", Newton lists the universal qualities of matter as "extension, hardness, impenetrability, mobility, and inertia". Similarly in *Optics* he conjectures that God created matter as "solid, massy, hard, impenetrable, movable particles", which were "...even so very hard as never to wear or break in pieces". The "primary" properties of matter were amenable to mathematical description, unlike "secondary" qualities such as color or taste. Like Descartes, Newton rejected the essential nature of secondary qualities.

Newton developed Descartes' notion of matter by restoring to matter intrinsic properties in addition to extension (at least on a limited basis), such as mass. Newton's use of gravitational force, which worked "at a distance", effectively repudiated Descartes' mechanics, in which interactions happened exclusively by contact.

Though Newton's gravity would seem to be a *power* of bodies, Newton himself did not admit it to be an *essential* property of matter. Carrying the logic forward more consistently, Joseph Priestley (1733-1804) argued that corporeal properties transcend contact mechanics: chemical properties require the *capacity* for attraction. He argued matter has other inherent powers besides the so-called primary qualities of Descartes, et al.

Nineteenth and Twentieth Centuries

Since Priestley's time, there has been a massive expansion in knowledge of the constituents of the material world (viz., molecules, atoms, subatomic particles), but there has been no further development in the *definition* of matter. Rather the question has been set aside. Noam Chomsky (born 1928) summarizes the situation that has prevailed since that time:

What is the concept of body that finally emerged?[...] The answer is that there is no clear and definite conception of body.[...] Rather, the material world is whatever we discover it to be, with whatever properties it must be assumed to have for the purposes of explanatory theory. Any intelligible theory that offers genuine explanations and that can be assimilated to the core notions of physics becomes part of the theory of the material world, part of our account of body. If we have such a theory in some domain, we seek to assimilate it to the core notions of physics, perhaps modifying these notions as we carry out this enterprise.

— *Noam Chomsky, Language and problems of knowledge: the Managua lectures, p. 144*

So matter is whatever physics studies and the object of study of physics is matter: there is no independent general definition of matter, apart from its fitting into the methodology of measurement and controlled experimentation. In sum, the boundaries between what constitutes matter and everything else remains as vague as the demarcation problem of delimiting science from everything else.

In the 19th century, following the development of the periodic table, and of atomic theory, atoms were seen as being the fundamental constituents of matter; atoms formed molecules and compounds.

The common definition in terms of occupying space and having mass is in contrast with most physical and chemical definitions of matter, which rely instead upon its structure and upon attributes not necessarily related to volume and mass. At the turn of the nineteenth century, the knowledge of matter began a rapid evolution.

Aspects of the Newtonian view still held sway. James Clerk Maxwell discussed matter in his work *Matter and Motion*. He carefully separates "matter" from space and time, and defines it in terms of the object referred to in Newton's first law of motion.

However, the Newtonian picture was not the whole story. In the 19th century, the term "matter" was actively discussed by a host of scientists and philosophers, and a brief outline can be found in Levere. A textbook discussion from 1870 suggests matter is what is made up of atoms:

Three divisions of matter are recognized in science: masses, molecules and atoms. A Mass of matter is any portion of matter appreciable by the senses. A Molecule is the smallest particle of matter into which a body can be divided without losing its identity. An Atom is a still smaller particle produced by division of a molecule.

Rather than simply having the attributes of mass and occupying space, matter was held to have chemical and electrical properties. In 1909 the famous physicist J. J. Thomson (1856-1940) wrote about the "constitution of matter" and was concerned with the possible connection between matter and electrical charge.

There is an entire literature concerning the "structure of matter", ranging from the "electrical structure" in the early 20th century, to the more recent "quark structure of matter", introduced today with the remark: *Understanding the quark structure of matter has been one of the most important advances in contemporary physics*. In this connection, physicists speak of *matter fields*, and speak of particles as "quantum excitations of a mode of the matter field". And here is a quote from de Sabbata and Gasperini: "With the word "matter" we denote, in this context, the sources of the interactions, that is spinor fields (like quarks and leptons), which are believed to be the fundamental components of matter, or scalar fields, like the Higgs particles, which are used to introduced mass in a gauge theory (and that, however, could be composed of more fundamental fermion fields)."

In the late 19th century with the discovery of the electron, and in the early 20th century, with the discovery of the atomic nucleus, and the birth of particle physics, matter was seen as made up of electrons, protons and neutrons interacting to form atoms. Today, we know that even protons and neutrons are not indivisible, they can be divided into quarks, while electrons are part of a particle family called leptons. Both quarks and leptons are elementary particles, and are currently seen as being the fundamental constituents of matter.

These quarks and leptons interact through four fundamental forces: gravity, electromagnetism, weak interactions, and strong interactions. The Standard Model of particle physics is currently the best explanation for all of physics, but despite decades of efforts, gravity cannot yet be accounted for at the quantum level; it is only described by classical physics. Interactions between quarks and leptons are the result of an exchange of force-carrying particles (such as photons) between quarks and leptons. The force-carrying particles are not themselves building blocks. As one consequence, mass and energy (which cannot be created or destroyed) cannot always be related to matter (which can be created out of non-matter particles such as photons, or even out of pure energy, such as kinetic energy). Force carriers are usually not considered matter: the carriers of the electric force (photons) possess energy and the carriers of the weak force (W and Z bosons) are massive, but neither are considered matter either. However, while these particles are not considered matter, they do contribute to the total mass of atoms, subatomic particles, and all systems that contain them.

Summary

The modern conception of matter has been refined many times in history, in light of the improvement in knowledge of just *what* the basic building blocks are, and in how they interact. The term "matter" is used throughout physics in a bewildering variety of contexts: for example, one refers to "condensed matter physics", "elementary matter", "partonic" matter, "dark" matter, "anti"-matter, "strange" matter, and "nuclear" matter. In discussions of matter and antimatter, normal matter has been referred to by Alfvén as *koinomatter* (Gk. *common matter*). It is fair to say that in physics, there is no broad consensus as to a general definition of matter, and the term "matter" usually is used in conjunction with a specifying modifier.

The history of the concept of matter is a history of the fundamental *length scales* used to define matter. Different building blocks apply depending upon whether one defines matter on an atomic or elementary particle level. One may use a definition that matter is atoms, or that matter is hadrons, or that matter is leptons and quarks depending upon the scale at which one wishes to define matter.

These quarks and leptons interact through four fundamental forces: gravity, electromagnetism, weak interactions, and strong interactions. The Standard Model of particle physics is currently the best explanation for all of physics, but despite decades of efforts, gravity cannot yet be accounted for at the quantum level; it is only described by classical physics.

State of Matter

In physics, a state of matter is one of the distinct forms that matter takes on. Four states of matter are observable in everyday life: solid, liquid, gas, and plasma. Many other states are known to exist only in extreme situations, such as Bose–Einstein condensates, neutron-degenerate matter and quark-gluon plasma, which occur in situations of extreme cold, extreme density and extremely high-energy color-charged matter respectively. Some other states are believed to be possible but remain theoretical for now. For a complete list of all exotic states of matter.

Historically, the distinction is made based on qualitative differences in properties. Matter in the solid state maintains a fixed volume and shape, with component particles (atoms, molecules or ions) close together and fixed into place. Matter in the liquid state maintains a fixed volume, but has a variable shape that adapts to fit its container. Its particles are still close together but move freely. Matter in the gaseous state has both variable volume and shape, adapting both to fit its container. Its particles are neither close together nor fixed in place. Matter in the plasma state has variable volume and shape, but as well as neutral atoms, it contains a significant number of ions and electrons, both of which can move around freely. Plasma is the most common form of visible matter in the universe.

The four fundamental states of matter. Clockwise from top left, they are solid, liquid, plasma, and gas, represented by an ice sculpture, a drop of water, electrical arcing from a tesla coil, and the air around clouds, respectively.

The term phase is sometimes used as a synonym for state of matter, but a system can contain several immiscible phases of the same state of matter.

The Four Fundamental States

Solid

In a solid the particles (ions, atoms or molecules) are closely packed together. The forces between particles are strong so that the particles cannot move freely but can only vibrate. As a result, a solid has a stable, definite shape, and a definite volume. Solids can only change their shape by force, as when broken or cut.

A crystalline solid: atomic resolution image of strontium titanate. Brighter atoms are Sr and darker ones are Ti.

In crystalline solids, the particles (atoms, molecules, or ions) are packed in a regularly ordered, repeating pattern. There are various different crystal structures, and the same substance can have more than one structure (or solid phase). For example, iron has a body-centred cubic structure at temperatures below 912 °C, and a face-centred cubic structure between 912 and 1394 °C. Ice has fifteen known crystal structures, or fifteen solid phases, which exist at various temperatures and pressures.

Glasses and other non-crystalline, amorphous solids without long-range order are not thermal equilibrium ground states; therefore they are described below as nonclassical states of matter.

Solids can be transformed into liquids by melting, and liquids can be transformed into solids by freezing. Solids can also change directly into gases through the process of sublimation, and gases can likewise change directly into solids through deposition.

Liquid

Structure of a classical monatomic liquid. Atoms have many nearest neighbors in contact, yet no long-range order is present.

A liquid is a nearly incompressible fluid that conforms to the shape of its container but retains a (nearly) constant volume independent of pressure. The volume is definite if the temperature and pressure are constant. When a solid is heated above its melting point, it becomes liquid, given that the pressure is higher than the triple point of the substance. Intermolecular (or interatomic or interionic) forces are still important, but the molecules have enough energy to move relative to each other and the structure is mobile. This means that the shape of a liquid is not definite but is determined by its container. The volume is usually greater than that of the corresponding solid, the best known exception being water, H_2O. The highest temperature at which a given liquid can exist is its critical temperature.

Gas

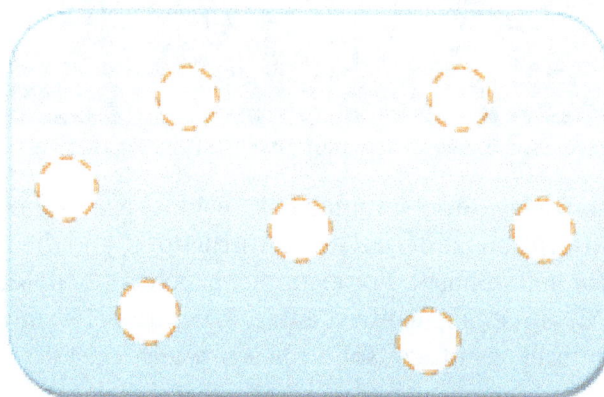

The spaces between gas molecules are very big. Gas molecules have very weak or no bonds at all.
The molecules in "gas" can move freely and fast.

A gas is a compressible fluid. Not only will a gas conform to the shape of its container but it will also expand to fill the container.

In a gas, the molecules have enough kinetic energy so that the effect of intermolecular forces is small (or zero for an ideal gas), and the typical distance between neighboring molecules is much greater than the molecular size. A gas has no definite shape or volume, but occupies the entire container in which it is confined. A liquid may be converted to a gas by heating at constant pressure to the boiling point, or else by reducing the pressure at constant temperature.

At temperatures below its critical temperature, a gas is also called a vapor, and can be liquefied by compression alone without cooling. A vapor can exist in equilibrium with a liquid (or solid), in which case the gas pressure equals the vapor pressure of the liquid (or solid).

A supercritical fluid (SCF) is a gas whose temperature and pressure are above the critical temperature and critical pressure respectively. In this state, the distinction between liquid and gas disappears. A supercritical fluid has the physical properties of a gas, but its high density confers solvent properties in some cases, which leads to useful applications. For example, supercritical carbon dioxide is used to extract caffeine in the manufacture of decaffeinated coffee.

Plasma

Like a gas, plasma does not have definite shape or volume. Unlike gases, plasmas are electrically conductive, produce magnetic fields and electric currents, and respond strongly to electromagnetic forces. Positively charged nuclei swim in a "sea" of freely-moving disassociated electrons, similar to the way such charges exist in conductive metal. In fact it is this electron "sea" that allows matter in the plasma state to conduct electricity.

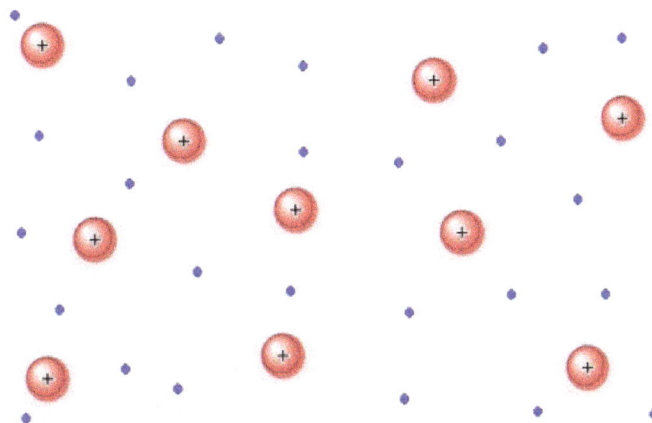

In a plasma, electrons are ripped away from their nuclei, forming an electron "sea".
This gives it the ability to conduct electricity.

The plasma state is often misunderstood, but it is actually quite common on Earth, and the majority of people observe it on a regular basis without even realizing it. Lightning, electric sparks, fluorescent lights, neon lights, plasma televisions, some types of flame and the stars are all examples of illuminated matter in the plasma state.

A gas is usually converted to a plasma in one of two ways, either from a huge voltage difference between two points, or by exposing it to extremely high temperatures.

Heating matter to high temperatures causes electrons to leave the atoms, resulting in the presence of free electrons. At very high temperatures, such as those present in stars, it is assumed that essentially all electrons are "free", and that a very high-energy plasma is essentially bare nuclei swimming in a sea of electrons.

Phase Transitions

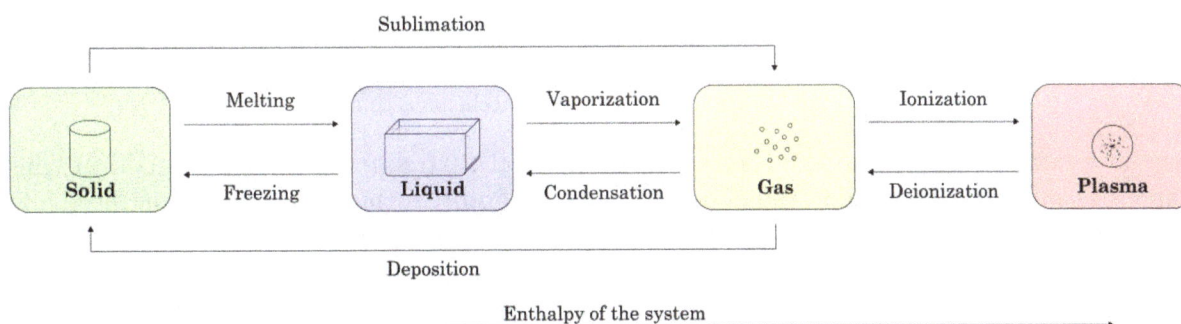

This diagram illustrates transitions between the four fundamental states of matter.

A state of matter is also characterized by phase transitions. A phase transition indicates a change in structure and can be recognized by an abrupt change in properties. A distinct state of matter can be defined as any set of states distinguished from any other set of states by a phase transition. Water can be said to have several distinct solid states. The appearance of superconductivity is associated with a phase transition, so there are superconductive states. Likewise, ferromagnetic states are demarcated by phase transitions and have distinctive properties. When the change of state occurs in stages the intermediate steps are called mesophases. Such phases have been exploited by the introduction of liquid crystal technology.

The state or *phase* of a given set of matter can change depending on pressure and temperature conditions, transitioning to other phases as these conditions change to favor their existence; for example, solid transitions to liquid with an increase in temperature. Near absolute zero, a substance exists as a solid. As heat is added to this substance it melts into a liquid at its melting point, boils into a gas at its boiling point, and if heated high enough would enter a plasma state in which the electrons are so energized that they leave their parent atoms.

Forms of matter that are not composed of molecules and are organized by different forces can also be considered different states of matter. Superfluids (like Fermionic condensate) and the quark–gluon plasma are examples.

In a chemical equation, the state of matter of the chemicals may be shown as (s) for solid, (l) for liquid, and (g) for gas. An aqueous solution is denoted (aq). Matter in the plasma state is seldom used (if at all) in chemical equations, so there is no standard symbol to denote it. In the rare equations that plasma is used in plasma is symbolized as (p).

Non-classical States

Glass

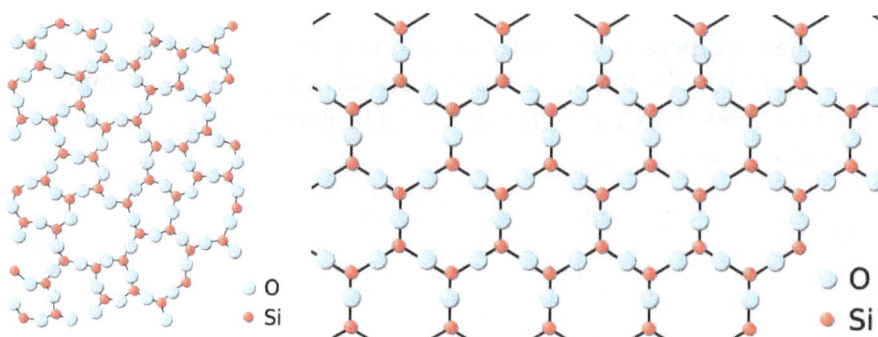

Schematic representation of a random-network glassy form (left) and ordered crystalline lattice (right) of identical chemical composition.

Glass is a non-crystalline or amorphous solid material that exhibits a glass transition when heated towards the liquid state. Glasses can be made of quite different classes of materials: inorganic networks (such as window glass, made of silicate plus additives), metallic alloys, ionic melts, aqueous solutions, molecular liquids, and polymers. Thermodynamically, a glass is in a metastable state with respect to its crystalline counterpart. The conversion rate, however, is practically zero.

Crystals with Some Degree of Disorder

A plastic crystal is a molecular solid with long-range positional order but with constituent molecules retaining rotational freedom; in an orientational glass this degree of freedom is frozen in a quenched disordered state.

Similarly, in a spin glass magnetic disorder is frozen.

Liquid Crystal States

Liquid crystal states have properties intermediate between mobile liquids and ordered solids. Generally, they are able to flow like a liquid, but exhibiting long-range order. For example, the nematic phase consists of long rod-like molecules such as para-azoxyanisole, which is nematic in the temperature range 118–136 °C. In this state the molecules flow as in a liquid, but they all point in the same direction (within each domain) and cannot rotate freely. Like a crystalline solid, but unlike a liquid, liquid crystals react to polarized light.

Other types of liquid crystals are described in the main article on these states. Several types have technological importance, for example, in liquid crystal displays.

Magnetically Ordered

Transition metal atoms often have magnetic moments due to the net spin of electrons that remain unpaired and do not form chemical bonds. In some solids the magnetic moments on different atoms are ordered and can form a ferromagnet, an antiferromagnet or a ferrimagnet.

In a ferromagnet—for instance, solid iron—the magnetic moment on each atom is aligned in the same direction (within a magnetic domain). If the domains are also aligned, the solid is a permanent magnet, which is magnetic even in the absence of an external magnetic field. The magnetization disappears when the magnet is heated to the Curie point, which for iron is 768 °C.

An antiferromagnet has two networks of equal and opposite magnetic moments, which cancel each other out so that the net magnetization is zero. For example, in nickel(II) oxide (NiO), half the nickel atoms have moments aligned in one direction and half in the opposite direction.

In a ferrimagnet, the two networks of magnetic moments are opposite but unequal, so that cancellation is incomplete and there is a non-zero net magnetization. An example is magnetite (Fe_3O_4), which contains Fe^{2+} and Fe^{3+} ions with different magnetic moments.

A quantum spin liquid (QSL) is a disordered state in a system of interacting quantum spins which preserves its disorder to very low temperatures, unlike other disordered states. It is not a liquid in physical sense, but a solid whose magnetic order is inherently disordered. The name "liquid" is due to an analogy with the molecular disorder in a conventional liquid. A QSL is neither a ferromagnet, where magnetic domains are parallel, nor an antiferromagnet, where the magnetic domains are antiparallel; instead, the magnetic domains are randomly oriented. This can be realized e.g. by geometrically frustrated magnetic moments that cannot point uniformly parallel or antiparallel. When cooling down and settling to a state, the domain must "choose" an orientation, but if the possible states are similar in energy, one will be chosen randomly. Consequently, despite strong short-range order, there is no long-range magnetic order.

Microphase-separated

Copolymers can undergo microphase separation to form a diverse array of periodic nanostructures, as shown in the example of the styrene-butadiene-styrene block copolymer shown at right. Microphase separation can be understood by analogy to the phase separation between oil and water. Due to chemical incompatibility between the blocks, block copolymers undergo a similar phase separation. However, because the blocks are covalently bonded to each other, they cannot demix macroscopically as water and oil can, and so instead the blocks form nanometer-sized structures. Depending on the relative lengths of each block and the overall block topology of the polymer, many morphologies can be obtained, each its own phase of matter.

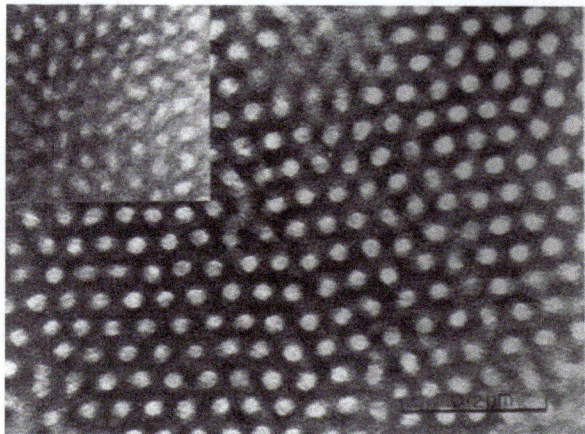

SBS block copolymer in TEM

Ionic liquids also display microphase separation. The anion and cation are not necessarily compatible and would demix otherwise, but electric charge attraction prevents them from separating. Their anions and cations appear to diffuse within compartmentalized layers or micelles instead of freely as in a uniform liquid.

Low-temperature States

Superfluid

Liquid helium in a superfluid phase creeps up on the walls of the cup in a Rollin film, eventually dripping out from the cup.

Close to absolute zero, some liquids form a second liquid state described as superfluid because it has zero viscosity (or infinite fluidity; i.e., flowing without friction). This was discovered in 1937 for helium, which forms a superfluid below the lambda temperature of 2.17 K. In this state it will attempt to "climb" out of its container. It also has infinite thermal conductivity so that no temperature gradient can form in a superfluid. Placing a superfluid in a spinning container will result in quantized vortices.

These properties are explained by the theory that the common isotope helium-4 forms a Bose–Einstein condensate in the superfluid state. More recently, Fermionic condensate superfluids have been formed at even lower temperatures by the rare isotope helium-3 and by lithium-6.

Bose–Einstein Condensate

In 1924, Albert Einstein and Satyendra Nath Bose predicted the "Bose–Einstein condensate" (BEC), sometimes referred to as the fifth state of matter. In a BEC, matter stops behaving as independent particles, and collapses into a single quantum state that can be described with a single, uniform wavefunction.

In the gas phase, the Bose–Einstein condensate remained an unverified theoretical prediction for many years. In 1995, the research groups of Eric Cornell and Carl Wieman, of JILA at the University of Colorado at Boulder, produced the first such condensate experimentally. A Bose–Einstein condensate is "colder" than a solid. It may occur when atoms have very similar (or the same) quantum levels, at temperatures very close to absolute zero (−273.15 °C).

Fermionic Condensate

A *fermionic condensate* is similar to the Bose–Einstein condensate but composed of fermions. The Pauli exclusion principle prevents fermions from entering the same quantum state, but a pair of fermions can behave as a boson, and multiple such pairs can then enter the same quantum state without restriction.

Rydberg Molecule

One of the metastable states of strongly non-ideal plasma is Rydberg matter, which forms upon condensation of excited atoms. These atoms can also turn into ions and electrons if they reach a certain temperature. In April 2009, *Nature* reported the creation of Rydberg molecules from a Rydberg atom and a ground state atom, confirming that such a state of matter could exist. The experiment was performed using ultracold rubidium atoms.

Quantum Hall State

A *quantum Hall state* gives rise to quantized Hall voltage measured in the direction perpendicular to the current flow. A *quantum spin Hall state* is a theoretical phase that may pave the way for the development of electronic devices that dissipate less energy and generate less heat. This is a derivation of the Quantum Hall state of matter.

Photonic Matter

Photonic matter is a phenomenon where photons interacting with a gas develop apparent mass, and can interact with each other, even forming photonic "molecules". The source of mass is the gas, which is massive. This is in contrast to photons moving in empty space, which have no rest mass, and cannot interact.

Dropleton

A "quantum fog" of electrons and holes that flow around each other and even ripple like a liquid, rather than existing as discrete pairs.

High-energy States

Degenerate Matter

Under extremely high pressure, as in the cores of dead stars, ordinary matter undergoes a transition to a series of exotic states of matter collectively known as degenerate matter, which are supported mainly by quantum mechanical effects. In physics, "degenerate" refers to two states that have the same energy and are thus interchangeable. Degenerate matter is supported by the Pauli exclusion principle, which prevents two fermionic particles from occupying the same quantum state. Unlike regular plasma, degenerate plasma expands little when heated, because there are simply no momentum states left. Consequently, degenerate stars collapse into very high densities. More massive degenerate stars are smaller, because the gravitational force increases, but pressure does not increase proportionally.

Electron-degenerate matter is found inside white dwarf stars. Electrons remain bound to atoms but are able to transfer to adjacent atoms. Neutron-degenerate matter is found in neutron stars. Vast gravitational pressure compresses atoms so strongly that the electrons are forced to combine with protons via inverse beta-decay, resulting in a superdense conglomeration of neutrons. Normally free neutrons outside an atomic nucleus will decay with a half life of just under 15 minutes, but in a neutron star, the decay is overtaken by inverse decay. Cold degenerate matter is also present in planets such as Jupiter and in the even more massive brown dwarfs, which are expected to have a core with metallic hydrogen. Because of the degeneracy, more massive brown dwarfs are not significantly larger. In metals, the electrons can be modeled as a degenerate gas moving in a lattice of non-degenerate positive ions.

Quark Matter

In regular cold matter, quarks, fundamental particles of nuclear matter, are confined by the strong force into hadrons that consist of 2–4 quarks, such as protons and neutrons. Quark matter or quantum chromodynanamical (QCD) matter is a group of phases where the strong force is overcome and quarks are deconfined and free to move. Quark matter phases occur at extremely high densities or temperatures.

Strange matter is a type of quark matter that may exist inside some neutron stars close to the Tolman–Oppenheimer–Volkoff limit (approximately 2–3 solar masses). It may be stable at lower energy states once formed, although this is not known.

Quark–gluon plasma is a very high-temperature phase in which quarks become free and able to move independently, rather than being perpetually bound into particles, in a sea of gluons, subatomic particles that transmit the strong force that binds quarks together. This is analogous to the liberation of electrons from atoms in a plasma. This state is briefly attainable in extremely high-energy heavy ion collisions in particle accelerators, and allows scientists to observe the properties of individual quarks, and not just theorize. Quark–gluon plasma was discovered at CERN in 2000. Unlike plasma, which flows like a gas, interactions within QGP are strong and it flows like a liquid.

At high densities but relatively low temperatures, quarks are theorized to for a quark liquid whose nature is presently unknown. It form a distinct color-flavor locked (CFL) phase at even higher densities. This phase is superconductive for color charge. These phases may occur in neutron stars but they are presently theoretical.

Color-glass Condensate

Color-glass condensate is a type of matter theorized to exist in atomic nuclei traveling near the speed of light. According to Einstein's theory of relativity, a high-energy nucleus appears length contracted, or compressed, along its direction of motion. As a result, the gluons inside the nucleus appear to a stationary observer as a "gluonic wall" traveling near the speed of light. At very high energies, the density of the gluons in this wall is seen to increase greatly. Unlike the quark–gluon plasma produced in the collision of such walls, the color-glass condensate describes the walls themselves, and is an intrinsic property of the particles that can only be observed under high-energy conditions such as those at RHIC and possibly at the Large Hadron Collider as well.

Very High Energy States

The gravitational singularity predicted by general relativity to exist at the center of a black hole is *not* a phase of matter; it is not a material object at all (although the mass-energy of matter contributed to its creation) but rather a property of spacetime at a location. It could be argued, of course, that all particles are properties of spacetime at a location, leaving a half-note of controversy on the subject.

Other Proposed States
Supersolid

A supersolid is a spatially ordered material (that is, a solid or crystal) with superfluid properties. Similar to a superfluid, a supersolid is able to move without friction but retains a rigid shape. Although a supersolid is a solid, it exhibits so many characteristic properties different from other solids that many argue it is another state of matter.

String-net Liquid

In a string-net liquid, atoms have apparently unstable arrangement, like a liquid, but are still consistent in overall pattern, like a solid. When in a normal solid state, the atoms of matter align themselves in a grid pattern, so that the spin of any electron is the opposite of the spin of all electrons touching it. But in a string-net liquid, atoms are arranged in some pattern that requires some electrons to have neighbors with the same spin. This gives rise to curious properties, as well as supporting some unusual proposals about the fundamental conditions of the universe itself.

Superglass

A superglass is a phase of matter characterized, at the same time, by superfluidity and a frozen amorphous structure.

Dark Matter

While dark matter is estimated to comprise 83% of the mass of matter in the universe, most of its properties remain a mystery due to the fact that it neither absorbs nor emits electromagnetic radiation, and there are many competing theories regarding what dark matter is actually made of. Thus, while it is hypothesized to exist and comprise the vast majority of matter in the universe, almost all of its properties are unknown and a matter of speculation, because it has only been observed through its gravitational effects.

Phase (Matter)

In the physical sciences, a phase is a region of space (a thermodynamic system), throughout which all physical properties of a material are essentially uniform. Examples of physical properties include density, index of refraction, magnetization and chemical composition. A simple description is that a phase is a region of material that is chemically uniform, physically distinct, and (often) mechanically separable. In a system consisting of ice and water in a glass jar, the ice cubes are one phase, the water is a second phase, and the humid air over the water is a third phase. The glass of the jar is another separate phase.

The term *phase* is sometimes used as a synonym for state of matter, but there can be several immiscible phases of the same state of matter. Also, the term *phase* is sometimes used to refer to a set of equilibrium states demarcated in terms of state variables such as pressure and temperature by a phase boundary on a phase diagram. Because phase boundaries relate to changes in the organization of matter, such as a change from liquid to solid or a more subtle change from one crystal structure to another, this latter usage is similar to the use of "phase" as a synonym for state of matter. However, the state of matter and phase diagram usages are not commensurate with the formal definition given above and the intended meaning must be determined in part from the context in which the term is used.

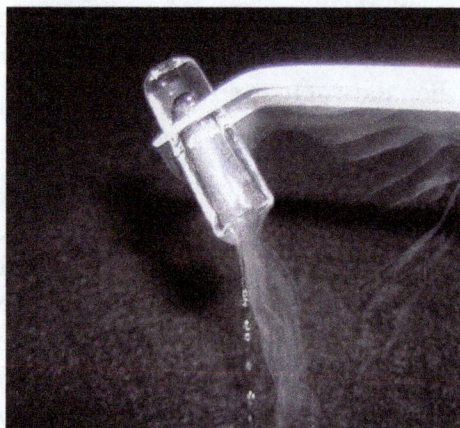

A small piece of rapidly melting argon ice shows the transition from solid to liquid.

Types of Phases

Temperature (°C)

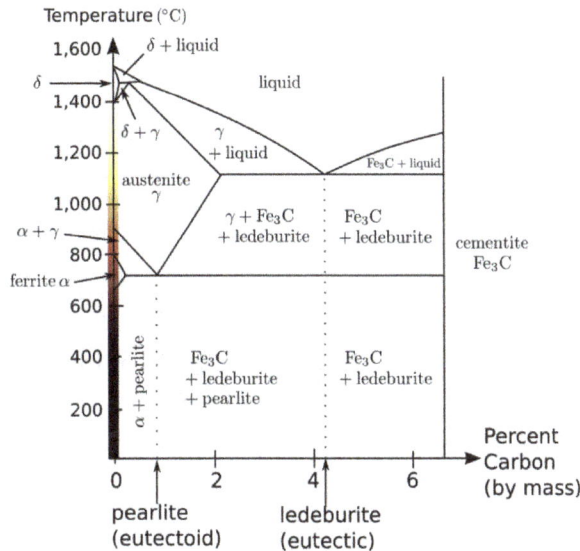

Iron-carbon phase diagram, showing the conditions necessary to form different phases

Distinct phases may be described as different states of matter such as gas, liquid, solid, plasma or Bose–Einstein condensate. Useful mesophases between solid and liquid form other states of matter.

Distinct phases may also exist within a given state of matter. As shown in the diagram for iron alloys, several phases exist for both the solid and liquid states. Phases may also be differentiated based on solubility as in polar (hydrophilic) or non-polar (hydrophobic). A mixture of water (a polar liquid) and oil (a non-polar liquid) will spontaneously separate into two phases. Water has a very low solubility (is insoluble) in oil, and oil has a low solubility in water. Solubility is the maximum amount of a solute that can dissolve in a solvent before the solute ceases to dissolve and remains in a separate phase. A mixture can separate into more than two liquid phases and the concept of phase separation extends to solids, i.e., solids can form solid solutions or crystallize into distinct crystal phases. Metal pairs that are mutually soluble can form alloys, whereas metal pairs that are mutually insoluble cannot.

As many as eight immiscible liquid phases have been observed. Mutually immiscible liquid phases are formed from water (aqueous phase), hydrophobic organic solvents, perfluorocarbons (fluorous phase), silicones, several different metals, and also from molten phosphorus. Not all organic solvents are completely miscible, e.g. a mixture of ethylene glycol and toluene may separate into two distinct organic phases.

Phases do not need to macroscopically separate spontaneously. Emulsions and colloids are examples of immiscible phase pair combinations that do not physically separate.

Phase Equilibrium

Left to equilibration, many compositions will form a uniform single phase, but depending on the temperature and pressure even a single substance may separate into two or more distinct phases. Within each phase, the properties are uniform but between the two phases properties differ.

Water in a closed jar with an air space over it forms a two phase system. Most of the water is in the liquid phase, where it is held by the mutual attraction of water molecules. Even at equilibrium molecules are constantly in motion and, once in a while, a molecule in the liquid phase gains enough kinetic energy to break away from the liquid phase and enter the gas phase. Likewise, every once in a while a vapor molecule collides with the liquid surface and condenses into the liquid. At equilibrium, evaporation and condensation processes exactly balance and there is no net change in the volume of either phase.

At room temperature and pressure, the water jar reaches equilibrium when the air over the water has a humidity of about 3%. This percentage increases as the temperature goes up. At 100 °C and atmospheric pressure, equilibrium is not reached until the air is 100% water. If the liquid is heated a little over 100 °C, the transition from liquid to gas will occur not only at the surface, but throughout the liquid volume: the water boils.

Number of Phases

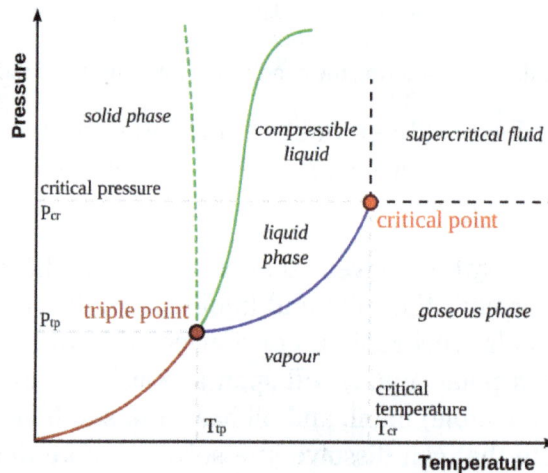

A typical phase diagram for a single-component material, exhibiting solid, liquid and gaseous phases. The solid green line shows the usual shape of the liquid–solid phase line. The dotted green line shows the anomalous behavior of water when the pressure increases. The triple point and the critical point are shown as red dots.

For a given composition, only certain phases are possible at a given temperature and pressure. The number and type of phases that will form is hard to predict and is usually determined by experiment. The results of such experiments can be plotted in phase diagrams.

The phase diagram shown here is for a single component system. In this simple system, which phases that are possible depends only on pressure and temperature. The markings show points where two or more phases can co-exist in equilibrium. At temperatures and pressures away from the markings, there will be only one phase at equilibrium.

In the diagram, the blue line marking the boundary between liquid and gas does not continue indefinitely, but terminates at a point called the critical point. As the temperature and pressure approach the critical point, the properties of the liquid and gas become progressively more similar. At the critical point, the liquid and gas become indistinguishable. Above the critical point, there are no longer separate liquid and gas phases: there is only a generic fluid phase referred to as a supercritical fluid. In water, the critical point occurs at around 647 K (374 °C or 705 °F) and 22.064 MPa.

An unusual feature of the water phase diagram is that the solid–liquid phase line (illustrated by the dotted green line) has a negative slope. For most substances, the slope is positive as exemplified by the dark green line. This unusual feature of water is related to ice having a lower density than liquid water. Increasing the pressure drives the water into the higher density phase, which causes melting.

Another interesting though not unusual feature of the phase diagram is the point where the solid–liquid phase line meets the liquid–gas phase line. The intersection is referred to as the triple point. At the triple point, all three phases can coexist.

Experimentally, the phase lines are relatively easy to map due to the interdependence of temperature and pressure that develops when multiple phases forms. Consider a test apparatus consisting of a closed and well insulated cylinder equipped with a piston. By charging the right amount of water and applying heat, the system can be brought to any point in the gas region of the phase diagram. If the piston is slowly lowered, the system will trace a curve of increasing temperature and pressure within the gas region of the phase diagram. At the point where gas begins to condense to liquid, the direction of the temperature and pressure curve will abruptly change to trace along the phase line until all of the water has condensed.

Interfacial Phenomena

Between two phases in equilibrium there is a narrow region where the properties are not that of either phase. Although this region may be very thin, it can have significant and easily observable effects, such as causing a liquid to exhibit surface tension. In mixtures, some components may preferentially move toward the interface. In terms of modeling, describing, or understanding the behavior of a particular system, it may be efficacious to treat the interfacial region as a separate phase.

Crystal Phases

A single material may have several distinct solid states capable of forming separate phases. Water is a well-known example of such a material. For example, water ice is ordinarily found in the hexagonal form ice Ih, but can also exist as the cubic ice Ic, the rhombohedral ice II, and many other forms. Polymorphism is the ability of a solid to exist in more than one crystal form. For pure chemical elements, polymorphism is known as allotropy. For example, diamond, graphite, and fullerenes are different allotropes of carbon.

Phase Transitions

When a substance undergoes a phase transition (changes from one state of matter to another) it usually either takes up or releases energy. For example, when water evaporates, the increase in kinetic energy as the evaporating molecules escape the attractive forces of the liquid is reflected in a decrease in temperature. The energy required to induce the phase transition is taken from the internal thermal energy of the water, which cools the liquid to a lower temperature; hence evaporation is useful for cooling. The reverse process, condensation, releases heat. The heat energy, or enthalpy, associated with a solid to liquid transition is the enthalpy of fusion and that associated with a solid to gas transition is the enthalpy of sublimation.

Solid

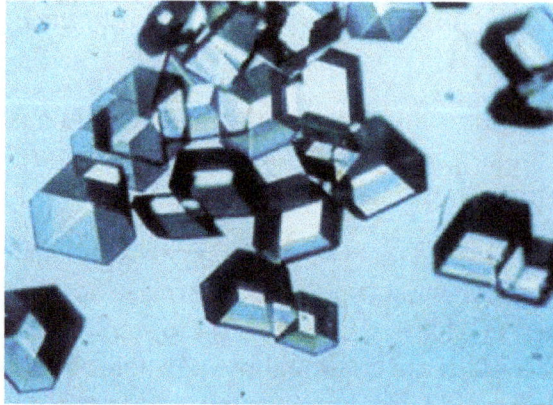

Single crystalline form of solid insulin.

Solid is one of the four fundamental states of matter (the others being liquid, gas, and plasma). It is characterized by structural rigidity and resistance to changes of shape or volume. Unlike a liquid, a solid object does not flow to take on the shape of its container, nor does it expand to fill the entire volume available to it like a gas does. The atoms in a solid are tightly bound to each other, either in a regular geometric lattice (crystalline solids, which include metals and ordinary ice) or irregularly (an amorphous solid such as common window glass).

The branch of physics that deals with solids is called solid-state physics, and is the main branch of condensed matter physics (which also includes liquids). Materials science is primarily concerned with the physical and chemical properties of solids. Solid-state chemistry is especially concerned with the synthesis of novel materials, as well as the science of identification and chemical composition.

Microscopic Description

Model of closely packed atoms within a crystalline solid.

The atoms, molecules or ions which make up solids may be arranged in an orderly repeating pattern, or irregularly. Materials whose constituents are arranged in a regular pattern are known as crystals. In some cases, the regular ordering can continue unbroken over a large scale, for example diamonds, where each diamond is a single crystal. Solid objects that are large enough to see and handle are rarely composed of a single crystal, but instead are made of a large number of single crystals, known as crystallites, whose size can vary from a few nanometers to several meters. Such materials are called polycrystalline. Almost all common metals, and many ceramics, are polycrystalline.

In other materials, there is no long-range order in the position of the atoms. These solids are known as amorphous solids; examples include polystyrene and glass.

Whether a solid is crystalline or amorphous depends on the material involved, and the conditions in which it was formed. Solids which are formed by slow cooling will tend to be crystalline, while solids which are frozen rapidly are more likely to be amorphous. Likewise, the specific crystal structure adopted by a crystalline solid depends on the material involved and on how it was formed.

While many common objects, such as an ice cube or a coin, are chemically identical throughout, many other common materials comprise a number of different substances packed together. For example, a typical rock is an aggregate of several different minerals and mineraloids, with no specific chemical composition. Wood is a natural organic material consisting primarily of cellulose fibers embedded in a matrix of organic lignin. In materials science, composites of more than one constituent material can be designed to have desired properties.

Classes of Solids

The forces between the atoms in a solid can take a variety of forms. For example, a crystal of sodium chloride (common salt) is made up of ionic sodium and chlorine, which are held together by ionic bonds. In diamond or silicon, the atoms share electrons and form covalent bonds. In metals, electrons are shared in metallic bonding. Some solids, particularly moWst organic compounds, are held together with van der Waals forces resulting from the polarization of the electronic charge cloud on each molecule. The dissimilarities between the types of solid result from the differences between their bonding.

Metals

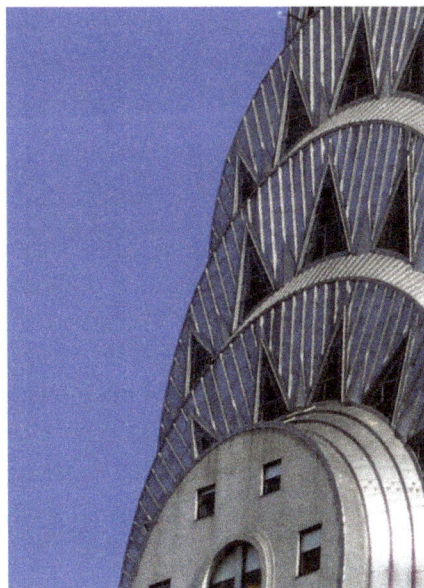

The pinnacle of New York's Chrysler Building, the world's tallest steel-supported brick building, is clad with stainless steel.

Metals typically are strong, dense, and good conductors of both electricity and heat. The bulk of the elements in the periodic table, those to the left of a diagonal line drawn from boron to polonium, are metals. Mixtures of two or more elements in which the major component is a metal are known as alloys.

People have been using metals for a variety of purposes since prehistoric times. The strength and reliability of metals has led to their widespread use in construction of buildings and other structures, as well as in most vehicles, many appliances and tools, pipes, road signs and railroad tracks. Iron and aluminium are the two most commonly used structural metals, and they are also the most abundant metals in the Earth's crust. Iron is most commonly used in the form of an alloy, steel, which contains up to 2.1% carbon, making it much harder than pure iron.

Because metals are good conductors of electricity, they are valuable in electrical appliances and for carrying an electric current over long distances with little energy loss or dissipation. Thus, electrical power grids rely on metal cables to distribute electricity. Home electrical systems, for example, are wired with copper for its good conducting properties and easy machinability. The high thermal conductivity of most metals also makes them useful for stovetop cooking utensils.

The study of metallic elements and their alloys makes up a significant portion of the fields of solid-state chemistry, physics, materials science and engineering.

Metallic solids are held together by a high density of shared, delocalized electrons, known as "metallic bonding". In a metal, atoms readily lose their outermost ("valence") electrons, forming positive ions. The free electrons are spread over the entire solid, which is held together firmly by electrostatic interactions between the ions and the electron cloud. The large number of free electrons gives metals their high values of electrical and thermal conductivity. The free electrons also prevent transmission of visible light, making metals opaque, shiny and lustrous.

More advanced models of metal properties consider the effect of the positive ions cores on the delocalised electrons. As most metals have crystalline structure, those ions are usually arranged into a periodic lattice. Mathematically, the potential of the ion cores can be treated by various models, the simplest being the nearly free electron model.

Minerals

Minerals are naturally occurring solids formed through various geological processes under high pressures. To be classified as a true mineral, a substance must have a crystal structure with uniform physical properties throughout. Minerals range in composition from pure elements and simple salts to very complex silicates with thousands of known forms. In contrast, a rock sample is a random aggregate of minerals and/or mineraloids, and has no specific chemical composition. The vast majority of the rocks of the Earth's crust consist of quartz (crystalline SiO_2), feldspar, mica, chlorite, kaolin, calcite, epidote, olivine, augite, hornblende, magnetite, hematite, limonite and a few other minerals. Some minerals, like quartz, mica or feldspar are common, while others have been found in only a few locations worldwide. The largest group of minerals by far is the silicates (most rocks are $\geq 95\%$ silicates), which are composed largely of silicon and oxygen, with the addition of ions of aluminium, magnesium, iron, calcium and other metals.

A collection of various minerals.

Ceramics

Si_3N_4 ceramic bearing parts

Ceramic solids are composed of inorganic compounds, usually oxides of chemical elements. They are chemically inert, and often are capable of withstanding chemical erosion that occurs in an acidic or caustic environment. Ceramics generally can withstand high temperatures ranging from 1000 to 1600 °C (1800 to 3000 °F). Exceptions include non-oxide inorganic materials, such as nitrides, borides and carbides.

Traditional ceramic raw materials include clay minerals such as kaolinite, more recent materials include aluminium oxide (alumina). The modern ceramic materials, which are classified as advanced ceramics, include silicon carbide and tungsten carbide. Both are valued for their abrasion resistance, and hence find use in such applications as the wear plates of crushing equipment in mining operations.

Most ceramic materials, such as alumina and its compounds, are formed from fine powders, yielding a fine grained polycrystalline microstructure which is filled with light scattering centers comparable to the wavelength of visible light. Thus, they are generally opaque materials, as opposed to transparent materials. Recent nanoscale (e.g. sol-gel) technology has, however, made possible the production of polycrystalline transparent ceramics such as transparent alumina and alumina compounds for such applications as high-power lasers. Advanced ceramics are also used in the medicine, electrical and electronics industries.

Ceramic engineering is the science and technology of creating solid-state ceramic materials, parts and devices. This is done either by the action of heat, or, at lower temperatures, using precipitation reactions from chemical solutions. The term includes the purification of raw materials, the study and production of the chemical compounds concerned, their formation into components, and the study of their structure, composition and properties.

Mechanically speaking, ceramic materials are brittle, hard, strong in compression and weak in shearing and tension. Brittle materials may exhibit significant tensile strength by supporting a static load. Toughness indicates how much energy a material can absorb before mechanical failure, while fracture toughness (denoted K_{Ic}) describes the ability of a material with inherent micro-structural flaws to resist fracture via crack growth and propagation. If a material has a large value of fracture toughness, the basic principles of fracture mechanics suggest that it will most likely undergo ductile fracture. Brittle fracture is very characteristic of most ceramic and glass-ceramic materials which typically exhibit low (and inconsistent) values of K_{Ic}.

For an example of applications of ceramics, the extreme hardness of zirconia is utilized in the manufacture of knife blades, as well as other industrial cutting tools. Ceramics such as alumina, boron carbide and silicon carbide have been used in bulletproof vests to repel large-caliber rifle fire. Silicon nitride parts are used in ceramic ball bearings, where their high hardness makes them wear resistant. In general, ceramics are also chemically resistant and can be used in wet environments where steel bearings would be susceptible to oxidation (or rust).

As another example of ceramic applications, in the early 1980s, Toyota researched production of an adiabatic ceramic engine with an operating temperature of over 6000 °F (3300 °C). Ceramic engines do not require a cooling system and hence allow a major weight reduction and therefore greater fuel efficiency. In a conventional metallic engine, much of the energy released from the fuel must be dissipated as waste heat in order to prevent a meltdown of the metallic parts. Work is also being done in developing ceramic parts for gas turbine engines. Turbine engines made with ceramics could operate more efficiently, giving aircraft greater range and payload for a set amount of fuel. However, such engines are not in production because the manufacturing of ceramic parts in the sufficient precision and durability is difficult and costly. Processing methods often result in a wide distribution of microscopic flaws which frequently play a detrimental role in the sintering process, resulting in the proliferation of cracks, and ultimate mechanical failure.

Glass Ceramics

A high strength glass-ceramic cooktop with negligible thermal expansion.

Glass-ceramics are used to make cookware (originally known by the brand name CorningWare) and stovetops which have both high resistance to thermal shock and extremely low permeability to liquids. The negative coefficient of thermal expansion of the crystalline ceramic phase can be balanced with the positive coefficient of the glassy phase. At a certain point (~70% crystalline) the glass-ceramic has a net coefficient of thermal expansion close to zero. This type of glass-ceramic exhibits excellent mechanical properties and can sustain repeated and quick temperature changes up to 1000 °C.

Glass ceramics may also occur naturally when lightning strikes the crystalline (e.g. quartz) grains found in most beach sand. In this case, the extreme and immediate heat of the lightning (~2500 °C) creates hollow, branching rootlike structures called fulgurite via fusion.

Organic Solids

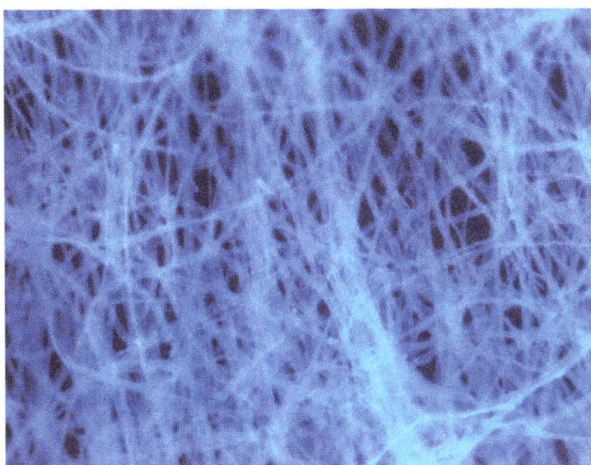

The individual wood pulp fibers in this sample are around 10 μm in diameter.

Organic chemistry studies the structure, properties, composition, reactions, and preparation by synthesis (or other means) of chemical compounds of carbon and hydrogen, which may contain any number of other elements such as nitrogen, oxygen and the halogens: fluorine, chlorine, bromine and iodine. Some organic compounds may also contain the elements phosphorus or sulfur. Examples of organic solids include wood, paraffin wax, naphthalene and a wide variety of polymers and plastics.

Wood

Wood is a natural organic material consisting primarily of cellulose fibers embedded in a matrix of lignin. Regarding mechanical properties, the fibers are strong in tension, and the lignin matrix resists compression. Thus wood has been an important construction material since humans began building shelters and using boats. Wood to be used for construction work is commonly known as *lumber* or *timber*. In construction, wood is not only a structural material, but is also used to form the mould for concrete.

Wood-based materials are also extensively used for packaging (e.g. cardboard) and paper which are both created from the refined pulp. The chemical pulping processes use a combination of high temperature and alkaline (kraft) or acidic (sulfite) chemicals to break the chemical bonds of the lignin before burning it out.

Polymers

STM image of self-assembled supramolecular chains of the organic semiconductor quinacridone on graphite.

One important property of carbon in organic chemistry is that it can form certain compounds, the individual molecules of which are capable of attaching themselves to one another, thereby forming a chain or a network. The process is called polymerization and the chains or networks polymers, while the source compound is a monomer. Two main groups of polymers exist: those artificially manufactured are referred to as industrial polymers or synthetic polymers (plastics) and those naturally occurring as biopolymers.

Monomers can have various chemical substituents, or functional groups, which can affect the chemical properties of organic compounds, such as solubility and chemical reactivity, as well as the physical properties, such as hardness, density, mechanical or tensile strength, abrasion resistance, heat resistance, transparency, color, etc.. In proteins, these differences give the polymer the ability to adopt a biologically active conformation in preference to others.

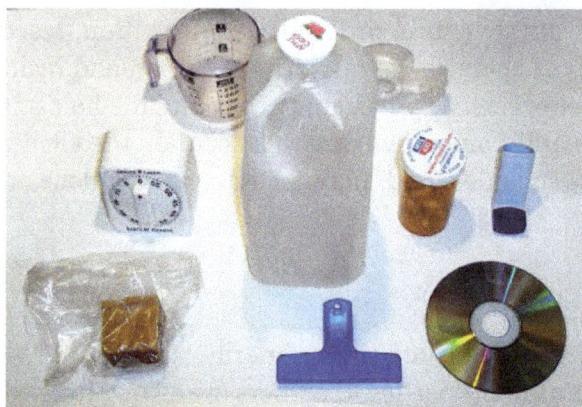

Household items made of various kinds of plastic.

People have been using natural organic polymers for centuries in the form of waxes and shellac which is classified as a thermoplastic polymer. A plant polymer named cellulose provided the tensile strength for natural fibers and ropes, and by the early 19th century natural rubber was in widespread use. Polymers are the raw materials (the resins) used to make what are commonly called plastics. Plastics are the final product, created after one or more polymers or additives have been added to a resin during processing, which is then shaped into a final form. Polymers which have been around, and which are in current widespread use, include carbon-based polyethylene, polypropylene, polyvinyl chloride, polystyrene, nylons, polyesters, acrylics, polyurethane, and polycarbonates, and silicon-based silicones. Plastics are generally classified as "commodity", "specialty" and "engineering" plastics.

Composite Materials

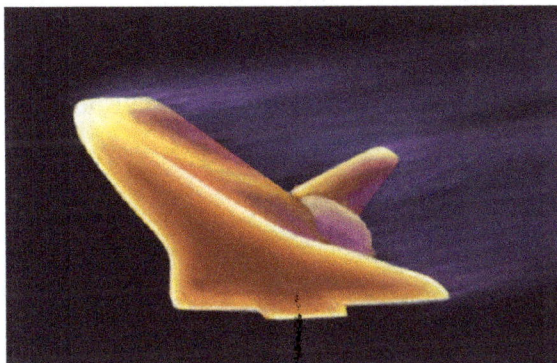

Simulation of the outside of the Space Shuttle as it heats up to over 1500 °C during re-entry

A cloth of woven carbon fiber filaments, a common element in composite materials

Composite materials contain two or more macroscopic phases, one of which is often ceramic. For example, a continuous matrix, and a dispersed phase of ceramic particles or fibers.

Applications of composite materials range from structural elements such as steel-reinforced concrete, to the thermally insulative tiles which play a key and integral role in NASA's Space Shuttle thermal protection system which is used to protect the surface of the shuttle from the heat of re-entry into the Earth's atmosphere. One example is Reinforced Carbon-Carbon (RCC), the light gray material which withstands reentry temperatures up to 1510 °C (2750 °F) and protects the nose cap and leading edges of Space Shuttle's wings. RCC is a laminated composite material made from graphite rayon cloth and impregnated with a phenolic resin. After curing at high temperature in an autoclave, the laminate is pyrolized to convert the resin to carbon, impregnated with furfural alcohol in a vacuum chamber, and cured/pyrolized to convert the furfural alcohol to carbon. In order to provide oxidation resistance for reuse capability, the outer layers of the RCC are converted to silicon carbide.

Domestic examples of composites can be seen in the "plastic" casings of television sets, cell-phones and so on. These plastic casings are usually a composite made up of a thermoplastic matrix such as acrylonitrile butadiene styrene (ABS) in which calcium carbonate chalk, talc, glass fibers or carbon fibers have been added for strength, bulk, or electro-static dispersion. These additions may be referred to as reinforcing fibers, or dispersants, depending on their purpose.

Thus, the matrix material surrounds and supports the reinforcement materials by maintaining their relative positions. The reinforcements impart their special mechanical and physical properties to enhance the matrix properties. A synergism produces material properties unavailable from the individual constituent materials, while the wide variety of matrix and strengthening materials provides the designer with the choice of an optimum combination.

Semiconductors

Semiconductor chip on crystalline silicon substrate.

Semiconductors are materials that have an electrical resistivity (and conductivity) between that of metallic conductors and non-metallic insulators. They can be found in the periodic table moving diagonally downward right from boron. They separate the electrical conductors (or metals, to the left) from the insulators (to the right).

Devices made from semiconductor materials are the foundation of modern electronics, including radio, computers, telephones, etc. Semiconductor devices include the transistor, solar cells, diodes and integrated circuits. Solar photovoltaic panels are large semiconductor devices that directly convert light into electrical energy.

In a metallic conductor, current is carried by the flow of electrons", but in semiconductors, current can be carried either by electrons or by the positively charged "holes" in the electronic band structure of the material. Common semiconductor materials include silicon, germanium and gallium arsenide.

Nanomaterials

Many traditional solids exhibit different properties when they shrink to nanometer sizes. For example, nanoparticles of usually yellow gold and gray silicon are red in color; gold nanoparticles melt at much lower temperatures (~300 °C for 2.5 nm size) than the gold slabs (1064 °C); and metallic nanowires are much stronger than the corresponding bulk metals. The high surface area of nanoparticles makes them extremely attractive for certain applications in the field of energy. For example, platinum metals may provide improvements as automotive fuel catalysts, as well as proton exchange membrane (PEM) fuel cells. Also, ceramic oxides (or cermets) of lanthanum, cerium, manganese and nickel are now being developed as solid oxide fuel cells (SOFC). Lithium, lithium–titanate and tantalum nanoparticles are being applied in lithium ion batteries. Silicon nanoparticles have been shown to dramatically expand the storage capacity of lithium ion batteries during the expansion/contraction cycle. Silicon nanowires cycle without significant degradation and present the potential for use in batteries with greatly expanded storage times. Silicon nanoparticles are also being used in new forms of solar energy cells. Thin film deposition of silicon quantum dots on the polycrystalline silicon substrate of a photovoltaic (solar) cell increases voltage output as much as 60% by fluorescing the incoming light prior to capture. Here again, surface area of the nanoparticles (and thin films) plays a critical role in maximizing the amount of absorbed radiation.

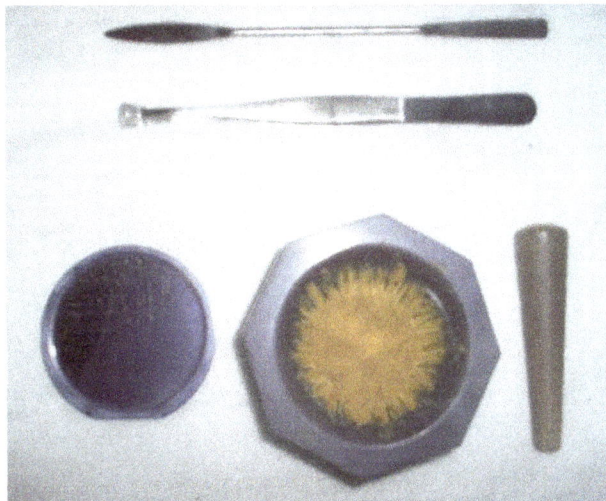

Bulk silicon (left) and silicon nanopowder (right)

Biomaterials

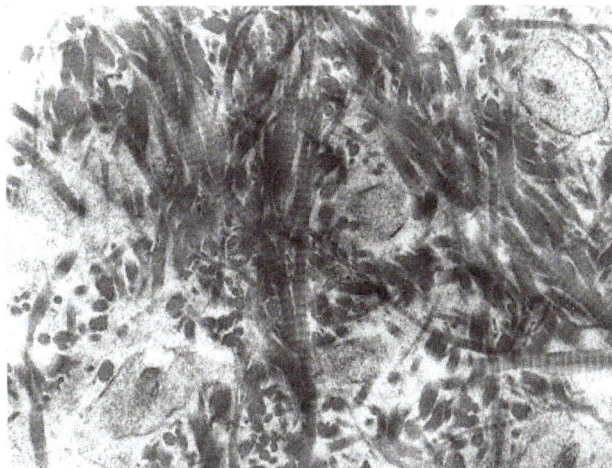

Collagen fibers of woven bone

Many natural (or biological) materials are complex composites with remarkable mechanical properties. These complex structures, which have risen from hundreds of million years of evolution, are inspiring materials scientists in the design of novel materials. Their defining characteristics include structural hierarchy, multifunctionality and self-healing capability. Self-organization is also a fundamental feature of many biological materials and the manner by which the structures are assembled from the molecular level up. Thus, self-assembly is emerging as a new strategy in the chemical synthesis of high performance biomaterials.

Physical Properties

Physical properties of elements and compounds which provide conclusive evidence of chemical composition include odor, color, volume, density (mass per unit volume), melting point, boiling point, heat capacity, physical form and shape at room temperature (solid, liquid or gas; cubic, trigonal crystals, etc.), hardness, porosity, index of refraction and many others. This section discusses some physical properties of materials in the solid state.

Mechanical

Granite rock formation in the Chilean Patagonia. Like most inorganic minerals formed by oxidation in the Earth's atmosphere, granite consists primarily of crystalline silica SiO_2 and alumina Al_2O_3.

The mechanical properties of materials describe characteristics such as their strength and resistance to deformation. For example, steel beams are used in construction because of their high strength, meaning that they neither break nor bend significantly under the applied load.

Mechanical properties include elasticity and plasticity, tensile strength, compressive strength, shear strength, fracture toughness, ductility (low in brittle materials), and indentation hardness. Solid mechanics is the study of the behavior of solid matter under external actions such as external forces and temperature changes.

A solid does not exhibit macroscopic flow, as fluids do. Any degree of departure from its original shape is called deformation. The proportion of deformation to original size is called strain. If the applied stress is sufficiently low, almost all solid materials behave in such a way that the strain is directly proportional to the stress (Hooke's law). The coefficient of the proportion is called the modulus of elasticity or Young's modulus. This region of deformation is known as the linearly elastic region. Three models can describe how a solid responds to an applied stress:

- Elasticity – When an applied stress is removed, the material returns to its undeformed state.

- Viscoelasticity – These are materials that behave elastically, but also have damping. When the applied stress is removed, work has to be done against the damping effects and is converted to heat within the material. This results in a hysteresis loop in the stress–strain curve. This implies that the mechanical response has a time-dependence.

- Plasticity – Materials that behave elastically generally do so when the applied stress is less than a yield value. When the stress is greater than the yield stress, the material behaves plastically and does not return to its previous state. That is, irreversible plastic deformation (or viscous flow) occurs after yield which is permanent.

Many materials become weaker at high temperatures. Materials which retain their strength at high temperatures, called refractory materials, are useful for many purposes. For example, glass-ceramics have become extremely useful for countertop cooking, as they exhibit excellent mechanical properties and can sustain repeated and quick temperature changes up to 1000 °C. In the aerospace industry, high performance materials used in the design of aircraft and/or spacecraft exteriors must have a high resistance to thermal shock. Thus, synthetic fibers spun out of organic polymers and polymer/ceramic/metal composite materials and fiber-reinforced polymers are now being designed with this purpose in mind.

Thermal

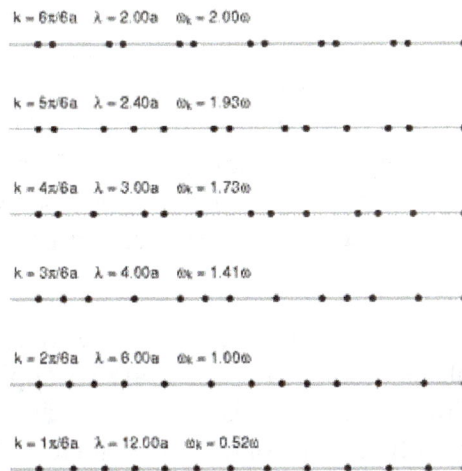

$$k = 6\pi/6a \quad \lambda = 2.00a \quad \omega_k = 2.00\omega$$

$$k = 5\pi/6a \quad \lambda = 2.40a \quad \omega_k = 1.93\omega$$

$$k = 4\pi/6a \quad \lambda = 3.00a \quad \omega_k = 1.73\omega$$

$$k = 3\pi/6a \quad \lambda = 4.00a \quad \omega_k = 1.41\omega$$

$$k = 2\pi/6a \quad \lambda = 6.00a \quad \omega_k = 1.00\omega$$

$$k = 1\pi/6a \quad \lambda = 12.00a \quad \omega_k = 0.52\omega$$

Normal modes of atomic vibration in a crystalline solid.

Because solids have thermal energy, their atoms vibrate about fixed mean positions within the ordered (or disordered) lattice. The spectrum of lattice vibrations in a crystalline or glassy network provides the foundation for the kinetic theory of solids. This motion occurs at the atomic level, and thus cannot be observed or detected without highly specialized equipment, such as that used in spectroscopy.

Thermal properties of solids include thermal conductivity, which is the property of a material that indicates its ability to conduct heat. Solids also have a specific heat capacity, which is the capacity of a material to store energy in the form of heat (or thermal lattice vibrations).

Electrical

Video of superconducting levitation of YBCO

Electrical properties include conductivity, resistance, impedance and capacitance. Electrical conductors such as metals and alloys are contrasted with electrical insulators such as glasses and ceramics. Semiconductors behave somewhere in between. Whereas conductivity in metals is caused by electrons, both electrons and holes contribute to current in semiconductors. Alternatively, ions support electric current in ionic conductors.

Many materials also exhibit superconductivity at low temperatures; they include metallic elements such as tin and aluminium, various metallic alloys, some heavily doped semiconductors, and certain ceramics. The electrical resistivity of most electrical (metallic) conductors generally decreases gradually as the temperature is lowered, but remains finite. In a superconductor however, the resistance drops abruptly to zero when the material is cooled below its critical temperature. An electric current flowing in a loop of superconducting wire can persist indefinitely with no power source.

A dielectric, or electrical insulator, is a substance that is highly resistant to the flow of electric current. A dielectric, such as plastic, tends to concentrate an applied electric field within itself which property is used in capacitors. A capacitor is an electrical device that can store energy in the electric field between a pair of closely spaced conductors (called 'plates'). When voltage is applied to the capacitor, electric charges of equal magnitude, but opposite polarity, build up on each plate. Capacitors are used in electrical circuits as energy-storage devices, as well as in electronic filters to differentiate between high-frequency and low-frequency signals.

Electro-mechanical

Piezoelectricity is the ability of crystals to generate a voltage in response to an applied mechanical stress. The piezoelectric effect is reversible in that piezoelectric crystals, when subjected to an externally applied voltage, can change shape by a small amount. Polymer materials like rubber, wool, hair, wood fiber, and silk often behave as electrets. For example, the polymer polyvinylidene fluoride (PVDF) exhibits a piezoelectric response several times larger than the traditional piezoelectric material quartz (crystalline SiO_2). The deformation (~0.1%) lends itself to useful technical applications such as high-voltage sources, loudspeakers, lasers, as well as chemical, biological, and acousto-optic sensors and/or transducers.

Optical

Materials can transmit (e.g. glass) or reflect (e.g. metals) visible light.

Many materials will transmit some wavelengths while blocking others. For example, window glass is transparent to visible light, but much less so to most of the frequencies of ultraviolet light that cause sunburn. This property is used for frequency-selective optical filters, which can alter the color of incident light.

For some purposes, both the optical and mechanical properties of a material can be of interest. For example, the sensors on an infrared homing ("heat-seeking") missile must be protected by a cover which is transparent to infrared radiation. The current material of choice for high-speed infrared-guided missile domes is single-crystal sapphire. The optical transmission of sapphire does not actually extend to cover the entire mid-infrared range (3–5 μm), but starts to drop off at wavelengths greater than approximately 4.5 μm at room temperature. While the strength of sapphire is better than that of other available mid-range infrared dome materials at room temperature, it weakens above 600 °C. A long-standing trade-off exists between optical bandpass and mechanical durability; new materials such as transparent ceramics or optical nanocomposites may provide improved performance.

Guided lightwave transmission involves the field of fiber optics and the ability of certain glasses to transmit, simultaneously and with low loss of intensity, a range of frequencies (multimode optical waveguides) with little interference between them. Optical waveguides are used as components in integrated optical circuits or as the transmission medium in optical communication systems.

Opto-electronic

A solar cell or photovoltaic cell is a device that converts light energy into electrical energy. Fundamentally, the device needs to fulfill only two functions: photo-generation of charge carriers (electrons and holes) in a light-absorbing material, and separation of the charge carriers to a conductive contact that will transmit the electricity (simply put, carrying electrons off through a metal contact into an external circuit). This conversion is called the photoelectric effect, and the field of research related to solar cells is known as photovoltaics.

Solar cells have many applications. They have long been used in situations where electrical power from the grid is unavailable, such as in remote area power systems, Earth-orbiting satellites and space probes, handheld calculators, wrist watches, remote radiotelephones and water pumping applications. More recently, they are starting to be used in assemblies of solar modules (photovoltaic arrays) connected to the electricity grid through an inverter, that is not to act as a sole supply but as an additional electricity source.

All solar cells require a light absorbing material contained within the cell structure to absorb photons and generate electrons via the photovoltaic effect. The materials used in solar cells tend to have the property of preferentially absorbing the wavelengths of solar light that reach the earth surface. However, some solar cells are optimized for light absorption beyond Earth's atmosphere as well.

Liquid

The formation of a spherical droplet of liquid water minimizes the surface area, which is the natural result of surface tension in liquids.

A liquid is a nearly incompressible fluid that conforms to the shape of its container but retains a (nearly) constant volume independent of pressure. As such, it is one of the four fundamental states of matter (the others being solid, gas, and plasma), and is the only state with a definite volume but no fixed shape. A liquid is made up of tiny vibrating particles of matter, such as atoms, held together by intermolecular bonds. Water is, by far, the most common liquid on Earth. Like a gas, a liquid is able to flow and take the shape of a container. Most liquids resist compression, although others can be compressed. Unlike a gas, a liquid does not disperse to fill every space of a container, and maintains a fairly constant density. A distinctive property of the liquid state is surface tension, leading to wetting phenomena.

The density of a liquid is usually close to that of a solid, and much higher than in a gas. Therefore, liquid and solid are both termed condensed matter. On the other hand, as liquids and gases share the ability to flow, they are both called fluids. Although liquid water is abundant on Earth, this state of matter is actually the least common in the known universe, because liquids require a relatively narrow temperature/pressure range to exist. Most known matter in the universe is in gaseous form (with traces of detectable solid matter) as interstellar clouds or in plasma form within stars.

Introduction

Liquid is one of the four primary states of matter, with the others being solid, gas and plasma. A liquid is a fluid. Unlike a solid, the molecules in a liquid have a much greater freedom to move. The forces that bind the molecules together in a solid are only temporary in a liquid, allowing a liquid to flow while a solid remains rigid.

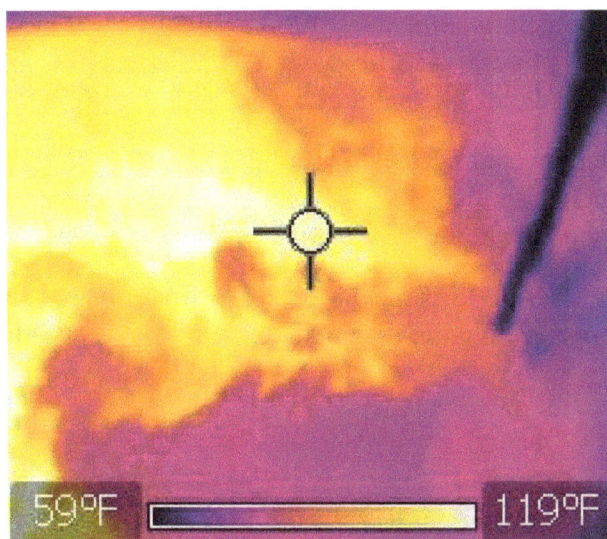

Thermal image of a sink full of hot water with cold water being added,
showing how the hot and the cold water flow into each other.

A liquid, like a gas, displays the properties of a fluid. A liquid can flow, assume the shape of a container, and, if placed in a sealed container, will distribute applied pressure evenly to every surface in the container. If you place the liquid in a bag, you can squeeze it into any shape you want. Unlike a gas, a liquid may not always mix readily with another liquid, will not always fill every space in the container, forming its own surface, and will not compress significantly, except under extremely high pressures. These properties make a liquid suitable for applications such as hydraulics.

Liquid particles are bound firmly but not rigidly. They are able to move around one another freely, resulting in a limited degree of particle mobility. As the temperature increases, the increased vibrations of the molecules causes distances between the molecules to increase. When a liquid reaches its boiling point, the cohesive forces that bind the molecules closely together break, and the liquid changes to its gaseous state (unless superheating occurs). If the temperature is decreased, the distances between the molecules become smaller. When the liquid reaches its freezing point the molecules will usually lock into a very specific order, called crystallizing, and the bonds between them become more rigid, changing the liquid into its solid state (unless supercooling occurs).

Examples

Only two elements are liquid at standard conditions for temperature and pressure: mercury and bromine. Four more elements have melting points slightly above room temperature: francium, caesium, gallium and rubidium. Metal alloys that are liquid at room temperature include NaK, a sodium-potassium metal alloy, galinstan, a fusible alloy liquid, and some amalgams (alloys involving mercury).

Pure substances that are liquid under normal conditions include water, ethanol and many other organic solvents. Liquid water is of vital importance in chemistry and biology; it is believed to be a necessity for the existence of life.

Inorganic liquids include water, magma, inorganic nonaqueous solvents and many acids.

Important everyday liquids include aqueous solutions like household bleach, other mixtures of different substances such as mineral oil and gasoline, emulsions like vinaigrette or mayonnaise, suspensions like blood, and colloids like paint and milk.

Many gases can be liquefied by cooling, producing liquids such as liquid oxygen, liquid nitrogen, liquid hydrogen and liquid helium. Not all gases can be liquified at atmospheric pressure, for example carbon dioxide can only be liquified at pressures above 5.1 atm.

Some materials cannot be classified within the classical three states of matter; they possess solid-like and liquid-like properties. Examples include liquid crystals, used in LCD displays, and biological membranes.

Applications

Liquids have a variety of uses, as lubricants, solvents, and coolants. In hydraulic systems, liquid is used to transmit power.

In tribology, liquids are studied for their properties as lubricants. Lubricants such as oil are chosen for viscosity and flow characteristics that are suitable throughout the operating temperature range of the component. Oils are often used in engines, gear boxes, metalworking, and hydraulic systems for their good lubrication properties.

Many liquids are used as solvents, to dissolve other liquids or solids. Solutions are found in a wide variety of applications, including paints, sealants, and adhesives. Naphtha and acetone are used frequently in industry to clean oil, grease, and tar from parts and machinery. Body fluids are water based solutions.

Surfactants are commonly found in soaps and detergents. Solvents like alcohol are often used as antimicrobials. They are found in cosmetics, inks, and liquid dye lasers. They are used in the food industry, in processes such as the extraction of vegetable oil.

Liquids tend to have better thermal conductivity than gases, and the ability to flow makes a liquid suitable for removing excess heat from mechanical components. The heat can be removed by channeling the liquid through a heat exchanger, such as a radiator, or the heat can be removed with the liquid during evaporation. Water or glycol coolants are used to keep engines from overheating. The coolants used in nuclear reactors include water or liquid metals, such as sodium or bismuth. Liquid propellant films are used to cool the thrust chambers of rockets. In machining, water and oils are used to remove the excess heat generated, which can quickly ruin both the work piece and the tooling. During perspiration, sweat removes heat from the human body by evaporating. In the heating, ventilation, and air-conditioning industry (HVAC), liquids such as water are used to transfer heat from one area to another.

Liquid is the primary component of hydraulic systems, which take advantage of Pascal's law to provide fluid power. Devices such as pumps and waterwheels have been used to change liquid motion into mechanical work since ancient times. Oils are forced through hydraulic pumps, which transmit this force to hydraulic cylinders. Hydraulics can be found in many applications, such as automotive brakes and transmissions, heavy equipment, and airplane control systems. Various hydraulic presses are used extensively in repair and manufacturing, for lifting, pressing, clamping and forming.

Liquids are sometimes used in measuring devices. A thermometer often uses the thermal expansion of liquids, such as mercury, combined with their ability to flow to indicate temperature. A manometer uses the weight of the liquid to indicate air pressure.

Mechanical Properties

Volume

Quantities of liquids are commonly measured in units of volume. These include the SI unit cubic metre (m^3) and its divisions, in particular the cubic decimeter, more commonly called the litre (1 dm^3 = 1 L = 0.001 m^3), and the cubic centimetre, also called millilitre (1 cm^3 = 1 mL = 0.001 L = 10^{-6} m^3).

The volume of a quantity of liquid is fixed by its temperature and pressure. Liquids generally expand when heated, and contract when cooled. Water between 0 °C and 4 °C is a notable exception. Liquids have little compressibility. Water, for example, will compress by only 46.4 parts per million for every unit increase in atmospheric pressure (bar). At around 4000 bar (58,000 psi) of pressure, at room temperature, water only experiences an 11% decrease in volume. In the study of fluid dynamics, liquids are often treated as incompressible, especially when studying incompressible flow. This incompressible nature makes a liquid suitable for transmitting hydraulic power, because very little of the energy is lost in the form of compression. However, the very slight compressibility does lead to other phenomena. The banging of pipes, called water hammer, occurs when a valve is suddenly closed, creating a huge pressure-spike at the valve that travels backward through the system at just under the speed of sound. Another phenomenon caused by liquid's incompressibility is cavitation. Because liquids have little elasticity they can literally be pulled apart in areas of high turbulence or a dramatic change in direction, such as the trailing edge of a boat propeller or a sharp corner in a pipe. A liquid in an area of low pressure (vacuum) vaporizes and forms bubbles, which then collapse as they enter high pressure areas. This causes liquid to fill the cavities left by the bubbles with tremendous localized force, eroding any adjacent solid surface.

Pressure and Buoyancy

In a gravitational field, liquids exert pressure on the sides of a container as well as on anything within the liquid itself. This pressure is transmitted in all directions and increases with depth. If a liquid is at rest in a uniform gravitational field, the pressure, p, at any depth, z, is given by

$$p = \rho g z$$

where:

ρ is the density of the liquid (assumed constant)

g is the gravitational acceleration.

Note that this formula assumes that the pressure *at* the free surface is zero, and that surface tension effects may be neglected.

Objects immersed in liquids are subject to the phenomenon of buoyancy. (Buoyancy is also observed in other fluids, but is especially strong in liquids due to their high density.)

Surfaces

Surface waves in water

Unless the volume of a liquid exactly matches the volume of its container, one or more surfaces are observed. The surface of a liquid behaves like an elastic membrane in which surface tension appears, allowing the formation of drops and bubbles. Surface waves, capillary action, wetting, and ripples are other consequences of surface tension. In a confined liquid, defined by geometric constraints on a nanoscopic scale, most molecules sense some surface effects, which can result in physical properties grossly deviating from those of the bulk liquid.

Free Surface

A free surface is the surface of a fluid that is subject to both zero perpendicular normal stress and parallel shear stress, such as the boundary between, e.g., liquid water and the air in the Earth's atmosphere.

Level

The liquid level (as in, e.g., water level) is the height associated with the liquid free surface, especially when it's the top-most surface. It may be measured with a level sensor.

Flow

Viscosity measures the resistance of a liquid which is being deformed by either shear stress or extensional stress.

When a liquid is supercooled towards the glass transition, the viscosity increases dramatically. The liquid then becomes a viscoelastic medium that shows both the elasticity of a solid and the fluidity of a liquid, depending on the time scale of observation or on the frequency of perturbation.

Sound Propagation

The speed of sound in a fluid is given by $c = \sqrt{K/\rho}$ where K is the bulk modulus of the fluid, and ρ the density. To give a typical value, in fresh water $c=1497$ m/s at 25 °C.

Thermodynamics

Phase Transitions

At a temperature below the boiling point, any matter in liquid form will evaporate until the condensation of gas above reach an equilibrium. At this point the gas will condense at the same rate as the liquid evaporates. Thus, a liquid cannot exist permanently if the evaporated liquid is continually removed. A liquid at its boiling point will evaporate more quickly than the gas can condense at the current pressure. A liquid at or above its boiling point will normally boil, though superheating can prevent this in certain circumstances.

At a temperature below the freezing point, a liquid will tend to crystallize, changing to its solid form. Unlike the transition to gas, there is no equilibrium at this transition under constant pressure, so unless supercooling occurs, the liquid will eventually completely crystallize. Note that this is only true under constant pressure, so e.g. water and ice in a closed, strong container might reach an equilibrium where both phases coexist.

Liquids in Space

The phase diagram explains why liquids do not exist in space or any other vacuum. Since the pressure is zero (except on surfaces or interiors of planets and moons) water and other liquids exposed to space will either immediately boil or freeze depending on the temperature. In regions of space near the earth, water will freeze if the sun is not shining directly on it and vapourize (sublime) as soon as it is in sunlight. If water exists as ice on the moon, it can only exist in shadowed holes where the sun never shines and where the surrounding rock doesn't heat it up too much. At some point near the orbit of Saturn, the light from the sun is too faint to sublime ice to water vapour. This is evident from the longevity of the ice that composes Saturn's rings.

Solutions

Liquids can display immiscibility. The most familiar mixture of two immiscible liquids in everyday life is the vegetable oil and water in Italian salad dressing. A familiar set of miscible liquids is water and alcohol. Liquid components in a mixture can often be separated from one another via fractional distillation.

Microscopic Properties

Static Structure Factor

In a liquid, atoms do not form a crystalline lattice, nor do they show any other form of long-range order. This is evidenced by the absence of Bragg peaks in X-ray and neutron diffraction. Under normal conditions, the diffraction pattern has circular symmetry, expressing the isotropy of the liquid. In radial direction, the diffraction intensity smoothly oscillates. This is usually described by the static structure factor $S(q)$, with wavenumber $q=(4\pi/\lambda)\sin\theta$ given by the wavelength λ of the probe (photon or neutron) and the Bragg angle θ. The oscillations of $S(q)$ express the *near order* of the liquid, i.e. the correlations between an atom and a few shells of nearest, second nearest, ... neighbors.

A more intuitive description of these correlations is given by the radial distribution function $g(r)$, which is basically the Fourier transform of $S(q)$. It represents a spatial average of a temporal snapshot of pair correlations in the liquid.

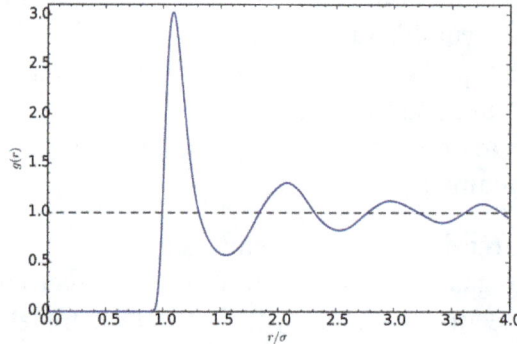

Radial distribution function of the Lennard-Jones model fluid.

Sound Dispersion and Structural Relaxation

The above expression for the sound velocity $c = \sqrt{K / \rho}$ contains the bulk modulus K. If K is frequency independent then the liquid behaves as a linear medium, so that sound propagates without dissipation and without mode coupling. In reality, any liquid shows some dispersion: with increasing frequency, K crosses over from the low-frequency, liquid-like limit K_∞ to the high-frequency, solid-like limit K_0. In normal liquids, most of this cross over takes place at frequencies between GHz and THz, sometimes called hypersound.

At sub-GHz frequencies, a normal liquid cannot sustain shear waves: the zero-frequency limit of the shear modulus is $G_0 = 0$. This is sometimes seen as the defining property of a liquid. However, just as the bulk modulus K, the shear modulus G is frequency dependent, and at hypersound frequencies it shows a similar cross over from the liquid-like limit G_0 to a solid-like, non-zero limit G_∞.

According to the Kramers-Kronig relation, the dispersion in the sound velocity (given by the real part of K or G) goes along with a maximum in the sound attenuation (dissipation, given by the imaginary part of K or G). According to linear response theory, the Fourier transform of K or G describes how the system returns to equilibrium after an external perturbation; for this reason, the dispersion step in the GHz..THz region is also called structural relaxation. According to the fluctuation-dissipation theorem, relaxation *towards* equilibrium is intimately connected to fluctuations *in* equilibrium. The density fluctuations associated with sound waves can be experimentally observed by Brillouin scattering.

On supercooling a liquid towards the glass transition, the crossover from liquid-like to solid-like response moves from GHz to MHz, kHz, Hz, ...; equivalently, the characteristic time of structural relaxation increases from ns to µs, ms, s, ... This is the microscopic explanation for the above-mentioned viscoelastic behaviour of glass-forming liquids.

Effects of Association

The mechanisms of atomic/molecular diffusion (or particle displacement) in solids are closely related to the mechanisms of viscous flow and solidification in liquid materials. Descriptions of viscosity in terms of molecular "free space" within the liquid were modified as needed in order to account for liquids whose molecules are known to be "associated" in the liquid state at ordinary temperatures. When various molecules combine together to form an associated molecule, they enclose within a semi-rigid system a certain amount of space which before was available as free space for mobile molecules. Thus, increase in viscosity upon cooling due to the tendency of most substances to become *associated* on cooling.

Similar arguments could be used to describe the effects of pressure on viscosity, where it may be assumed that the viscosity is chiefly a function of the volume for liquids with a finite compressibility. An increasing viscosity with rise of pressure is therefore expected. In addition, if the volume is expanded by heat but reduced again by pressure, the viscosity remains the same.

The local tendency to orientation of molecules in small groups lends the liquid (as referred to previously) a certain degree of association. This association results in a considerable "internal pressure" within a liquid, which is due almost entirely to those molecules which, on account of their temporary low velocities (following the Maxwell distribution) have coalesced with other molecules. The internal pressure between several such molecules might correspond to that between a group of molecules in the solid form.

Gas

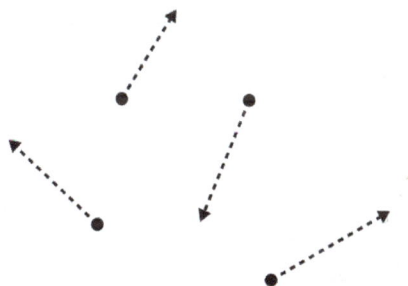

Gas phase particles (atoms, molecules, or ions) move around freely in the absence of an applied electric field.

Gas is one of the four fundamental states of matter (the others being solid, liquid, and plasma). A pure gas may be made up of individual atoms (e.g. a noble gas like neon), elemental molecules made from one type of atom (e.g. oxygen), or compound molecules made from a variety of atoms (e.g. carbon dioxide). A gas mixture would contain a variety of pure gases much like the air. What distinguishes a gas from liquids and solids is the vast separation of the individual gas particles. This separation usually makes a colorless gas invisible to the human observer. The interaction of gas particles in the presence of electric and gravitational fields are considered negligible as indicated by the constant velocity vectors in the image. One type of commonly known gas is steam.

The gaseous state of matter is found between the liquid and plasma states, the latter of which provides the upper temperature boundary for gases. Bounding the lower end of the temperature scale lie degenerative quantum gases which are gaining increasing attention. High-density atomic gases super cooled to incredibly low temperatures are classified by their statistical behavior as either a Bose gas or a Fermi gas.

Elemental Gases

The only chemical elements which are stable multi atom homonuclear molecules at standard temperature and pressure (STP), are hydrogen (H_2), nitrogen (N_2) and oxygen (O_2); plus two halogens, fluorine (F_2) and chlorine (Cl_2). These gases, when grouped together with the monatomic noble gases; which are helium (He), neon (Ne), argon (Ar), krypton (Kr), xenon (Xe) and radon (Rn) ; are called "elemental gases". Alternatively they are sometimes known as "molecular gases" to distinguish them from molecules that are also chemical compounds.

Etymology

An alternative story is that Van Helmont's word is corrupted from *gahst* (or *geist*), signifying a ghost or spirit. This was because certain gases suggested a supernatural origin, such as from their ability to cause death, extinguish flames, and to occur in "mines, bottom of wells, churchyards and other lonely places".

Physical Characteristics

Drifting smoke particles provide clues to the movement of the surrounding gas.

Because most gases are difficult to observe directly, they are described through the use of four physical properties or macroscopic characteristics: pressure, volume, number of particles (chemists group them by moles) and temperature. These four characteristics were repeatedly observed by scientists such as Robert Boyle, Jacques Charles, John Dalton, Joseph Gay-Lussac and Amedeo Avogadro for a variety of gases in various settings. Their detailed studies ultimately led to a mathematical relationship among these properties expressed by the ideal gas law

Gas particles are widely separated from one another, and consequently have weaker intermolecular bonds than liquids or solids. These intermolecular forces result from electrostatic interactions between gas particles. Like-charged areas of different gas particles repel, while oppositely charged regions of different gas particles attract one another; gases that contain permanently charged ions are known as plasmas. Gaseous compounds with polar covalent bonds contain permanent charge imbalances and so experience relatively strong intermolecular forces, although the molecule while the compound's net charge remains neutral. Transient, randomly induced charges exist across non-polar covalent bonds of molecules and electrostatic interactions caused by them are referred to as Van der Waals forces. The interaction of these intermolecular forces varies within a substance which determines many of the physical properties unique to each gas. A comparison of *boiling points* for compounds formed by ionic and covalent bonds leads us to this conclusion. The drifting smoke particles in the image provides some insight into low pressure gas behavior.

Compared to the other states of matter, gases have low density and viscosity. Pressure and temperature influence the particles within a certain volume. This variation in particle separation and speed is referred to as *compressibility*. This particle separation and size influences optical properties of gases as can be found in the following list of refractive indices. Finally, gas particles spread apart or diffuse in order to homogeneously distribute themselves throughout any container.

Macroscopic

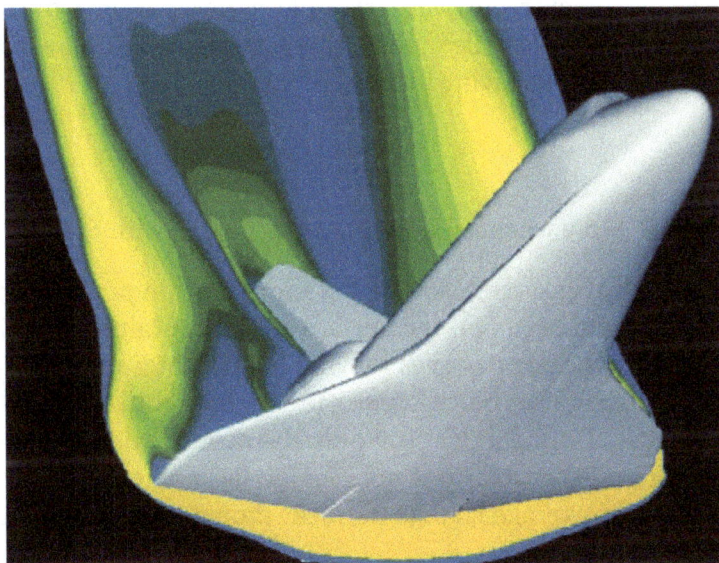

Shuttle imagery of re-entry phase.

When observing a gas, it is typical to specify a frame of reference or length scale. A *larger* length scale corresponds to a macroscopic or global point of view of the gas. This region (referred to as a volume) must be sufficient in size to contain a large sampling of gas particles. The resulting statistical analysis of this sample size produces the "average" behavior (i.e. velocity, temperature or pressure) of all the gas particles within the region. In contrast, a *smaller* length scale corresponds to a microscopic or particle point of view.

Macroscopically, the gas characteristics measured are either in terms of the gas particles themselves (velocity, pressure, or temperature) or their surroundings (volume). For example, Robert Boyle studied pneumatic chemistry for a small portion of his career. One of his experiments related the macroscopic properties of pressure and volume of a gas. His experiment used a J-tube manometer which looks like a test tube in the shape of the letter J. Boyle trapped an inert gas in the closed end of the test tube with a column of mercury, thereby making the number of particles and the temperature constant. He observed that when the pressure was increased in the gas, by adding more mercury to the column, the trapped gas' volume decreased (this is known as an inverse relationship). Furthermore, when Boyle multiplied the pressure and volume of each observation, the product was constant. This relationship held for every gas that Boyle observed leading to the law, (PV=k), named to honor his work in this field.

There are many mathematical tools available for analyzing gas properties. As gases are subjected to extreme conditions, these tools become a bit more complex, from the Euler equations for inviscid flow to the Navier–Stokes equations that fully account for viscous effects. These equations are adapted to the conditions of the gas system in question. Boyle's lab equipment allowed the use of algebra to obtain his analytical results. His results were possible because he was studying gases in relatively low pressure situations where they behaved in an "ideal" manner. These ideal relationships apply to safety calculations for a variety of flight conditions on the materials in use. The high technology equipment in use today was designed to help us safely explore the more exotic operating environments where the gases no longer behave in an "ideal" manner. This advanced math, including statistics and multivariable calculus, makes possible the solution to such complex dynamic situations as space vehicle reentry. An example is the analysis of the space shuttle reentry pictured to ensure the material properties under this loading condition are appropriate. In this flight regime, the gas is no longer behaving ideally.

Pressure

The symbol used to represent *pressure* in equations is "p" or "P" with SI units of pascals.

When describing a container of gas, the term pressure (or absolute pressure) refers to the average force per unit area that the gas exerts on the surface of the container. Within this volume, it is sometimes easier to visualize the gas particles moving in straight lines until they collide with the container. The force imparted by a gas particle into the container during this collision is the change in momentum of the particle. During a collision only the normal component of velocity changes. A particle traveling parallel to the wall does not change its momentum. Therefore, the average force on a surface must be the average change in linear momentum from all of these gas particle collisions.

Pressure is the sum of all the normal components of force exerted by the particles impacting the walls of the container divided by the surface area of the wall.

Temperature

The symbol used to represent *temperature* in equations is T with SI units of kelvins.

The speed of a gas particle is proportional to its absolute temperature. The volume of the balloon in the video shrinks when the trapped gas particles slow down with the addition of extremely cold nitrogen. The temperature of any physical system is related to the motions of the particles (molecules and atoms) which make up the [gas] system. In statistical mechanics, temperature is the measure of the average kinetic energy stored in a particle. The methods of storing this energy are dictated by the degrees of freedom of the particle itself (energy modes). Kinetic energy added (endothermic process) to gas particles by way of collisions produces linear, rotational, and vibrational motion. In contrast, a molecule in a solid can only increase its vibrational modes with the addition of heat as the lattice crystal structure prevents both linear and rotational motions. These heated gas molecules have a greater speed range which constantly varies due to constant collisions with other particles. The speed range can be described by the Maxwell–Boltzmann distribution. Use of this distribution implies ideal gases near thermodynamic equilibrium for the system of particles being considered.

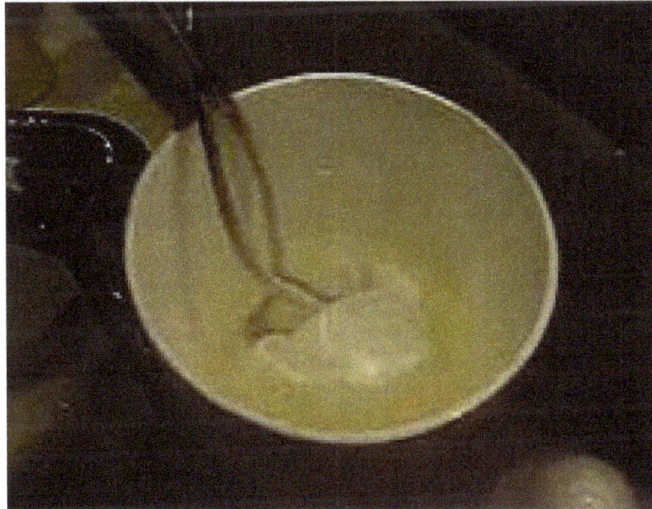

Air balloon shrinks after submersion in liquid nitrogen

Specific Volume

The symbol used to represent *specific volume* in equations is "v" with SI units of cubic meters per kilogram.

The symbol used to represent volume in equations is "V" with SI units of cubic meters.

When performing a thermodynamic analysis, it is typical to speak of intensive and extensive properties. Properties which depend on the amount of gas (either by mass or volume) are called *extensive* properties, while properties that do not depend on the amount of gas are called *intensive* properties. Specific volume is an example of an *intensive* property because it is the ratio of volume occupied by a *unit of mass* of a gas that is identical throughout a system at equilibrium. 1000 atoms a gas occupy the same space as any other 1000 atoms for any given temperature and pressure. This concept is easier to visualize for solids such as iron which are incompressible compared to gases. Since a gas fills any container in which it is placed, volume is an *extensive property*.

Density

The symbol used to represent density in equations is ρ (rho) with SI units of kilograms per cubic meter. This term is the reciprocal of specific volume.

Since gas molecules can move freely within a container, their mass is normally characterized by density. Density is the amount of mass per unit volume of a substance, or the inverse of specific volume. For gases, the density can vary over a wide range because the particles are free to move closer together when constrained by pressure or volume. This variation of density is referred to as compressibility. Like pressure and temperature, density is a state variable of a gas and the change in density during any process is governed by the laws of thermodynamics. For a static gas, the density is the same throughout the entire container. Density is therefore a scalar quantity. It can be shown by kinetic theory that the density is *inversely* proportional to the size of the container in which a fixed mass of gas is confined. In this case of a fixed mass, the density decreases as the volume increases.

Microscopic

If one could observe a gas under a powerful microscope, one would see a collection of particles (molecules, atoms, ions, electrons, etc.) without any definite shape or volume that are in more or less random motion. These neutral gas particles only change direction when they collide with another particle or with the sides of the container. In an ideal gas, these collisions are perfectly elastic. This particle or microscopic view of a gas is described by the Kinetic-molecular theory. The assumptions behind this theory can be found in the postulates section of Kinetic Theory.

Kinetic Theory

Kinetic theory provides insight into the macroscopic properties of gases by considering their molecular composition and motion. Starting with the definitions of momentum and kinetic energy, one can use the conservation of momentum and geometric relationships of a cube to relate macroscopic system properties of temperature and pressure to the microscopic property of kinetic energy per molecule. The theory provides averaged values for these two properties.

The theory also explains how the gas system responds to change. For example, as a gas is heated from absolute zero, when it is (in theory) perfectly still, its internal energy (temperature) is increased. As a gas is heated, the particles speed up and its temperature rises. This results in greater numbers of collisions with the container per unit time due to the higher particle speeds associated with elevated temperatures. The pressure increases in proportion to the number of collisions per unit time.

Brownian Motion

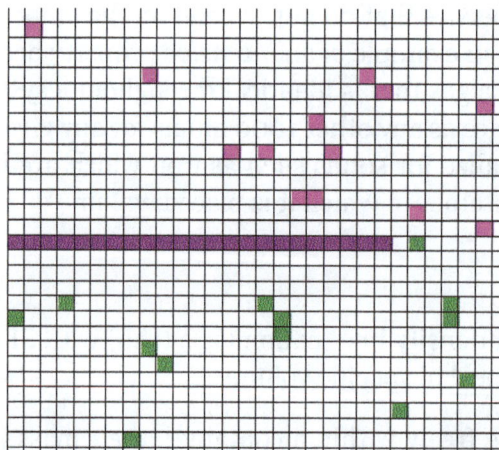

Random motion of gas particles results in diffusion.

Brownian motion is the mathematical model used to describe the random movement of particles suspended in a fluid. The gas particle animation, using pink and green particles, illustrates how this behavior results in the spreading out of gases (entropy). These events are also described by particle theory.

Since it is at the limit of (or beyond) current technology to observe individual gas particles (atoms or molecules), only theoretical calculations give suggestions about how they move, but their motion is different from Brownian motion because Brownian motion involves a smooth drag due to the frictional force of many gas molecules, punctuated by violent collisions of an individual (or several) gas molecule(s) with the particle. The particle (generally consisting of millions or billions of atoms) thus moves in a jagged course, yet not so jagged as would be expected if an individual gas molecule were examined.

Intermolecular Forces

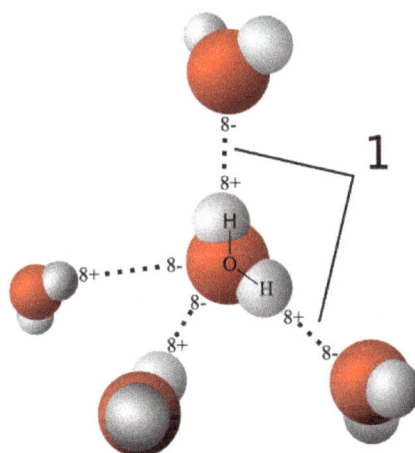

When gases are compressed, intermolecular forces like those shown here start to play a more active role.

As discussed earlier, momentary attractions (or repulsions) between particles have an effect on gas dynamics. In physical chemistry, the name given to these intermolecular forces is *van der Waals force*. These forces play a key role in determining physical properties of a gas such as viscosity and flow rate. Ignoring these forces in certain conditions allows a real gas to be treated like an ideal gas. This assumption allows the use of ideal gas laws which greatly simplifies calculations.

Proper use of these gas relationships requires the Kinetic-molecular theory (KMT). When gas particles possess a magnetic charge or Intermolecular force they gradually influence one another as the spacing between them is reduced (the hydrogen bond model illustrates one example). In the absence of any charge, at some point when the spacing between gas particles is greatly reduced they can no longer avoid collisions between themselves at normal gas temperatures. Another case for increased collisions among gas particles would include a fixed volume of gas, which upon heating would contain very fast particles. *This means that these ideal equations provide reasonable results* except *for extremely high pressure (compressible) or high temperature (ionized) conditions*. Notice that all of these excepted conditions allow energy transfer to take place within the gas system. The absence of these internal transfers is what is referred to as ideal conditions in which the energy exchange occurs only at the boundaries of the system. Real gases experience some of these collisions and intermolecular forces. When these collisions are statistically negligible (incompressible), results from these ideal equations are still meaningful. If the gas particles are compressed into close proximity they behave more like a liquid.

Simplified Models

An *equation of state* (for gases) is a mathematical model used to roughly describe or predict the state properties of a gas. At present, there is no single equation of state that accurately predicts the properties of all gases under all conditions. Therefore, a number of much more accurate equations of state have been developed for gases in specific temperature and pressure ranges. The "gas models" that are most widely discussed are "perfect gas", "ideal gas" and "real gas". Each of these models has its own set of assumptions to facilitate the analysis of a given thermodynamic system. Each successive model expands the temperature range of coverage to which it applies.

Ideal and Perfect Gas Models

The equation of state for an ideal or perfect gas is the ideal gas law and reads

$$PV = nRT,$$

where P is the pressure, V is the volume, n is amount of gas (in mol units), R is the universal gas constant, 8.314 J/(mol K), and T is the temperature. Written this way, it is sometimes called the "chemist's version", since it emphasizes the number of molecules n. It can also be written as

$$P = \rho R_s T,$$

where R_s is the specific gas constant for a particular gas, in units J/(kg K), and $\rho = m/V$ is density. This notation is the "gas dynamicist's" version, which is more practical in modeling of gas flows involving acceleration without chemical reactions.

The ideal gas law does not make an assumption about the specific heat of a gas. In the most general case, the specific heat is a function of both temperature and pressure. If the pressure-dependence is neglected (and possibly the temperature-dependence as well) in a particular application, sometimes the gas is said to be a perfect gas, although the exact assumptions may vary depending on the author and/or field of science.

For an ideal gas, the ideal gas law applies without restrictions on the specific heat. An ideal gas is a simplified "real gas" with the assumption that the compressibility factor Z is set to 1 meaning that this pneumatic ratio remains constant. A compressibility factor of one also requires the four state variables to follow the ideal gas law.

This approximation is more suitable for applications in engineering although simpler models can be used to produce a "ball-park" range as to where the real solution should lie. An example where the "ideal gas approximation" would be suitable would be inside a combustion chamber of a jet engine. It may also be useful to keep the elementary reactions and chemical dissociations for calculating emissions.

Real Gas

21 April 1990 eruption of Mount Redoubt, Alaska,
illustrating real gases not in thermodynamic equilibrium.

Each one of the assumptions listed below adds to the complexity of the problem's solution. As the density of a gas increases with rising pressure, the intermolecular forces play a more substantial role in gas behavior which results in the ideal gas law no longer providing "reasonable" results. At the upper end of the engine temperature ranges (e.g. combustor sections – 1300 K), the complex fuel particles absorb internal energy by means of rotations and vibrations that cause their specific heats to vary from those of diatomic molecules and noble gases. At more than double that temperature, electronic excitation and dissociation of the gas particles begins to occur causing the pressure to adjust to a greater number of particles (transition from gas to plasma). Finally, all of the thermodynamic processes were presumed to describe uniform gases whose velocities varied according to a fixed distribution. Using a non-equilibrium situation implies the flow field must be characterized in some manner to enable a solution. One of the first attempts to expand the boundaries of the ideal gas law was to include coverage for different thermodynamic processes by adjusting the equation to read $pV^n = constant$ and then varying the n through different values such as the specific heat ratio, γ.

Real gas effects include those adjustments made to account for a greater range of gas behavior:

- Compressibility effects (Z allowed to vary from 1.0)

- Variable heat capacity (specific heats vary with temperature)

- Van der Waals forces (related to compressibility, can substitute other equations of state)

- Non-equilibrium thermodynamic effects

- Issues with molecular dissociation and elementary reactions with variable composition.

For most applications, such a detailed analysis is excessive. Examples where "Real Gas effects" would have a significant impact would be on the Space Shuttle re-entry where extremely high temperatures and pressures are present or the gases produced during geological events as in the image of the 1990 eruption of Mount Redoubt.

Historical Synthesis
Boyle's Law

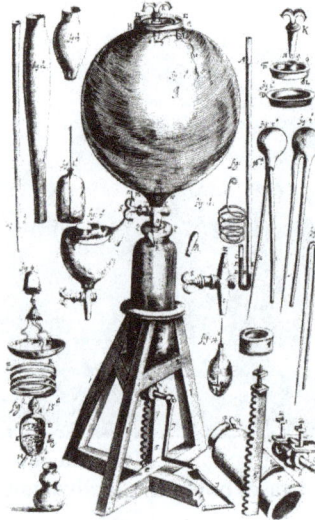

Boyle's equipment.

Boyle's Law was perhaps the first expression of an equation of state. In 1662 Robert Boyle performed a series of experiments employing a J-shaped glass tube, which was sealed on one end. Mercury was added to the tube, trapping a fixed quantity of air in the short, sealed end of the tube. Then the volume of gas was carefully measured as additional mercury was added to the tube. The pressure of the gas could be determined by the difference between the mercury level in the short end of the tube and that in the long, open end. The image of Boyle's Equipment shows some of the exotic tools used by Boyle during his study of gases.

Through these experiments, Boyle noted that the pressure exerted by a gas held at a constant temperature varies inversely with the volume of the gas. For example, if the volume is halved, the pressure is doubled; and if the volume is doubled, the pressure is halved. Given the inverse relationship between pressure and volume, the product of pressure (P) and volume (V) is a constant (k) for a given mass of confined gas as long as the temperature is constant. Stated as a formula, thus is:

$$PV = k$$

Because the before and after volumes and pressures of the fixed amount of gas, where the before and after temperatures are the same both equal the constant k, they can be related by the equation:

$$P_1V_1 = P_2V_2.$$

Charles's Law

In 1787, the French physicist and balloon pioneer, Jacques Charles, found that oxygen, nitrogen, hydrogen, carbon dioxide, and air expand to the same extent over the same 80 kelvin interval. He noted that, for an ideal gas at constant pressure, the volume is directly proportional to its temperature:

$$\frac{V_1}{T_1} = \frac{V_2}{T_2}$$

Gay-lussac's Law

In 1802, Joseph Louis Gay-Lussac published results of similar, though more extensive experiments. Gay-Lussac credited Charle's earlier work by naming the law in his honor. Gay-Lussac himself is credited with the law describing pressure, which he found in 1809. It states that the pressure exerted on a container's sides by an ideal gas is proportional to its temperature.

$$\frac{P_1}{T_1} = \frac{P_2}{T_2}$$

Avogadro's Law

In 1811, Amedeo Avogadro verified that equal volumes of pure gases contain the same number of particles. His theory was not generally accepted until 1858 when another Italian chemist Stanislao Cannizzaro was able to explain non-ideal exceptions. For his work with gases a century prior, the number that bears his name Avogadro's constant represents the number of atoms found in 12 grams of elemental carbon-12 (6.022×10^{23} mol^{-1}). This specific number of gas particles, at standard temperature and pressure (ideal gas law) occupies 22.40 liters, which is referred to as the molar volume.

Avogadro's law states that the volume occupied by an ideal gas is proportional to the number of moles (or molecules) present in the container. This gives rise to the molar volume of a gas, which at STP is 22.4 dm^3 (or litres). The relation is given by

$$\frac{V_1}{n_1} = \frac{V_2}{n_2}$$

where n is equal to the number of moles of gas (the number of molecules divided by Avogadro's Number).

Dalton's Law

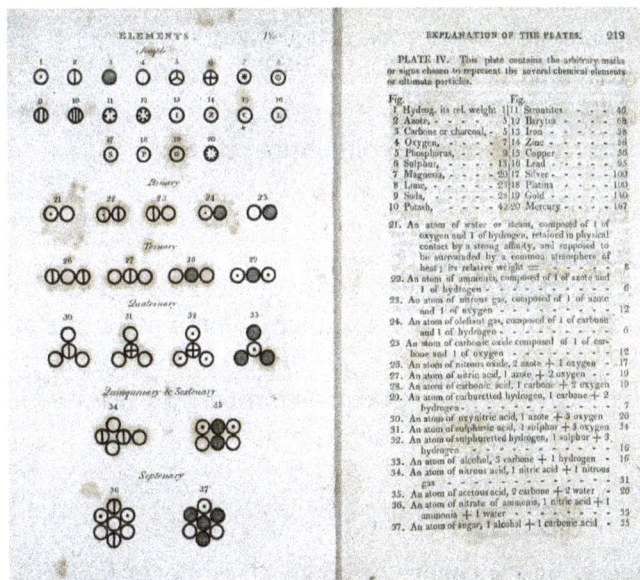

Dalton's notation.

In 1801, John Dalton published the Law of Partial Pressures from his work with ideal gas law relationship: The pressure of a mixture of non reactive gases is equal to the sum of the pressures of all of the constituent gases alone. Mathematically, this can be represented for n species as:

$$Pressure_{total} = Pressure_1 + Pressure_2 + ... + Pressure_n$$

The image of Dalton's journal depicts symbology he used as shorthand to record the path he followed. Among his key journal observations upon mixing unreactive "elastic fluids" (gases) were the following:

- Unlike liquids, heavier gases did not drift to the bottom upon mixing.

- Gas particle identity played no role in determining final pressure (they behaved as if their size was negligible).

Special Topics

Compressibility

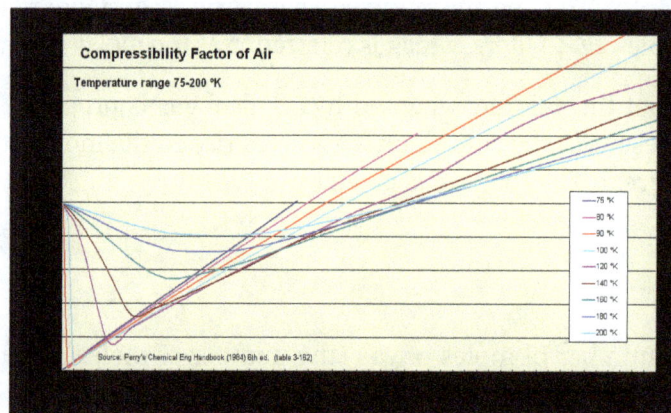

Compressibility factors for air.

Thermodynamicists use this factor (Z) to alter the ideal gas equation to account for compressibility effects of real gases. This factor represents the ratio of actual to ideal specific volumes. It is sometimes referred to as a "fudge-factor" or correction to expand the useful range of the ideal gas law for design purposes. *Usually* this Z value is very close to unity. The compressibility factor image illustrates how Z varies over a range of very cold temperatures.

Reynolds Number

In fluid mechanics, the Reynolds number is the ratio of inertial forces ($v_s\rho$) to viscous forces (μ/L). It is one of the most important dimensionless numbers in fluid dynamics and is used, usually along with other dimensionless numbers, to provide a criterion for determining dynamic similitude. As such, the Reynolds number provides the link between modeling results (design) and the full-scale actual conditions. It can also be used to characterize the flow.

Viscosity

Viscosity, a physical property, is a measure of how well adjacent molecules stick to one another. A solid can withstand a shearing force due to the strength of these sticky intermolecular forces. A fluid will continuously deform when subjected to a similar load. While a gas has a lower value of viscosity than a liquid, it is still an observable property. If gases had no viscosity, then they would not stick to the surface of a wing and form a boundary layer. A study of the delta wing in the Schlieren image reveals that the gas particles stick to one another.

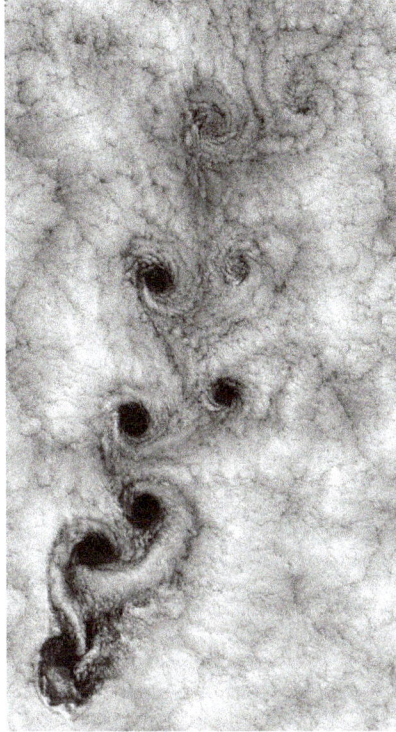

Satellite view of weather pattern in vicinity of Robinson Crusoe Islands on 15 September 1999, shows a unique turbulent cloud pattern called a Kármán vortex street

Turbulence

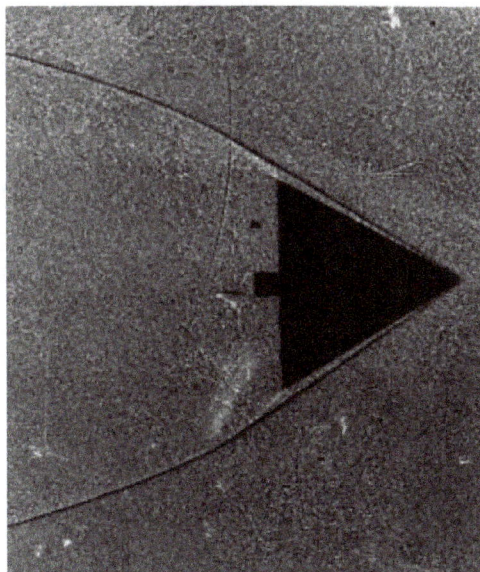

Delta wing in wind tunnel. The shadows form as the indices of refraction change within the gas as it compresses on the leading edge of this wing.

In fluid dynamics, turbulence or turbulent flow is a flow regime characterized by chaotic, stochastic property changes. This includes low momentum diffusion, high momentum convection, and rapid variation of pressure and velocity in space and time. The Satellite view of weather around Robinson Crusoe Islands illustrates just one example.

Boundary Layer

Particles will, in effect, "stick" to the surface of an object moving through it. This layer of particles is called the boundary layer. At the surface of the object, it is essentially static due to the friction of the surface. The object, with its boundary layer is effectively the new shape of the object that the rest of the molecules "see" as the object approaches. This boundary layer *can* separate from the surface, essentially creating a new surface and completely changing the flow path. The classical example of this is a stalling airfoil. The delta wing image clearly shows the boundary layer thickening as the gas flows from right to left along the leading edge.

Maximum Entropy Principle

As the total number of degrees of freedom approaches infinity, the system will be found in the macrostate that corresponds to the highest multiplicity. In order to illustrate this principle, observe the skin temperature of a frozen metal bar. Using a thermal image of the skin temperature, note the temperature distribution on the surface. This initial observation of temperature represents a "microstate." At some future time, a second observation of the skin temperature produces a second microstate. By continuing this observation process, it is possible to produce a series of microstates that illustrate the thermal history of the bar's surface. Characterization of this historical series of microstates is possible by choosing the macrostate that successfully classifies them all into a single grouping.

Thermodynamic Equilibrium

When energy transfer ceases from a system, this condition is referred to as thermodynamic equilibrium. Usually this condition implies the system and surroundings are at the same temperature so that heat no longer transfers between them. It also implies that external forces are balanced (volume does not change), and all chemical reactions within the system are complete. The timeline varies for these events depending on the system in question. A container of ice allowed to melt at room temperature takes hours, while in semiconductors the heat transfer that occurs in the device transition from an on to off state could be on the order of a few nanoseconds.

Plasma (Physics)

Plasma is one of the four fundamental states of matter, the others being solid, liquid, and gas. A plasma has properties unlike those of the other states.

A plasma can be created by heating a gas or subjecting it to a strong electromagnetic field, applied with a laser or microwave generator at temperatures above 5000 °C. This decreases or increases the number of electrons in the atoms or molecules, creating positive or negative charged particles called ions, and is accompanied by the dissociation of molecular bonds, if present.

The presence of a significant number of charge carriers makes plasma electrically conductive so that it responds strongly to electromagnetic fields. Like gas, plasma does not have a definite shape or a definite volume unless enclosed in a container. Unlike gas, under the influence of a magnetic field, it may form structures such as filaments, beams and double layers.

Plasma is the most abundant form of ordinary matter in the universe (the properties of dark matter are still mostly unknown; whether it can be equated to ordinary matter has yet to be determined), most of which is in the rarefied intergalactic regions, particularly the intracluster medium, and in stars, including the Sun. A common form of plasma on Earth is produced in neon signs.

Much of the understanding of plasma has come from the pursuit of controlled nuclear fusion and fusion power, for which plasma physics provides the scientific foundation.

Properties and Parameters

Artist's rendition of the Earth's plasma fountain, showing oxygen, helium, and hydrogen ions that gush into space from regions near the Earth's poles. The faint yellow area shown above the north pole represents gas lost from Earth into space; the green area is the aurora borealis, where plasma energy pours back into the atmosphere.

Definition

Plasma is an electrically neutral medium of unbound positive and negative particles (i.e. the over-all charge of a plasma is roughly zero). It is important to note that although the particles are un-bound, they are not 'free' in the sense of not experiencing forces. When a charged particle moves, it generates an electric current with magnetic fields; in plasma, the movement of a charged particle affects and is affected by the general field created by the movement of other charges. This governs collective behavior with many degrees of variation. Three factors are listed in the definition of a plasma stream:

1. The plasma approximation: Charged particles must be close enough together that each particle influences many nearby charged particles, rather than just interacting with the

closest particle (these collective effects are a distinguishing feature of a plasma). The plasma approximation is valid when the number of charge carriers within the sphere of influence (called the *Debye sphere* whose radius is the Debye screening length) of a particular particle is higher than unity to provide collective behavior of the charged particles. The average number of particles in the Debye sphere is given by the plasma parameter.

2. Bulk interactions: The Debye screening length (defined above) is short compared to the physical size of the plasma. This criterion means that interactions in the bulk of the plasma are more important than those at its edges, where boundary effects may take place. When this criterion is satisfied, the plasma is quasineutral.

3. Plasma frequency: The electron plasma frequency (measuring plasma oscillations of the electrons) is large compared to the electron-neutral collision frequency (measuring frequency of collisions between electrons and neutral particles). When this condition is valid, electrostatic interactions dominate over the processes of ordinary gas kinetics.

Ranges of Parameters

The factors of a plasma stream can vary by many orders of magnitude, but the properties of plasmas with apparently disparate parameters may be very similar. The following chart considers only conventional atomic plasmas and not exotic phenomena like quark gluon plasmas:

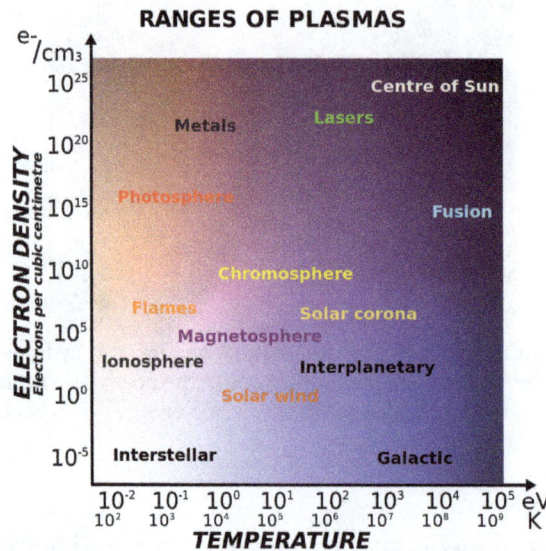

Range of plasmas. Density increases upwards, temperature increases towards the right.
The free electrons in a metal may be considered an electron plasma.

Typical ranges of plasma parameters: orders of magnitude (OOM)		
Characteristic	**Terrestrial plasmas**	**Cosmic plasmas**
Size in meters	10^{-6} m (lab plasmas) to 10^2 m (lightning) (~8 OOM)	10^{-6} m (spacecraft sheath) to 10^{25} m (intergalactic nebula) (~31 OOM)
Lifetime in seconds	10^{-12} s (laser-produced plasma) to 10^7 s (fluorescent lights) (~19 OOM)	10^1 s (solar flares) to 10^{17} s (intergalactic plasma) (~16 OOM)

Density in particles per cubic meter	10^7 m^{-3} to 10^{32} m^{-3} (inertial confinement plasma)	1 m^{-3} (intergalactic medium) to 10^{30} m^{-3} (stellar core)
Temperature in Kelvin	~0 K (crystalline non-neutral plasma) to 10^8 K (magnetic fusion plasma)	10^2 K (aurora) to 10^7 K (solar core)
Magnetic fields in teslas	10^{-4} T (lab plasma) to 10^3 T (pulsed-power plasma)	10^{-12} T (intergalactic medium) to 10^{11} T (near neutron stars)

Degree of Ionization

For plasma to exist, ionization is necessary. The term "plasma density" by itself usually refers to the "electron density", that is, the number of free electrons per unit volume. The degree of ionization of a plasma is the proportion of atoms that have lost or gained electrons, and is controlled mostly by the temperature. Even a partially ionized gas in which as little as 1% of the particles are ionized can have the characteristics of a plasma (i.e., response to magnetic fields and high electrical conductivity). The degree of ionization, α, is defined as $\alpha = \dfrac{n_i}{n_i + n_n}$,, where n_i is the number density of ions and n_n is the number density of neutral atoms. The *electron density* is related to this by the average charge state $\langle Z \rangle$ of the ions through $n_e = \langle Z \rangle n_i$,, where n_e is the number density of electrons.

Temperatures

Plasma temperature is commonly measured in kelvins or electronvolts and is, informally, a measure of the thermal kinetic energy per particle. High temperatures are usually needed to sustain ionization, which is a defining feature of a plasma. The degree of plasma ionization is determined by the electron temperature relative to the ionization energy (and more weakly by the density), in a relationship called the Saha equation. At low temperatures, ions and electrons tend to recombine into bound states—atoms—and the plasma will eventually become a gas.

In most cases the electrons are close enough to thermal equilibrium that their temperature is relatively well-defined, even when there is a significant deviation from a Maxwellian energy distribution function, for example, due to UV radiation, energetic particles, or strong electric fields. Because of the large difference in mass, the electrons come to thermodynamic equilibrium amongst themselves much faster than they come into equilibrium with the ions or neutral atoms. For this reason, the ion temperature may be very different from (usually lower than) the electron temperature. This is especially common in weakly ionized technological plasmas, where the ions are often near the ambient temperature.

Thermal Vs. Nonthermal Plasmas

Based on the relative temperatures of the electrons, ions and neutrals, plasmas are classified as "thermal" or "non-thermal". Thermal plasmas have electrons and the heavy particles at the same temperature, i.e. they are in thermal equilibrium with each other. Nonthermal plasmas on the other hand have the ions and neutrals at a much lower temperature (sometimes room temperature), whereas electrons are much "hotter" ($T_e T_n$).

Complete Vs. Incomplete Ionization

A plasma is sometimes referred to as being "hot" if it is nearly fully ionized, or "cold" if only a small fraction (for example 1%) of the gas molecules are ionized, but other definitions of the terms "hot plasma" and "cold plasma" are common. Even in a "cold" plasma, the electron temperature is still typically several thousand degrees Celsius. Plasmas utilized in "plasma technology" ("technological plasmas") are usually cold plasmas in the sense that only a small fraction of the gas molecules are ionized.

Plasma Potential

Lightning is an example of plasma present at Earth's surface. Typically, lightning discharges 30,000 amperes at up to 100 million volts, and emits light, radio waves, X-rays and even gamma rays. Plasma temperatures in lightning can approach 28,000 K (28,000 °C; 50,000 °F) and electron densities may exceed 10^{24} m^{-3}.

Since plasmas are very good electrical conductors, electric potentials play an important role. The potential as it exists on average in the space between charged particles, independent of the question of how it can be measured, is called the "plasma potential", or the "space potential". If an electrode is inserted into a plasma, its potential will generally lie considerably below the plasma potential due to what is termed a Debye sheath. The good electrical conductivity of plasmas makes their electric fields very small. This results in the important concept of "quasineutrality", which says the density of negative charges is approximately equal to the density of positive charges over large volumes of the plasma ($n_e = \langle Z \rangle n_i$,), but on the scale of the Debye length there can be charge imbalance. In the special case that *double layers* are formed, the charge separation can extend some tens of Debye lengths.

The magnitude of the potentials and electric fields must be determined by means other than simply finding the net charge density. A common example is to assume that the electrons satisfy the Boltzmann relation:

$$n_e \propto e^{e\Phi/k_B T_e}.$$

Differentiating this relation provides a means to calculate the electric field from the density:

$$\vec{E} = (k_B T_e / e)(\nabla n_e / n_e).$$

It is possible to produce a plasma that is not quasineutral. An electron beam, for example, has only negative charges. The density of a non-neutral plasma must generally be very low, or it must be very small, otherwise it will be dissipated by the repulsive electrostatic force.

In astrophysical plasmas, Debye screening prevents electric fields from directly affecting the plasma over large distances, i.e., greater than the Debye length. However, the existence of charged particles causes the plasma to generate, and be affected by, magnetic fields. This can and does cause extremely complex behavior, such as the generation of plasma double layers, an object that separates charge over a few tens of Debye lengths. The dynamics of plasmas interacting with external and self-generated magnetic fields are studied in the academic discipline of magnetohydrodynamics.

Magnetization

Plasma with a magnetic field strong enough to influence the motion of the charged particles is said to be magnetized. A common quantitative criterion is that a particle on average completes at least one gyration around the magnetic field before making a collision, i.e., $\omega_{ce} / v_{coll} > 1$, where ω_{ce} is the "electron gyrofrequency" and v_{coll} is the "electron collision rate". It is often the case that the electrons are magnetized while the ions are not. Magnetized plasmas are *anisotropic*, meaning that their properties in the direction parallel to the magnetic field are different from those perpendicular to it. While electric fields in plasmas are usually small due to the high conductivity, the electric field associated with a plasma moving in a magnetic field is given by $\mathbf{E} = -v \times \mathbf{B}$ (where $\mathbf{E}$ is the electric field, is the velocity, and $\mathbf{B}$ is the magnetic field), and is not affected by Debye shielding.

Comparison of Plasma and Gas Phases

Plasma is often called the *fourth state of matter* after solid, liquids and gases, despite plasma typically being an ionized gas. It is distinct from these and other lower-energy states of matter. Although it is closely related to the gas phase in that it also has no definite form or volume, it differs in a number of ways, including the following:

Common Plasmas

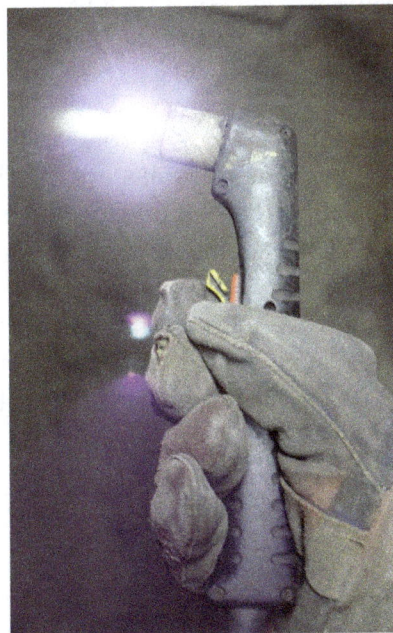

A handheld plasma cutter used to cut steel plates.
A jet of gas is heated with an electric arc into a plasma.

Plasmas are by far the most common phase of ordinary matter in the universe, both by mass and by volume. Essentially, all of the visible light from space comes from stars, which are plasmas with a temperature such that they radiate strongly at visible wavelengths. Most of the ordinary (or baryonic) matter in the universe, however, is found in the intergalactic medium, which is also a plasma, but much hotter, so that it radiates primarily as X-rays.

In 1937, Hannes Alfvén argued that if plasma pervaded the universe, it could then carry electric currents capable of generating a galactic magnetic field. After winning the Nobel Prize, he emphasized that:

In order to understand the phenomena in a certain plasma region, it is necessary to map not only the magnetic but also the electric field and the electric currents. Space is filled with a network of currents which transfer energy and momentum over large or very large distances. The currents often pinch to filamentary or surface currents. The latter are likely to give space, as also interstellar and intergalactic space, a cellular structure.

By contrast the current scientific consensus is that about 96% of the total energy density in the universe is not plasma or any other form of ordinary matter, but a combination of cold dark matter and dark energy. Our Sun, and all of the other stars, are made of plasma, much of interstellar space is filled with a plasma, albeit a very sparse one, and intergalactic space too. Even black holes, which are not directly visible, are thought to be fuelled by accreting ionising matter (i.e. plasma), and they are associated with astrophysical jets of luminous ejected plasma, such as M87's jet that extends 5,000 light-years.

In our solar system, interplanetary space is filled with the plasma of the Solar Wind that extends from the Sun out to the heliopause. However, the density of ordinary matter is much higher than average and much higher than that of either dark matter or dark energy. The planet Jupiter accounts for most of the *non*-plasma within the orbit of Pluto (about 0.1% by mass, or 10^{-15}% by volume).

Dust and small grains within a plasma will also pick up a net negative charge, so that they in turn may act like a very heavy negative ion component of the plasma.

Complex Plasma Phenomena

Although the underlying equations governing plasmas are relatively simple, plasma behavior is extraordinarily varied and subtle: the emergence of unexpected behavior from a simple model is a typical feature of a complex system. Such systems lie in some sense on the boundary between ordered and disordered behavior and cannot typically be described either by simple, smooth, mathematical functions, or by pure randomness. The spontaneous formation of interesting spatial features on a wide range of length scales is one manifestation of plasma complexity. The features are interesting, for example, because they are very sharp, spatially intermittent (the distance between features is much larger than the features themselves), or have a fractal form. Many of these features were first studied in the laboratory, and have subsequently been recognized throughout the universe. Examples of complexity and complex structures in plasmas include:

Filamentation

Striations or string-like structures, also known as Birkeland currents, are seen in many plasmas, like the plasma ball, the aurora, lightning, electric arcs, solar flares, and supernova remnants. They arc sometimes associated with larger current densities, and the interaction with the magnetic field can form a magnetic rope structure. High power microwave breakdown at atmospheric pressure also leads to the formation of filamentary structures.

Filamentation also refers to the self-focusing of a high power laser pulse. At high powers, the nonlinear part of the index of refraction becomes important and causes a higher index of refraction in the center of the laser beam, where the laser is brighter than at the edges, causing a feedback that focuses the laser even more. The tighter focused laser has a higher peak brightness (irradiance) that forms a plasma. The plasma has an index of refraction lower than one, and causes a defocusing of the laser beam. The interplay of the focusing index of refraction, and the defocusing plasma makes the formation of a long filament of plasma that can be micrometers to kilometers in length. One interesting aspect of the filamentation generated plasma is the relatively low ion density due to defocusing effects of the ionized electrons.

Shocks or Double Layers

Plasma properties change rapidly (within a few Debye lengths) across a two-dimensional sheet in the presence of a (moving) shock or (stationary) double layer. Double layers involve localized charge separation, which causes a large potential difference across the layer, but does not generate an electric field outside the layer. Double layers separate adjacent plasma regions with different physical characteristics, and are often found in current carrying plasmas. They accelerate both ions and electrons.

Electric Fields and Circuits

Quasineutrality of a plasma requires that plasma currents close on themselves in electric circuits. Such circuits follow Kirchhoff's circuit laws and possess a resistance and inductance. These circuits must generally be treated as a strongly coupled system, with the behavior in each plasma region dependent on the entire circuit. It is this strong coupling between system elements, together with nonlinearity, which may lead to complex behavior. Electrical circuits in plasmas store inductive (magnetic) energy, and should the circuit be disrupted, for example, by a plasma instability, the inductive energy will be released as plasma heating and acceleration. This is a common explanation for the heating that takes place in the solar corona. Electric currents, and in particular, magnetic-field-aligned electric currents (which are sometimes generically referred to as "Birkeland currents"), are also observed in the Earth's aurora, and in plasma filaments.

Cellular Structure

Narrow sheets with sharp gradients may separate regions with different properties such as magnetization, density and temperature, resulting in cell-like regions. Examples include the magnetosphere, heliosphere, and heliospheric current sheet. Hannes Alfvén wrote: "From the cosmological point of view, the most important new space research discovery is probably the cellular structure of space. As has been seen in every region of space accessible to in situ measurements, there are a number of 'cell walls', sheets of electric currents, which divide space into compartments with different magnetization, temperature, density, etc."

Critical Ionization Velocity

The critical ionization velocity is the relative velocity between an ionized plasma and a neutral gas, above which a runaway ionization process takes place. The critical ionization process is a quite general mechanism for the conversion of the kinetic energy of a rapidly streaming gas into ionization and plasma thermal energy. Critical phenomena in general are typical of complex systems, and may lead to sharp spatial or temporal features.

Ultracold Plasma

Ultracold plasmas are created in a magneto-optical trap (MOT) by trapping and cooling neutral atoms, to temperatures of 1 mK or lower, and then using another laser to ionize the atoms by giving each of the outermost electrons just enough energy to escape the electrical attraction of its parent ion.

One advantage of ultracold plasmas are their well characterized and tunable initial conditions, including their size and electron temperature. By adjusting the wavelength of the ionizing laser, the kinetic energy of the liberated electrons can be tuned as low as 0.1 K, a limit set by the frequency bandwidth of the laser pulse. The ions inherit the millikelvin temperatures of the neutral atoms, but are quickly heated through a process known as disorder induced heating (DIH). This type of non-equilibrium ultracold plasma evolves rapidly, and displays many other interesting phenomena.

One of the metastable states of a strongly nonideal plasma is Rydberg matter, which forms upon condensation of excited atoms.

Non-neutral Plasma

The strength and range of the electric force and the good conductivity of plasmas usually ensure that the densities of positive and negative charges in any sizeable region are equal ("quasineutrality"). A plasma with a significant excess of charge density, or, in the extreme case, is composed of a single species, is called a non-neutral plasma. In such a plasma, electric fields play a dominant role. Examples are charged particle beams, an electron cloud in a Penning trap and positron plasmas.

Dusty Plasma/Grain Plasma

A dusty plasma contains tiny charged particles of dust (typically found in space). The dust particles acquire high charges and interact with each other. A plasma that contains larger particles is called grain plasma. Under laboratory conditions, dusty plasmas are also called *complex plasmas*.

Impermeable Plasma

Impermeable plasma is a type of thermal plasma which acts like an impermeable solid with respect to gas or cold plasma and can be physically pushed. Interaction of cold gas and thermal plasma was briefly studied by a group led by Hannes Alfvén in 1960s and 1970s for its possible applications in insulation of fusion plasma from the reactor walls. However, later it was found that the external magnetic fields in this configuration could induce kink instabilities in the plasma and subsequently lead to an unexpectedly high heat loss to the walls. In 2013, a group of materials scientists reported that they have successfully generated stable impermeable plasma with no magnetic confinement using only an ultrahigh-pressure blanket of cold gas. While spectroscopic data on the characteristics of plasma were claimed to be difficult to obtain due to the high pressure, the passive effect of plasma on synthesis of different nanostructures clearly suggested the effective confinement. They also showed that upon maintaining the impermeability for a few tens of seconds, screening of ions at the plasma-gas interface could give rise to a strong secondary mode of heating (known as viscous heating) leading to different kinetics of reactions and formation of complex nanomaterials.

Mathematical Descriptions

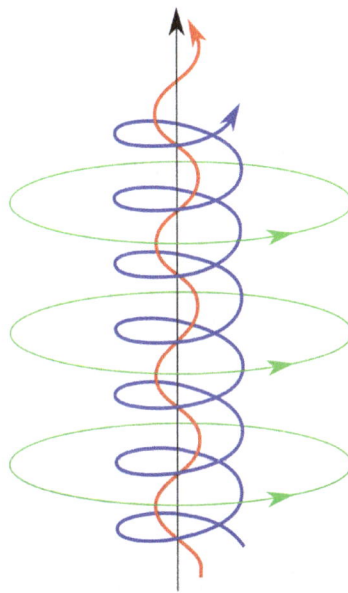

The complex self-constricting magnetic field lines and current paths in a field-aligned Birkeland current that can develop in a plasma.

To completely describe the state of a plasma, we would need to write down all the particle locations and velocities and describe the electromagnetic field in the plasma region. However, it is generally not practical or necessary to keep track of all the particles in a plasma. Therefore, plasma physicists commonly use less detailed descriptions, of which there are two main types:

Fluid Model

Fluid models describe plasmas in terms of smoothed quantities, like density and averaged velocity around each position. One simple fluid model, magnetohydrodynamics, treats the plasma as a single fluid governed by a combination of Maxwell's equations and the Navier–Stokes equations. A more general description is the two-fluid plasma picture, where the ions and electrons are described separately. Fluid models are often accurate when collisionality is sufficiently high to keep the plasma velocity distribution close to a Maxwell–Boltzmann distribution. Because fluid models usually describe the plasma in terms of a single flow at a certain temperature at each spatial location, they can neither capture velocity space structures like beams or double layers, nor resolve wave-particle effects.

Kinetic Model

Kinetic models describe the particle velocity distribution function at each point in the plasma and therefore do not need to assume a Maxwell–Boltzmann distribution. A kinetic description is often necessary for collisionless plasmas. There are two common approaches to kinetic description of a plasma. One is based on representing the smoothed distribution function on a grid in velocity and position. The other, known as the particle-in-cell (PIC) technique, includes kinetic information by following the trajectories of a large number of individual particles. Kinetic models are generally more computationally intensive than fluid models. The Vlasov equation may be used to describe the dynamics of a system of charged particles interacting with an electromagnetic field. In magnetized plasmas, a gyrokinetic approach can substantially reduce the computational expense of a fully kinetic simulation.

Artificial Plasmas

Most artificial plasmas are generated by the application of electric and/or magnetic fields through a gas. Plasma generated in a laboratory setting and for industrial use can be generally categorized by:

- The type of power source used to generate the plasma—DC, RF and microwave

- The pressure they operate at—vacuum pressure (< 10 mTorr or 1 Pa), moderate pressure (~ 1 Torr or 100 Pa), atmospheric pressure (760 Torr or 100 kPa)

- The degree of ionization within the plasma—fully, partially, or weakly ionized

- The temperature relationships within the plasma—thermal plasma $(T_e = T_i = T_{gas})$, non-thermal or "cold" plasma $(T_e T_i = T_{gas})$)

- The electrode configuration used to generate the plasma

- The magnetization of the particles within the plasma—magnetized (both ion and electrons are trapped in Larmor orbits by the magnetic field), partially magnetized (the electrons but not the ions are trapped by the magnetic field), non-magnetized (the magnetic field is too weak to trap the particles in orbits but may generate Lorentz forces)

Generation of Artificial Plasma

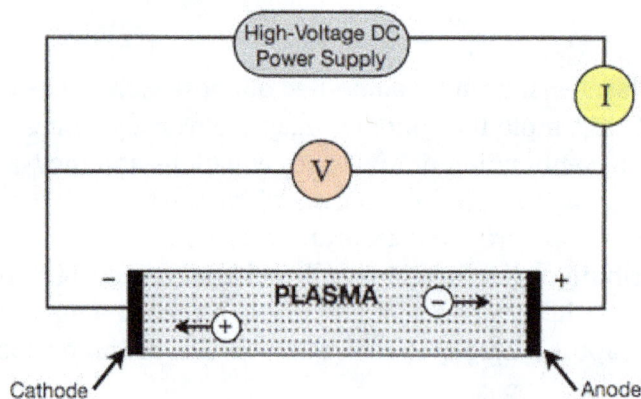

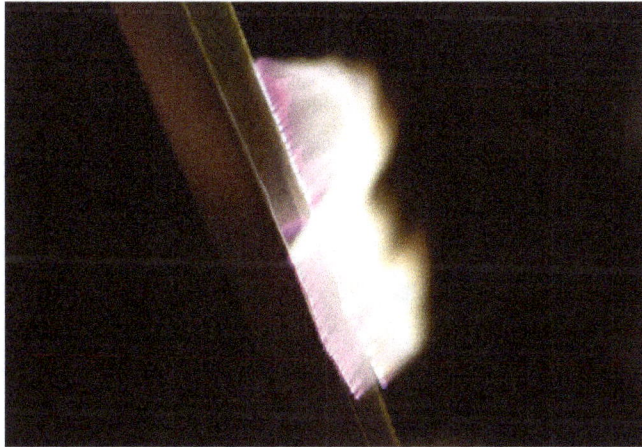

Artificial plasma produced in air by a Jacob's Ladder

Just like the many uses of plasma, there are several means for its generation, however, one principle is common to all of them: there must be energy input to produce and sustain it. For this case, plasma is generated when an electric current is applied across a dielectric gas or fluid (an electrically non-conducting material) as can be seen in the adjacent image, which shows a discharge tube as a simple example (DC used for simplicity).

The potential difference and subsequent electric field pull the bound electrons (negative) toward the anode (positive electrode) while the cathode (negative electrode) pulls the nucleus. As the voltage increases, the current stresses the material (by electric polarization) beyond its dielectric limit (termed strength) into a stage of electrical breakdown, marked by an electric spark, where the material transforms from being an insulator into a conductor (as it becomes increasingly ionized). The underlying process is the Townsend avalanche, where collisions between electrons and neutral gas atoms create more ions and electrons. The first impact of an electron on an atom results in one ion and two electrons. Therefore, the number of charged particles increases rapidly (in the millions) only "after about 20 successive sets of collisions", mainly due to a small mean free path (average distance travelled between collisions).

Electric Arc

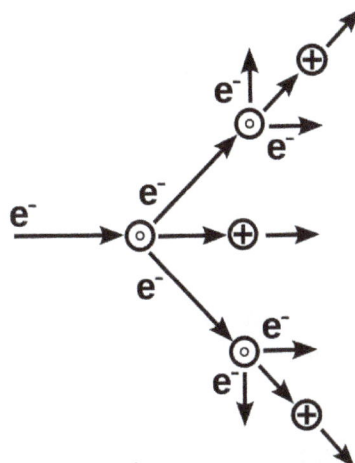

Cascade process of ionization. Electrons are 'e−', neutral atoms 'o', and cations '+'.

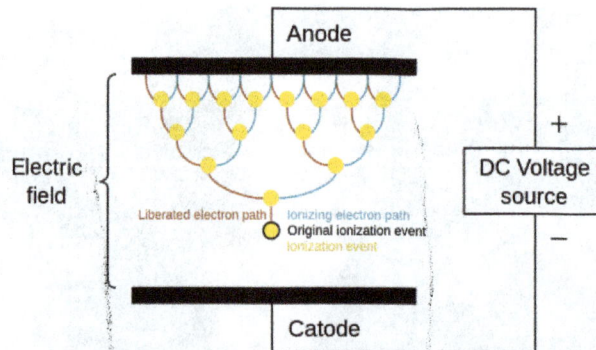

Avalanche effect between two electrodes. The original ionisation event liberates one electron, and each subsequent collision liberates a further electron, so two electrons emerge from each collision: the ionising electron and the liberated electron.

With ample current density and ionization, this forms a luminous electric arc (a continuous electric discharge similar to lightning) between the electrodes. Electrical resistance along the continuous electric arc creates heat, which dissociates more gas molecules and ionizes the resulting atoms (where degree of ionization is determined by temperature), and as per the sequence: solid-liquid-gas-plasma, the gas is gradually turned into a thermal plasma. A thermal plasma is in thermal equilibrium, which is to say that the temperature is relatively homogeneous throughout the heavy particles (i.e. atoms, molecules and ions) and electrons. This is so because when thermal plasmas are generated, electrical energy is given to electrons, which, due to their great mobility and large numbers, are able to disperse it rapidly and by elastic collision (without energy loss) to the heavy particles.

Examples of Industrial/Commercial Plasma

Because of their sizable temperature and density ranges, plasmas find applications in many fields of research, technology and industry. For example, in: industrial and extractive metallurgy, surface treatments such as plasma spraying (coating), etching in microelectronics, metal cutting and welding; as well as in everyday vehicle exhaust cleanup and fluorescent/luminescent lamps, while even playing a part in supersonic combustion engines for aerospace engineering.

Low-pressure Discharges

- *Glow discharge plasmas*: non-thermal plasmas generated by the application of DC or low frequency RF (<100 kHz) electric field to the gap between two metal electrodes. Probably the most common plasma; this is the type of plasma generated within fluorescent light tubes.

- *Capacitively coupled plasma (CCP)*: similar to glow discharge plasmas, but generated with high frequency RF electric fields, typically 13.56 MHz. These differ from glow discharges in that the sheaths are much less intense. These are widely used in the microfabrication and integrated circuit manufacturing industries for plasma etching and plasma enhanced chemical vapor deposition.

- *Cascaded Arc Plasma Source*: a device to produce low temperature (~1eV) high density plasmas (HDP).

- *Inductively coupled plasma (ICP)*: similar to a CCP and with similar applications but the electrode consists of a coil wrapped around the chamber where plasma is formed.

- *Wave heated plasma*: similar to CCP and ICP in that it is typically RF (or microwave). Examples include helicon discharge and electron cyclotron resonance (ECR).

Atmospheric Pressure

- *Arc discharge:* this is a high power thermal discharge of very high temperature (~10,000 K). It can be generated using various power supplies. It is commonly used in metallurgical processes. For example, it is used to smelt minerals containing Al_2O_3 to produce aluminium.

- *Corona discharge:* this is a non-thermal discharge generated by the application of high voltage to sharp electrode tips. It is commonly used in ozone generators and particle precipitators.

- *Dielectric barrier discharge (DBD):* this is a non-thermal discharge generated by the application of high voltages across small gaps wherein a non-conducting coating prevents the transition of the plasma discharge into an arc. It is often mislabeled 'Corona' discharge in industry and has similar application to corona discharges. It is also widely used in the web treatment of fabrics. The application of the discharge to synthetic fabrics and plastics functionalizes the surface and allows for paints, glues and similar materials to adhere.

- *Capacitive discharge:* this is a nonthermal plasma generated by the application of RF power (e.g., 13.56 MHz) to one powered electrode, with a grounded electrode held at a small separation distance on the order of 1 cm. Such discharges are commonly stabilized using a noble gas such as helium or argon.

- "Piezoelectric direct discharge plasma:" is a nonthermal plasma generated at the high-side of a piezoelectric transformer (PT). This generation variant is particularly suited for high efficient and compact devices where a separate high voltage power supply is not desired.

History

Plasma was first identified in a Crookes tube, and so described by Sir William Crookes in 1879 (he called it "radiant matter"). The nature of the Crookes tube "cathode ray" matter was subsequently identified by British physicist Sir J.J. Thomson in 1897. The term "plasma" was coined by Irving Langmuir in 1928, perhaps because the glowing discharge molds itself to the shape of the Crookes tube (a thing moulded or formed). Langmuir described his observations as:

Except near the electrodes, where there are *sheaths* containing very few electrons, the ionized gas contains ions and electrons in about equal numbers so that the resultant space charge is very small. We shall use the name *plasma* to describe this region containing balanced charges of ions and electrons.

Research

Plasmas are the object of study of the academic field of *plasma science* or *plasma physics*, including sub-disciplines such as space plasma physics. There are multiple journals devoted to the subject. It involves the following fields of active research.

Hall effect thruster. The electric field in a plasma double layer is so effective at accelerating ions that electric fields are used in ion drives.

References

- R. Penrose (1991). "The mass of the classical vacuum". In S. Saunders, H.R. Brown. The Philosophy of Vacuum. Oxford University Press. p. 21. ISBN 0-19-824449-5.

- M.H. Krieger (1998). Constitutions of Matter: Mathematically Modeling the Most Everyday of Physical Phenomena. University of Chicago Press. p. 22. ISBN 0-226-45305-7.

- G. F. Barker (1870). "Divisions of matter". A text-book of elementary chemistry: theoretical and inorganic. John F Morton & Co. p. 2. ISBN 978-1-4460-2206-1.

- L. Smolin (2007). The Trouble with Physics: The Rise of String Theory, the Fall of a Science, and What Comes Next. Mariner Books. p. 67. ISBN 0-618-91868-X.

- B. Povh; K. Rith; C. Scholz; F. Zetsche; M. Lavelle (2004). Particles and Nuclei: An Introduction to the Physical Concepts. Springer. p. 103. ISBN 3-540-20168-8.

- T. Hatsuda (2008). "Quark–gluon plasma and QCD". In H. Akai. Condensed matter theories. 21. Nova Publishers. p. 296. ISBN 1-60021-501-7.

- K.W Staley (2004). "Origins of the Third Generation of Matter". The Evidence for the Top Quark. Cambridge University Press. p. 8. ISBN 0-521-82710-8.

- P.J. Collings (2002). "Chapter 1: States of Matter". Liquid Crystals: Nature's Delicate Phase of Matter. Princeton University Press. ISBN 0-691-08672-9.

- National Research Council (US) (2006). Revealing the hidden nature of space and time. National Academies Press. p. 46. ISBN 0-309-10194-8.

- K. Pretzl (2004). "Dark Matter, Massive Neutrinos and Susy Particles". Structure and Dynamics of Elementary Matter. Walter Greiner. p. 289. ISBN 1-4020-2446-0.

- K. Freeman; G. McNamara (2006). "What can the matter be?". In Search of Dark Matter. Birkhäuser Verlag. p. 105. ISBN 0-387-27616-5.

Permissions

All chapters in this book are published with permission under the Creative Commons Attribution Share Alike License or equivalent. Every chapter published in this book has been scrutinized by our experts. Their significance has been extensively debated. The topics covered herein carry significant information for a comprehensive understanding. They may even be implemented as practical applications or may be referred to as a beginning point for further studies.

We would like to thank the editorial team for lending their expertise to make the book truly unique. They have played a crucial role in the development of this book. Without their invaluable contributions this book wouldn't have been possible. They have made vital efforts to compile up to date information on the varied aspects of this subject to make this book a valuable addition to the collection of many professionals and students.

This book was conceptualized with the vision of imparting up-to-date and integrated information in this field. To ensure the same, a matchless editorial board was set up. Every individual on the board went through rigorous rounds of assessment to prove their worth. After which they invested a large part of their time researching and compiling the most relevant data for our readers.

The editorial board has been involved in producing this book since its inception. They have spent rigorous hours researching and exploring the diverse topics which have resulted in the successful publishing of this book. They have passed on their knowledge of decades through this book. To expedite this challenging task, the publisher supported the team at every step. A small team of assistant editors was also appointed to further simplify the editing procedure and attain best results for the readers.

Apart from the editorial board, the designing team has also invested a significant amount of their time in understanding the subject and creating the most relevant covers. They scrutinized every image to scout for the most suitable representation of the subject and create an appropriate cover for the book.

The publishing team has been an ardent support to the editorial, designing and production team. Their endless efforts to recruit the best for this project, has resulted in the accomplishment of this book. They are a veteran in the field of academics and their pool of knowledge is as vast as their experience in printing. Their expertise and guidance has proved useful at every step. Their uncompromising quality standards have made this book an exceptional effort. Their encouragement from time to time has been an inspiration for everyone.

The publisher and the editorial board hope that this book will prove to be a valuable piece of knowledge for students, practitioners and scholars across the globe.

Index